Nutrition and Feeding of Fish and Crustaceans

Springer
London
Berlin
Heidelberg
New York
Barcelona
Hong Kong
Milan
Paris
Santa Clara
Singapore
Tokyo

Jean Guillaume, Sadasivam Kaushik, Pierre Bergot
and Robert Métailler

Nutrition and Feeding of Fish and Crustaceans

Springer

Published in association with
Praxis Publishing
Chichester, UK

PRAXIS

Editeurs/Editors:
J. Guillaume, R. Métailler
Unité mixte de Nutrition des poissons INRA-IFREMER, BP 70, 29280 Plouzané, France
S. Kaushik, P. Bergot
Unité mixte de Nutrition des poissons INRA-IFREMER, BP 3, 64310 Saint-Pée-sur-Nivelle, France
Translator: Dr Jennifer Watson, Department of Zoology, Univ. of Aberdeen, UK
Translation Editor: Dr Lindsay Laird, Department of Zoology, Univ. of Aberdeen, UK
Original French edition, *Nutrition et alimentation des poissons et crustacés*

This work has been published with the help of the French Ministère de la Culture - Centre national du livre
ouvrage publié avec le concours du Ministère française chargé de la culture - Centre national du livre

---

SPRINGER-PRAXIS BOOKS IN AQUACULTURE AND FISHERIES
SUBJECT *ADVISORY EDITORS*: Lindsay Laird, M.A., Ph.D., F.I.F.M., University of Aberdeen, UK
Selina Stead, B.Sc., M.Sc., Ph.D., Land Economy, University of Aberdeen, UK

ISBN 1-85233-241-7 Springer-Verlag Berlin Heidelberg New York
British Library Cataloguing in Publication Data
Nutrition and feeding of fish and crustaceans. -
(Springer-Praxis books in aquaculture and fisheries)
1. Fishes - Nutrition 2. Fishes - Food 3. Fishes - Feeding and feeds 4. Crustacea - Nutrition 5. Crustacea - Food
6. Crustacea - Feeding and feeds
I. Guillaume, Jean
597.1'53
ISBN 1852332417
Library of Congress Cataloging-in-Publication Data
Nutrition et alimentation des poissons et crustacés. English.
Nutrition and feeding of fish and crustaceans / Jean Guillaume ... [et al.] ;
[translator, Jennifer Watson].
p. cm.—(Springer-Praxis books in aquaculture and fisheries)
Includes bibliographical references (p.).
ISBN 1-85233-241-7 (alk. paper)
1. Fishes—Feeding and feeds. 2. Fishes—Nutrition. 3. Crustacea—Feeding and feeds. 4. Crustacea—Nutrition. I. Guillaume, Jean, 1936- II. Title.
III. Series.
SH156 .N8913 2001
639.8—dc21 2001020836

Printed by MPG Books Ltd, Bodmin, Cornwall, UK

Project copy editor: Rachael Wilkie
Cover design: Jim Wilkie
Typesetting: Originator, Gt Yarmouth, Norfolk, UK
Printed on acid-free paper supplied by Precision Publishing Papers Ltd, UK

Contents

List of Authors

Jacqueline Arzel
Institut Français de Recherches pour l'Exploitation de la Mer
Unité Mixte INRA-IFREMER de Nutrition des Poissons
IFREMER, Centre de Brest BP 70
29280 Plouzané

Pierre Bergot
Institut National de la Recherche Agronomique
Unité Mixte INRA-IFREMER de Nutrition des Poissons
INRA, Station d'Hydrobiologie
64310 Saint-Pée-sur-Nivelle

Thierry Boujard
Institut National de la Recherche Agonomique
Unité Mixte INRA-IFREMER de Nutrition des Poissons
INRA, Station d'Hydrobiologie
64310 Saint-Pée-sur-Nivelle

Chantal Cahu
Institute Français de Recherches pour l'Exploitation de la Mer
Unité Mixte INRA-IFREMER de Nutrition des Poissons
IFREMER, Centre de Brest BP 70
29280 Plouzané

Hubert J. Ceccaldi
Centre d'Étude des Ressources Animales Marines
Université Aix-Marseille III
Faculté de Sciences et Techniques de Saint Jérôme
Avenue Normandie-Niemen
13397 Marseille cedex 13

Georges Choubert
Institut National de la Recherche Agronomique
Unité Mixte INRA-IFREMER de Nutrition des Poissons
INRA, Station d'Hydrobiologie
64310 Saint-Pée-sur-Nivelle

Geneviève Corraze
Institut National de la Recherche Agronomique
Unité Mixte INRA-IFREMER de Nutrition des Poissons
INRA, Station d'Hydrobiologie
64310 Saint-Pée-sur-Nivelle

Gérard Cuzon
Institut Français de Recherches pour l'Exploitation de la Mer
Centre Océanologique du Pacifique
BP 7004
Taravao, Tahiti
Polynésie Française

André Fauré
Institut National de la Recherche Agronomique
Salmoniculture expérimentale IFREMER-INRA
Le Drennec BP 17
29450 Sizun
Current address: INRA Centre de Tours-Nouzilly
Services Généraux
37380 Monnaie

Jean-Louis Gaignon
Institut Française de Recherches pour l'Exploitation de la Mer
Laboratoire de Physiologie des Poissons
IFREMER, Centre de Brest BP 70
29280 Plouzané

François-Joël Gatesoupe
Institut National de la Recherche Agronomique
Unité Mixte INRA-IFREMER de Nutrition des Poissons
IFREMER, Centre de Brest BP 70
29280 Plouzané

Marie-Françoise Gouillou-Coustans
Institut National de la Recherches pour l'Exploitation de la Mer
Unité Mixte INRA-IFREMER de Nutrition des Poissons
IFREMER, Centre de Brest BP 70
29280 Plouzané

Jean Guillaume
Institut National de la Recherche Agronomique
Unité Mixte INRA-IFREMER de Nutrition des Poissons
IFREMER, Centre de Brest BP 70
29280 Plouzané

Sadasivam J. Kaushik
Institut National de la Recherche Agronomique
Unité Mixte INRA-IFREMER de Nutrition des Poissons
INRA, Station d'Hydrobiologie
64310 Saint-Pée-sur-Nivelle

Laurent Labbé
Institut National de la Recherche Agronomique
Salmoniculture expéremental e
IFREMER-INRA
Le Drennec BP 17
29450 Sizun

Muriel Mambrini
Institut National de la Recherche Agronomique
Unité Mixte INRA-IFREMER de Nutrition des Poissons
Station d'Hydrobiologie
64310 Saint-Pée-sur-Nivelle
Current address:
INRA Laboratoire de Génétique des poissons
Domaine de Vilvert
78280 Jouy-en-Josas

Françoise Médale
Institut National de la Recherche Agronomique
Unité Mixte INRA-IFREMER
de Nutrition des Poissons
INRA, Station d'Hydrobiologie
64310 Saint-Pée-sur-Nivelle

Jean-Pierre Melcion
Institut National de la Recherche Agronomique
Laboratoire de Technologie appliquée à la Nutrition
BP 1627
44316 Nantes cedex 03

Robert Métailler
Institut Français de Recherches pour l'Exploitation de la Mer
Unité Mixte INRA-IFREMER de Nutrition des Poissons
IFREMER, Centre de Brest BP 70
29280 Plouzané

Jeanine Person-Le Ruyet
Institut Français de Recherches pour l'Exploitation de la Mer
Laboratoire de Physiologie des Poissons
IFREMER, Centre de Brest BP 70
29280 Plouzané

Loïc Quéméner
Institut Français de Recherches pour l'Exploitation de la Mer
Unité Mixte INRA-IFREMER
Laboratoire de Physiologie des Poissons
IFREMER, Centre de Brest BP 70
29280 Plouzané

Jean Henri Robin
Institut Français de Recherches pour l'Exploitation de la Mer
Unité Mixte INRA-IFREMER de Nutrition des Poissons
IFREMER, Centre de Brest BP 70
29280 Plouzané

José Luis Zambonino Infante
Institut Français de Recherches pour l'Exploitation de la Mer
Unité Mixte INRA-IFREMER de Nutrition des Poissons
IFREMER, Centre de Brest BP 70
29280 Plouzané

Acknowledgements

We give our most grateful thanks to all those who aided us in the conception and writing of this book. Bernard Chevassus-Au-Lois, past director general of INRA, convinced us of the usefulness of our work and gave us encouragement. Benoît Fauconneau, head of the Hydrobiology and Wildlife department at INRA, and Bernard Leclerq, director of the INRA Poultry Research Station at Nouzilly, devoted a great deal of time to reading the manuscript and provided us with positive criticism and suggestions. Sylvie Gros, Didier Bazin, Hervé Le Delliou, Alain Sire and Frédéric Vallée made a vital contribution in the form of about a hundred illustrations and figures. Joëlle Mehur carried out a significant amount of secretarial work. We thank collectively, rather than individually, our colleagues at the two laboratories at Saint-Pée-sur-Nivelle and Plouzané, who form the combined INRA-IFREMER Fish Nutrition group, for the help that they gave us, in particular in re-reading manuscripts, as well as the service personnel at INRA Editions for the quality of the finished product.

Foreword

Aquaculture now provides 22% of consumable aquatic products (FAO statistics, 1996). It is expanding rapidly, while fisheries production appears to have reached a plateau in recent years. Numerous forms of aquaculture currently exist worldwide: the enormous production of Chinese and Indian carp (70% of freshwater aquaculture production) results almost exclusively from extensive production, while almost all reared shrimps are produced using semi-intensive methods in which the role of natural food is significant. However, in developed as well as in many developing countries, intensive production using manufactured feeds is playing an increasing role. The dynamism of this aquaculture production, and more specifically that of intensive production, has created a requirement for increasingly precise and diverse scientific and technical data. There is not the same amount of knowledge available in aquaculture in the areas of greatest interest (physiology, genetics, pathology, as well as nutrition) as there is in poultry rearing or pig production, for example, although the studies carried out in various aquaculture nutrition laboratories generate an increasing amount of scientific and technical knowledge. A requirement thus exists for an up-to-date review of the information.

The aim of this book is to provide reviewed data in an easily accessible form to students, researchers and industrial technicians who are interested in aquaculture production. All these individuals have easy access to books about higher vertebrates. Therefore, the various aspects of fish and crustacean nutrition are presented in a comparative way, with frequent reference to mammals and birds. Scientific information related to metabolism and energy or nutrients are presented succinctly; the reader wishing to know more is referred to several books or review articles cited in the bibliography. Despite this, the field of fish and crustacean nutrition and feeding has been covered as widely as possible. Particular aspects, such as crustacean digestive physiology or fish larvae have been described in separate chapters, as have choice of raw materials, processing and even feed utilisation by aquatic animals. With the aim of rendering the book as useful as possible, descriptive chapters have been followed by tables and numerical data that can be used directly in the formulation of aquaculture feeds.

Since the appearance of bovine spongioform encephalopathy (BSE), diets containing animal meals have become suspect in the eyes of the European public. The discredit placed on meat meals and the possible confusion with other meals of animal origin has led us to include a brief statement concerning these raw materials, although the circumstantial nature of this statement contrasts with the general spirit of the book.

We hope that this book, written largely by members of the combined Fish Nutrition group of IFREMER-INRA, will bring to its readers a large amount of useful information which until now has remained scattered or unpublished.

Jean Guillaume, Sadasivam Kaushik,
Pierre Bergot, Robert Métailler

Part I

General

1

Introduction

It is only recently that scientists have become interested in fish nutrition. Until the middle of the twentieth century, the only studies carried out were on the anatomy of the digestive tract and some aspects of digestive physiology or feeding of animals in their natural environment. From the end of the nineteenth century, fish culture production extended to the production of carnivorous species in ponds and tanks. Since then, feeding farmed fish, which previously depended on natural foods, has become dependent on input of food by man. The first 'compound' feeds, made from various basic materials and meeting the animals' requirements as best they could, were the 'Oregon moist pellets' in the 1950s. Dried pellets appeared in the United States at the end of the same decade and at the start of the 1960s in Europe. Since then, fish culture has come into line with agriculture and studies of the nutritional requirements of fish have progressed. In practice, it started earlier than this, at the very beginning of the 1940s in Cortland (Idaho, United States). From the 1960s onwards, considerable effort has been devoted to nutritional studies.

More than 20 research teams currently carry out pure and applied studies of fish nutrition throughout the world. Despite a late start, they have all progressed rapidly owing to the body of acquired knowledge concerning terrestrial species. In some respects, the nutrition of some fish such as the rainbow trout (called simply 'trout' in this book) appears to be well understood. In general terms, however, the knowledge remains limited, sometimes very much so, especially in relation to the large number of fish species. Fish nutrition suffers not only from being a young subject, but also from specific handicaps: difficulties in study linked to the environment and the nutritional characteristics of fish.

Significant difficulties are encountered by nutritionists such as physiologists studying aquatic animals. It is easier to perform an analysis of body composition on a fish than on a mammal, but it is much more difficult to quantify ingestion and thus to estimate a nutritional budget for a fish and even more so for a shrimp than for a terrestrial animal. Specialised techniques have had to be developed, the accuracy of which remains lower than for techniques used on terrestrial animals. A series of even greater difficulties has also slowed down research on larvae. This stage involves dealing with the very small size of the organisms, their rapid development into juveniles and features which are very far removed from vertebrates.

1.1 ORIGIN OF THE NUTRITIONAL CHARACTERISTICS OF FISH AND CRUSTACEANS

As indicated in Table 1.1, the nutritional characteristics of fish are zoological, biological and ecological in origin. The fish class is immense. According to current

Table 1.1. Nutritional characteristics of fish.

Origin of peculiarities	Consequences
Zoological	
40,000 species (?)	Large interspecific diversity Variability of requirements
Biological	
Larval stages	Unknown factors (?)
Continuous growth	Different requirements with age (?)
Possible absence of stomach	Unique digestive process
Physiological	
Ectothermy	Effect of temperature on dynamics of digestive functions Low energy requirement which varies greatly with temperature
Ammonotelism	Relatively high net utilisation of protein energy
Ecological	
Low average body temperature	Limited role played by intestinal flora Difficulty in maintaining membrane fluidity Unique requirement for certain fatty acids
Relatively low oxygen content Relatively high environmental viscosity	Requirement for considerable 'ventilation' and frequent use of anaerobiosis – hypertrophy of white muscle – large quantity of proteins required for synthesis of this tissue (?)
High density of environment	Reduced importance of skeleton Low requirements for Ca and P
Slow molecular diffusion	Unique role of chemical attractants
Environment rich in certain elements	Supply of dissolved minerals directly from the environment
Unique aquatic food chains (dominance of carnivores)	Different efficiencies of energy sources (massive protein catabolism and limited use of carbohydrates)
Abundance of polyunsaturated fatty acids	Reduced capacity for bioconversion

estimates, teleosts alone account for almost as many species as all other vertebrates put together—amphibians, reptiles, birds and mammals. According to some taxonomists using molecular techniques, this number may have been underestimated by half. These species show variable degrees of evolution and are adapted to very different biotopes, and only a very few have been the subject of nutrition studies. It is certainly tempting to generalise using the knowledge acquired from the few tens of species already studied, but great care must be taken not to extrapolate excessively, at least not in every area.

Fish form the most primitive class of vertebrates and show archaic features: these include the absence of mandibular teeth, the absence of a stomach in several families, the non-differentiation of the small and large intestines, physiological characteristics such as the importance of the mechanisms of intestinal cell wall absorption by pinocytosis and even the imperfect regulation of glycaemia after ingestion of carbohydrates. Another characteristic which is common to invertebrates and lower vertebrates is continuous growth, or lack of an adult stage in the proper sense of the term. However, the primitive characteristic which is without doubt of the greatest practical importance is the existence of larval stages. Only the selachians and some teleosts are ovoviviparous and their fry are born at a more advanced stage. Almost all species of interest to aquaculture spawn small or tiny eggs, giving rise to larvae whose physiology still presents a large number of unknown factors and the rearing of which is often arduous. It was therefore not just by chance that the first fish used for intensive aquaculture were the salmonids, a family in which the fry pass easily from hatching to the final stages of their 'metamorphosis'. These archaic characteristics are not equal in importance, but the larvae's particular requirements continue to seriously slow down aquaculture development and, for many cultured aquatic species, feeding of larvae is still carried out according to more or less empirical methods.

In physiological terms, fish have a large number of peculiar characteristics, not only because their endocrine system is very different to that of mammals and birds. Two characteristics, poikilothermy and ammonotely, merit attention.

- Poikilothermy dramatically influences energy expenditure, and therefore requirements, of animals that are directly dependent on ambient temperature. This characteristic has a direct influence on the relationships between energy and nutrients. For poikilotherms, whose body temperature is always less than that of homeotherms, it is not possible to survive except under conditions where cell membranes remain fluid. These conditions are created by the incorporation of large amounts of polyunsaturated fatty acids which are characteristic of fish lipids. This is one of the main features of the nutrition, not only of fish, but also other aquatic poikilotherms.
- Not all fish are ammonotelic. In teleosts, however, the majority of nitrogenous waste is discharged in the form of ammonia. This phenomenon influences the net energy value of proteins as well as environmental pollution by nitrogenous waste from fish.

There are only a few similarities between the environment in which fish and other aquatic organisms live and the terrestrial environment. Low partial pressure of oxygen along with the higher density and viscosity of water have led to common characteristics in different taxonomic groups of motile animals, in particular frequent recourse to anaerobic metabolism and energy saving, using flotation. The secondary consequences of these two features include increased ventilation costs as well as hypertrophy of 'white' muscle fibres (for anaerobic metabolism) and skeletal atrophy. These peculiarities in the organism's composition have repercussions for feeding requirements. Fish, for example, require less calcium and phosphorus for the growth of their skeleton, which is much weaker than that of a quadruped of the same biomass.

The aquatic environment contains dissolved organic and inorganic molecules, some of which are essential to organisms living within it. Compared to terrestrial animals, 'primitive' aquatic animals have the enormous advantage of being able to extract at least some of the nutrients they require directly from their environment. A disadvantage, however, is that dissolved molecules that can guide animals towards their prey or food diffuse much more slowly than in air. Feeding attractants, or phagostimulants, play a much more important role than on land, at least under certain conditions, such as the transfer from 'natural' food to compound feed for larvae or small juveniles.

It is not possible to give a general description of the nutritional requirements of fish without mentioning their position in the aquatic food chain. Both on land and in the water, the food chain commences with photosynthesis. On land, higher plants are most abundant, while in the sea the base of the pyramid consists mainly of unicellular algae. On land, herbivorous vertebrates, some of which are consumed by man, are numerous, whereas in the sea, algae are not generally consumed by fish but by zooplankton, which in turn are eaten by fish larvae. About 85% of marine fish are first or even second order carnivores and adult herbivorous fish are very rare. In contrast to the classical animal husbandryman who is primarily interested in animals that obtain the majority of their energy from starch and cellulose, aquaculturists deal with animals adapted to transform proteins and lipids.

Crustaceans have many unique nutritional characteristics. The anatomy of the digestive tract is far removed from that of the vertebrates, although its main functions remain the same. The exoskeleton, or carapace, is very different in composition from a bony skeleton, and can reduce the capacity for absorbing minerals present in the water via the integument. Growth by moulting creates a discontinuity in nutrient flux which has no equivalent in vertebrates. However, many factors such as poikilothermy, ammonotely and even the nature of the food chain influence requirements of both fish and crustaceans in the same way.

The list of organic nutrients and mineral elements essential for life is practically the same for fish, crustaceans and higher vertebrates (Figure 1.1). This analogy extends to sometimes important quantitative requirements, such as that for essential amino acids or water soluble vitamins. This is not the case for all quantitative requirements: the differences in relation to mammals and birds will become evident, as will be made clear in the following chapters.

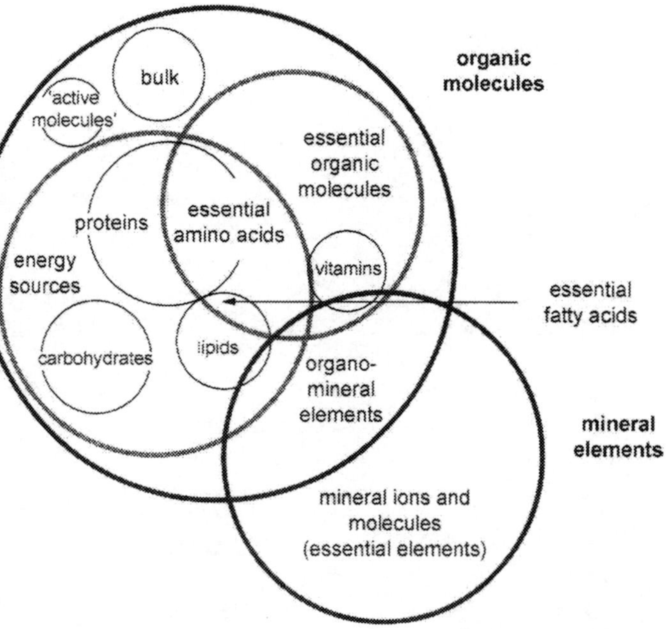

Figure 1.1. Diagram of the nutritional requirements of animals (not to scale). 'Active molecules': organic compounds in foods playing a positive (phagostimulant, growth factor, etc.) or negative (e.g. antinutritional factor) role (bulk: see Chapters 8 and 19).

1.2 BIBLIOGRAPHY

Guillaume, J. (1990) The nutritional characteristics and the formulation of diets for cultivated fish and crustaceans. In: Mellinger, J. (ed.), *Animal Nutrition and Transport Processes. Vol. 2. Nutrition in Wild and Domestic Animals*, Karger, Basle, pp. 203–214.

Love, R. M. (1980) *The Chemical Biology of Fishes*, Vol. 2. Academic Press, London.

NRC (National Research Council), (1993) *Nutrient Requirements of Fish*. National Academy Press, Washington.

2

Terminology and methodology

As with all scientific or technological methods, nutrition has its own terminology consisting of dedicated terms and common words to which specific meanings have been attributed. It is useful to define these terms in relation to the main principles.

Although often used to describe the action of feeding, here the word 'nutrition' is reserved for the 'science of nutrition', the branch of physiology that studies the whole series of actions which ensure the provision of the nutrients required for vital processes to an organism. 'Feeding', which is a vaguer term, will be employed here in the sense of nutrition applied to animal production and, more specifically, for everything involving the estimation of rations and their distribution. The methods for manufacture of compound feed will be subdivided into formulation (development of formulae) and feed technology (manufacturing processes).

In one sense, nutrition concerns nutrients (and energy) while feeding is concerned with raw materials. Nutrients are intermediate between food and metabolites; glucose, amino acids and vitamins are nutrients. Proteins, carbohydrates or lipids are often mistakenly termed nutrients. They are in fact macronutrients, a little-used term. The term 'energetic nutrients' is used for macronutrients that can provide energy, but food energy is not in itself a nutrient. Some utilisable products, such as trash fish, are undoubtedly food in the current sense of the word. However, ingredients or raw materials of compound feeds are being progressively defined (and increasingly called foods without any further definition), which provide almost the whole of the feed for fish in modern rearing units. Nutritionists use the term 'diet' to refer to experimental feeding. In nutrition, as in feeding, ration always indicates an amount distributed and not a 'diet' or a food. Ration expressed per unit body weight is the feeding rate. Ration can also be expressed as a percentage of maximal ration, i.e. the ration ingested when the animal is fed to satiation; this is the feeding level.

2.1 NUTRITION: MAIN BRANCHES AND FUNDAMENTAL IDEAS

Nutrition includes several stages: behaviour and intake of food, digestion and absorption (digestive phase), nutrient metabolism (metabolic phase) and excretion or elimination of wastes.

The first stage relates to ethology or the science of animal behaviour. The animal manifests appetence (or appetite) for a food, when this is characterised by a palatibility. It should be noted that, in nutrition, the taking of food (ingestion or consumption) cannot be called absorption. This term is reserved for the passage of food or non-food molecules from the intestinal lumen (which is on the exterior of the organism, in a biological sense) towards the interior. Digestion of macronutrients is only the first stage of food utilisation before absorption. This utilisation is often described as digestibility, a widely-used concept that can be quantified using an equation involving digestion and absorption. The concept of digestibility is only applied to dry and organic matter and to macronutrients. For nutrients in the true sense of the word, the term 'availability' is used. The different digestibility coefficients will be defined in Chapter 4.

The metabolic stage consists of a series of biochemical processes, the complex regulation of which is dependent on genetic and physiological factors. Assessment of metabolic processes is generally described by retention (the difference between absorption and excretion) of a compound or element in the body mass. When retention corresponds to increase in tissue or chemically well-defined mass (body proteins, for example), the term used is 'accretion'. The term 'assimilation' (no longer used in nutrition) must be avoided since it is ambiguous, often being applied to digestibility. Retention and excretion, like absorption, are always described by absolute values; their efficiency is calculated using retention or utilisation coefficients which will be defined in Chapters 5 and 11. Excretion relates exclusively to compounds produced by the internal environment, elimination of faeces being a form of egestion (a strict, but rarely-used term) or faecal discharge, or even faecal emission. Synthesised products such as mucus or sexual products are classed as production. Accretion of body tissues is also a form of production.

Nutritionists have very rapidly been led to study qualitative requirements, i.e. the necessary molecules or elements which are termed essential or indispensable or, when they are not absolutely indispensable, conditionally indispensible or semi-indispensable. The term 'semi-indispensable' has been rejected by the purists but will be employed here in one or other of its accepted forms: a molecule that can be synthesised from an essential molecule, a molecule that can be synthesised only under certain conditions, or a molecule that is difficult to synthesise. Quantitative requirements (quantities required of these molecules or elements) can be defined in two different ways: the absolute requirements are given in daily amounts (for example mg.animal^{-1} day or mg.100 g animal day^{-1}) and the relative requirements in content, or ratios (for example % of diet or amount/MJ energy). The requirements for every animal can be broken down into maintenance and production requirements. The state of maintenance corresponds to a stage where the food is providing sufficient nutrients to the organism to achieve a zero energy budget. The analytical approach, where total requirements are calculated from maintenance and growth requirements, is much less useful in the rearing of fish and crustaceans than for terrestrial mammals; it has a certain amount of use to calculate rations. In order to be useful in feeding, a requirement must first be converted to a standard that corresponds to a range of requirements with various safety margins.

2.2 SOME CRITERIA FOR FEEDING EFFICIENCY

Animal science experts often describe feed utilisation efficiency using the relationship between food ingested and gain in body mass (commonly called weight gain). This coefficient, which in fact does not describe an efficiency but the reverse, is called feed conversion ratio (FCR) or feed : gain ratio (FGR). It should be noted that this relationship is not only a function of the nature of the food and of the animal, but that it depends equally on the quantity of food ingested. In fact, gain in body mass is closely related (linear relationship) to the amount ingested, but is not directly proportional since it is proportional to the amount ingested above maintenance requirements. In the aquaculture profession, the FCR is sometimes used when adding the mass of dead fish to that of body weight gain. Feeding efficiency is the inverse of the FCR and is a more logical and more convenient criterion than the latter, perhaps because it is finite around maintenance. It would be desirable to substitute it for the consumption index. Other criteria, such as the Protein Efficiency Ratio (PER) can be used to describe the utilisation of proteins. This relates only to the gain in biomass, while others, such as the Net Protein Utilisation (NPU) or the Protein Utilisation Coefficient (PUC), or Protein Productive Value (PPV) describe the gain in body proteins. Yet more coefficients are calculated using the amino acid composition of proteins: the chemical Index and the Index of indispensable amino acids (or the Oser Index). The definitions of all these measures are given in Chapter 6.

Note that the availability of a nutrient, as described above, is sometimes used to describe its maximal utilisation when it is a limiting factor (i.e. when its scarcity slows the animal's performance). Availability can be a function not only of digestive processes, but also metabolic processes. It is therefore important to define the method used for its determination.

2.3 DESCRIPTION OF GROWTH

The main production in fish is that of flesh, which can be evaluated simply by monitoring growth in mass (commonly called weight). The instant criterion for measuring production is gain in mass (over a given time). Over short time periods, increase in biomass can be considered as a linear function of time, and Growth Rate is often defined as the increase in mass expressed as a percentage of the mean mass, while Daily Growth Rate is a growth rate per unit of time. Over longer periods, it is no longer possible to assume that growth is linear and the nutritionists very often use an exponential model, $M = M_0e^{kt}$ where M is mass and t, time. The exponent of the curve, k, multiplied by 100, is called the Specific Growth Rate (SGR) and describes the growth. However, the real growth curve is only exponential over a brief period, for example, during larval stages. In consequence, the best proposed adjustment corresponds to a power function (see the Muller-Feuga model, 1990, in Box 2.1). Instantaneous growth is proportional to the body mass M raised to the power r. Exponential growth is a particular case for this model, where $r = 1$. In general, when

Box 2.1 Description of growth using the Muller-Feuga (1990) model

General equation (in differential form)

$$\frac{dM}{dt} = \Gamma f_1(M) f_2(\theta) f_3(r)$$

where M = body mass
t = time
θ = temperature
r = feeding level = ration expressed as a percentage of maximal ration
Γ = correction factor, mainly a function of environmental quality, technical expertise of the farmer and genetic and physiological characteristics of the animals

Somatic function

$$f_1 = \Gamma M^r$$

Integration of this function gives a simple and general specific growth index (SGI):

$$\frac{M^{1-r} - M_0^{1-r}}{(1-r)(t-t_0)}$$

where M_0 and t_0 are the values of M and t at time 0.

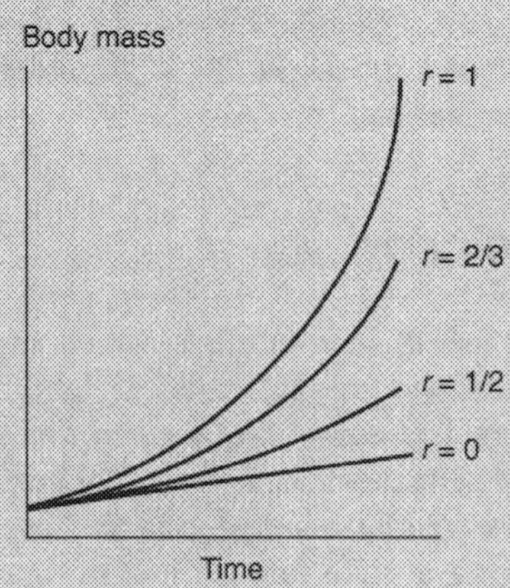

Temperature equation

$$f_2 = m^{\alpha(\theta_M - \theta)} - m^{\beta(\theta_M - \theta)}$$

where θ_M = maximum temperature tolerated by the animal
m = maintenance
α and β = coefficients

N.B. The equation gives realistic figures close to θ_M, but less realistic ones at low temperatures; below θ_1, the curve is close to a straight line joining the point of zero growth and the point θ_1 on the abscissa.

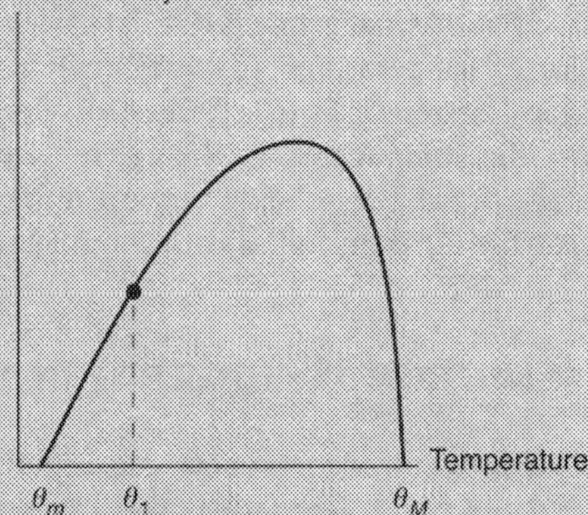

Ration equation

$$f_3 = \frac{Z(r)}{Z(r_s)}$$

where $Z(r)$ is growth rate expressed as a percentage of growth rate achieved with maximal ration. r_s is the level of ration giving the best compromise between maximal growth and FCR and m = maintenance ration.

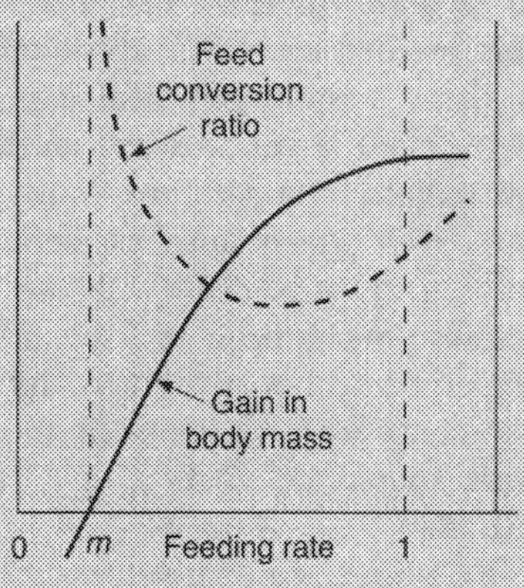

calculated by fit, and according to species and physiological stage, r lies between 0.5 and 0.9. Growth is also proportional to a thermal function (Chapter 5) and a food function and to a correction parameter. Instantaneous growth is proportional to body mass M raised to the power r with a value of between 0.5 and 0.9.

It has recently been proposed that a new parameter, Daily Growth Index (DGI), should replace SGR. It is derived from a simpler model than the Muller-Feuga model, but is relatively similar. It is assumed that growth is proportional to body mass raised to the power 2/3. This coefficient allows a better comparison of the growth rates of different groups than SGR, since it varies little whether the body mass is 1 g or several kilograms, even though, strictly speaking, the exponent should be adapted for every case. Although still not very well known, the DGI is already used by some researchers and aquaculture experts.

Another index, the Growth Coefficient (GC), has been proposed. This is the DGI divided by the 'sum of temperatures' (day-degrees), a concept widely used in plant physiology but which corresponds to a heresy in purely scientific terms. The growth coefficient allows the comparison, in theory, of groups reared at different temperatures, but it has the drawback of being very open to criticism in the way it is defined. The formulae for the different indices are summarised in Box 2.2.

2.4 SOME GUIDELINES FROM FISH NUTRITION EXPERIMENTS

It is not the aim of this book to consider statistics and experiments. Nevertheless, as experiments, often carried out in the laboratory, will frequently be referred to, the opportunity should be taken to make certain points.

The object of a nutrition experiment is to test a hypothesis; for example, is the addition of a certain amino acid beneficial to growth, or does the addition of a particular fatty acid modify the composition of body fats, or does the suppression of a certain vitamin affect survival? It can also be aimed at estimations of scale, such as the determination of the quantitative nutritional requirements or the estimation of a digestibility coefficient or an economic parameter. Hypothesis testing carries with it two types of risk: first and second type risks, called α and β respectively by the statisticians. The conclusions reached are never absolute and can be tested again with a different choice of risks, a more powerful experimental design or even modified conditions. No experiment is valid except under well-defined conditions (species, physiological stage, salinity, temperature, tank size, water turnover, test duration, nutritional context, etc.) which must be defined by the experimenter. Extrapolation to other conditions must be dealt with case by case.

The duration of aquaculture experiments is fixed in relation to to biological scale. Thus, for a growth experiment, it is advised to aim for a tripling in biomass during the trial. For experiments with vitamins to monitor the importance of reserves, very young animals should be used, with a tenfold increase in initial biomass sometimes being attained.

The number of fish to be used poses a completely different problem. It is rarely possible to rear fish individually. It is possible, however, to monitor the growth of

Box 2.2 Main coefficients used to estimate growth in fish and crustaceans

Daily growth rate:

$$DGR = \frac{Mf - Mi}{tMi} \times 100$$

Specific growth index*:

$$SGI = \frac{Mf^{1-r} - Mi^{1-r}}{t(1-r)} \times 100$$

Specific growth rate (where $m = 1$):

$$SGR = \frac{LnMf - LnMi}{t} \times 100$$

Daily growth index (where $m = 2/3$):

$$DGI = \frac{Mf^{1/3} - Mi^{1/3}}{t} \times 100$$

Growth coefficient:

$$GC = \frac{Mf^{1/3} - Mi^{1/3}}{\Sigma t\theta} \times 100$$

where Mi and Mf are the mean initial and final individual masses
t = duration of the period in days
Ln = Naperian logarithm
$\Sigma t\theta$ = 'sum of temperatures' = sum of average daily temperatures in °C.

* Muller-Feuga (1990). This index can only be applied after the value of r has been determined.

individual fish using marking, although this step cannot be extended to individual ingestion and hence utilisation of nutrients. Work is thus carried out on groups whose overall performances are recorded and analysed. It is these groups, and not the individual animals, which constitute the experimental units. While this method is entirely valid in itself, it requires certain reservations and precautions: thus feeding efficiency, a ration measured for a group, is not equal to the mean feeding efficiency of individuals in the group; for mortality, corrections are possible for biomass but difficult for feeding efficiency.

The number of replicates (groups per treatment) required can, in theory, be calculated fairly simply for first and second type risks, chosen *a priori*. If the order of magnitude of the difference (D) which is being sought for a given criterion is known, and also the coefficient of variation (cv) of the same criterion, one can refer to tables giving the number of repetitions required in relation to the ratio D/cv. This number will be high when the sought-after level of precision is high, or when the animal material is heterogeneous for the criterion being studied. In cultured fish populations, the cv of biomass is often twice as high as that for terrestrial vertebrates. It is possible to reduce it by grading and obtaining very homogeneous groups with a cv (intergroup) of 3, 2 or even 1%. If the formula indicated above is applied to these values it provides a greatly reduced number of repetitions, 4 or 3 for example. However, the cv to be taken into consideration is the final cv, which for artificially homogeneous groups is always much larger than that at the start as a result of the reappearance of the social hierarchy. In practice, this method is hardly ever used and the 3, 4 or 5 repetitions used more-or-less systematically by a large number of researchers are far from guaranteeing an ideally accurate test. The second type of risk (concluding that there is no difference when in fact a difference does exist) is very often underestimated. It should be remembered that cv varies not only in relation to populations and rearing methods, but also in relation to the studied criteria. These are generally classed in increasing order for the following categories: body protein content or digestibility coefficient, feeding efficiency, biomass gain and enzyme activity; the same protocol may appear insufficient for some categories and excessive for others. It is therefore necessary to take great care in the interpretation of negative results from certain experiments and *a fortiori* with the possibilities of extrapolating from the laboratory into actual practice. In contrast, observations made in the field, even though they are carried out over long time periods and on very large numbers of animals, very rarely correspond to rigorous comparisons allowing statistical tests and definitive conclusions.

2.5 BIBLIOGRAPHY

Altman, P. L. and Dittner, D. S. (1974) Nutrition, digestion and excretion. In: *Biology Data Books*, 2nd ed., pp 1433–1446, Bethesda, Maryland.

Castell, J. D. and Tiews, K. (1980) *Rapport du groupe de travail de la CECPI, de l'UISN et du CIEM sur la normalisation de la méthodologie dans la recherche sur la nutrition des poissons*. United Nations Organisation for Food and Agriculture.

Cowey, C. B. (1992) Estimating requirements of rainbow trout. *Aquaculture*, **100**, 177–189.

Muller-Feuga, A. (1990) *Modélisation de la croissance des poissons en élevage*. Scientific and Technical Report 21, IFREMER, Plouzané, France.

Philippeau, G. (1984) *Puissance d'une expérience. Nombre de répétitions nécessaires pour comparer deux ou plusieurs traitements*. ITCF, Paris.

Schumacher, A., Gropp, J., Tacon, A. G. J., Hilge, V. and Rosenthal, H. (eds) (1995) Proceedings of the EIFAC workshop on methodology for determination of nutrient requirements in fish. *J. Appl. Ichthyol.*, **11**, 129–400.

Part II

Nutrition of Aquatic Animals

3

Feeding behaviour and regulation of food intake

In aquaculture, good feeding management is demonstrated by a close correspondence between the amount of feed distributed and the amount ingested. However, it is not easy to attain this objective. In many cases, the amount of feed distributed exceeds the requirements, leading to excess expenditure by the producer and water pollution. In other situations, the amount distributed is lower than the demand, resulting in the limiting of fish growth and often increased heterogeneity in fish sizes. The amount of food likely to be ingested is influenced by numerous cues which can start, prolong or stop the ingestion of feed. These cues can result in significant changes in feeding behaviour in the short, medium or long term.

3.1 CENTRAL AND PERIPHERAL REGULATION OF FEED INTAKE

Although many studies have been carried out on fish nutrition, few have been concerned specifically with the mechanisms of voluntary intake. In homeothermic vertebrates, it has been clearly demonstrated that an organism integrates information from its immediate environment. The animal consequently modifies its voluntary intake via central or peripheral regulation.

- Central control is achieved by the hypothalamus which plays a primordial role in this area. With its appetite and satiation centres, this region of the brain integrates information provided by various physical, metabolic or endocrine signals. It is the origin of stimuli or inhibition of food intake by means of the central nervous system (Figure 3.1).
- Peripheral control encompasses several ways of controlling ingestion, such as humoral type signals which inhibit feeding. These signals, emitted by the digestive tract or the brain (peptides such as cholecystokinin, bombesine or somatostatin) can result in cessation of feed intake or food regurgitation. They form the main basis of short-term regulation.

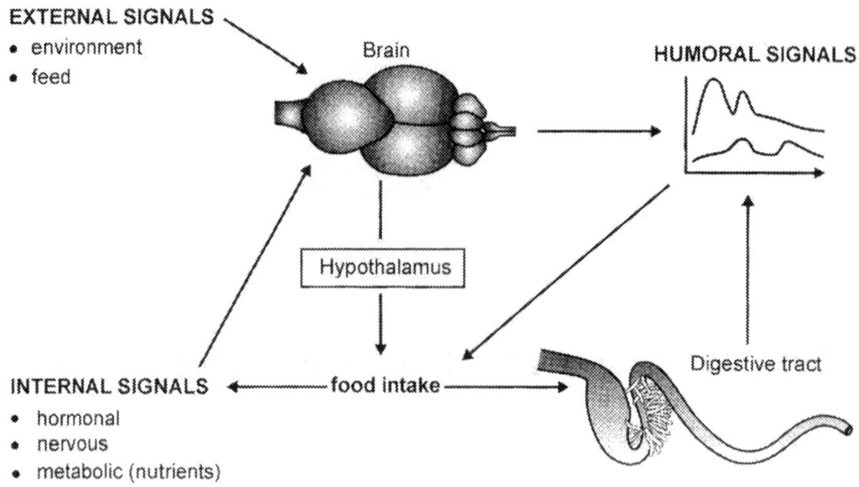

Figure 3.1. Simplified diagram of the neuro-hormonal regulation of appetite.

Some studies have shown that these complex systems act together to maintain body mass at an equilibrium, in relation to age, size, sex and season. The result of this regulatory process, and the data which is of interest to the fish farmer, is the amount of food voluntarily ingested per unit time and per unit biomass.

3.2 TECHNIQUES FOR MEASURING VOLUNTARY INTAKE

Manual feed distribution until satiation is reached has often been employed in order to measure objectively the amount of feed ingested voluntarily. This method appears to be simple but in fact depends on the unique perception of the experimenter as to when satiation is reached in a fish. It also requires the arbitrary imposition of food distribution patterns (frequency, timing and duration of meals) which inevitably are not in phase with the natural feeding rhythms unique to the species being studied. Another method consists of distributing an excess of feed containing a known proportion of a substance identifiable by X-ray (very small glass beads, for example), or a radioactive label. This method has its advantages: it eliminates errors arising from the loss or regurgitation of feed particles and allows the study of inter-individual variation in ingestion, but also requires either handling of sub-samples of fish after each meal (X-rays) or slaughter (radioactive labels). A final method consists of conditioning the animals to self-feed by manipulating a lever on a feed dispenser. These 'demand' feeders give total freedom to the fish in their choice of timing, frequency and amount of food. As a result, demand feeding is perhaps the only method of ensuring *ad libitum* feed intake. However, some species, such as sole or sturgeon, can be difficult to condition to the feeders. In addition, the amount of feed distributed in response to each demand must be carefully adjusted for each situation (species, density ...) and it is necessary to check regularly that all distributed feed is consumed.

Using different measurement techniques, it is possible to demonstrate the importance of different factors at the basis of variations in voluntary ingestion. Whether they are integrated by the organism centrally or peripherally, these factors can be grouped in relation to their nutritional, environmental or behavioural origin.

3.3 FACTORS AFFECTING VOLUNTARY INTAKE

3.3.1 Nutritional factors

Many experiments have shown that fish are capable of distinguishing between two different feeds. For example, using demand feeders, it has been shown that after two days of learning trout can differentiate between feeds based on casein and those based on fish meal, showing a marked preference for the latter. Under certain conditions, the fish can also differentiate between feeds that either have or lack an essential nutrient. Thus, after several days trout are capable of detecting the inversion of feeders containing feed with or without zinc. Similarly, sea bass can discriminate between feeds containing different levels of methionine. This capacity of fish to distinguish between two feeds and express a preference for the better balanced one has unfortunately only been observed in a few species, and it would be dangerous to make generalisations. Furthermore, this choice does not predict what might happen if only a deficient food is presented. There might be either an increase in voluntary intake or, in contrast, refusal to feed, depending on the degree of the deficiency and the nature of the nutrients.

When the distributed food is formulated so as to have no deficiencies, the amount of food ingested is regulated in relation to the energetic requirements of the animal, based on the digestible energy content of the food (Figure 3.2). This observation has been verified for several species, but must be modified in relation to the energy content of the diet and the nutritional history of the animals. In practice, as few data are available about this subject, it is not certain whether this regulation is very efficient the feed is particularly low in energy or when it is energy-rich (Chapter 22). It can also be observed, at least in species where the stomach is well defined, that following a short period of food deprivation to ensure the complete emptying of the stomach, the first meal taken corresponds to the volume of the stomach. Nerve receptors, sensitive to the pressure exerted on the stomach wall, limit maximal ingestion and contribute to the satiation mechanism. Regulation of ingestion on the basis of digestible energy only comes into play after the second or third meal. Some authors have also observed that, during the first days after a period of fasting, the energy ingested depends directly on the nutritional state of the fish: the greater the lipid reserves, the less voluntary ingestion occurs and vice versa.

Finally, the presence of attractants (Chapter 18) may overcome a lack of appetite or stimulate feeding. Such substances are frequently used in larval feeding since they can improve first-feeding (weaning) results (Chapter 13).

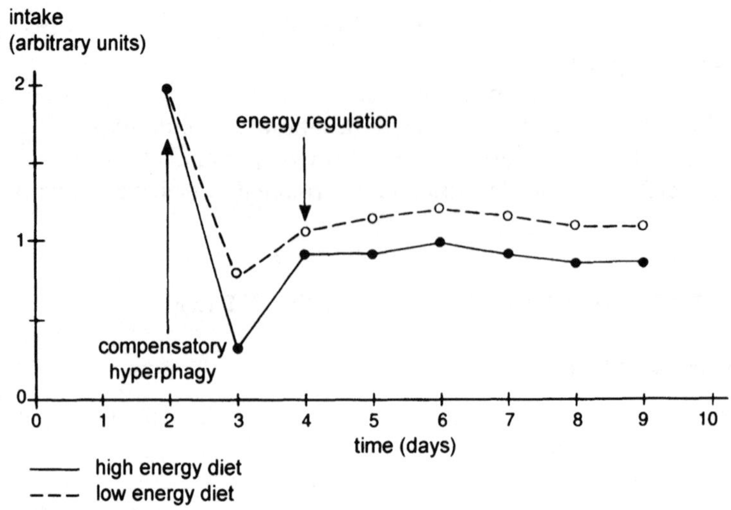

Figure 3.2. Change in intake of two diets with different energy contents, after a starvation period of several days.

3.3.2 Environmental factors

In ectotherms, and particularly in fish, temperature is the main environmental parameter influencing voluntary feed intake. For each species, there is a thermal optimum at which voluntary intake is maximal. In general, when the temperature exceeds this optimum, intake remains high for a while before falling sharply when a certain thermal threshold is reached. When temperature decreases, voluntary intake becomes progressively lower until another threshold is reached, below which the fish will not feed. In fish with a well-defined stomach, such as the salmonids and siluriforms, temperature also affects the time interval between feeds: there are longer intervals between feeding episodes at low temperatures than at high temperatures. This interval can be directly correlated with stomach emptying times, which is inversely proportional to temperature.

Fish are also very sensitive to the chemical characteristics of the water in which they are living. In this case, the behavioural response is usually 'all or nothing'. Thus, for the common carp, the presence of high levels of ammonia or nitrites blocks feeding well before causing fish mortalities. Similarly, in the trout, a dissolved oxygen level of less than 6 mg/l can stop feed consumption.

3.3.3 Behavioural factors

Temporary inhibition of feeding is often observed following various forms of stress such as handling the fish or the presence of predators in the direct vicinity of the cages. When the cause of stress is removed, feeding inhibition usually only lasts for a few hours. When the animals are fed *ad libitum*, this period of inhibition is followed

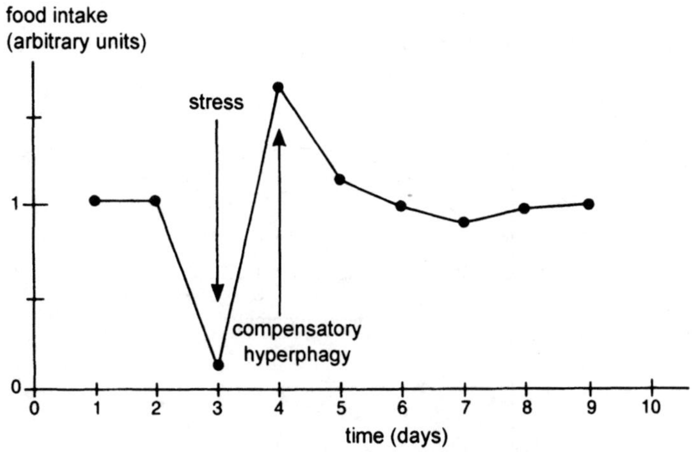

Figure 3.3. Change in voluntary food intake following a brief period of stress (e.g. capture–release).

by a short phase of compensatory increased intake, which cancels out or decreases the negative effect of stress on growth (Figure 3.3).

Some factors can have a long-term effect on feed consumption and in these cases the term 'chronic stress' would be used. For example, if the rearing density is too low, inter-individual relationships form between the fish and can lead to the formation of a hierarchy of individuals, resulting in the appearance of dominant and dominated relationships. This situation leads rapidly to high heterogeneity in growth rates. In contrast, in certain species such as sea bream and red porgy, if rearing density is too high the fish are forced to swim closer to their neighbours. In this situation, food consumption can decrease following the reduction in living space.

Independently of all apparent stress, temporal variations in voluntary intake of food in relation to the environment have been observed over different time scales (Figure. 3.4). The effect of temperature on variation in food consumption on a seasonal scale is well known. More controversial is the effect of photoperiod on food consumption, studied at a constant temperature. A consensus is developing, however, that it is not daylength in itself but change in daylength which affects food intake. Thus, for example, in the Atlantic salmon food intake is stimulated by increasing daylength (in spring). Significant variations over time periods of a fortnight or a month, linked to lunar cycles, have also been observed by several authors. Voluntary food intake also varies from day to day, even in stable laboratory conditions; an explanation is rarely found.

The final time scale for variation in food intake, which is by far the most thoroughly documented, is the diurnal pattern. The various studies on this subject allow the conclusion to be reached that most species show marked phases of feeding activity in relation to the day/night cycle. Some fish feed preferentially during the day (diurnal species, such as the trout). Their feeding activity is often characterised by a period of increased intake associated with the early hours of the day and occasionally a second period just before dusk. Other species, such as the sea bass or common

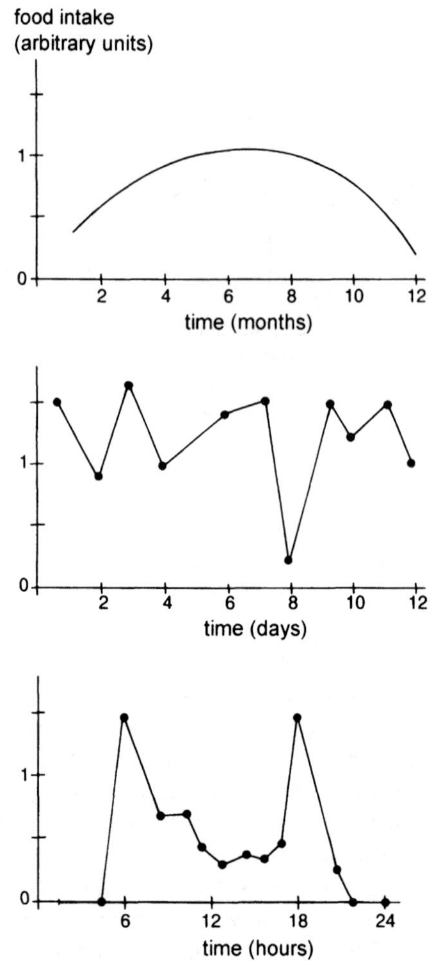

Figure 3.4. Temporal variations in voluntary food intake over different time scales.

carp, have the capacity to readily modify their feeding pattern and feed in the daytime at one time of the year and become nocturnal at another. It is usually difficult to detect the reason for these changes, which are observed even when food is always available. Some authors have shown that feeding activity in the Atlantic salmon, which is generally diurnal, becomes nocturnal when the temperature falls below 10°C.

Given the existence of feeding rhythms in fish over different time scales, questions must be asked about the control of these rhythms. Is control endogenous, or purely environmental? The data available are rather fragmented and it is difficult to prioritise clearly the significance of the various potential synchronisers. At the current state of knowledge, it appears that the main environmental synchronisers are the alternation between day and night, lunar cycles, photoperiod and temperature.

However, an internal clock, allowing at least partial endogenous control of these rhythms, does exist. Several authors have shown in the laboratory that the trout, sea bass and Channel catfish can develop their own feeding rhythm with a time period close to 24 h, when subjected to alternation in phases of light and dark with a periodicity of about 1 h.

The practical or physiological consequences of the imposition by the fish farmer of a more-or-less arbitrary feeding rhythm are still almost unknown. It appears, however, to be *a priori* advisable to base food distribution, as far as possible, on the species' own food consumption patterns or, preferably, to use demand feeding.

3.4 CONCLUSION

Many different factors affect voluntary feed intake, inducing responses which are also diverse, ranging from rejection of food through temporary cessation of ingestion to hypophagy or hyperphagy. Autoregulation of voluntary feed intake is influenced by nutritional factors among other things, and above all by the energetic content of the food. It also follows rhythms over different time scales.

These variations render the management of food distribution complex, and provide a good argument for the development of techniques for demand feeding.

3.5 BIBLIOGRAPHY

Ali, M. A. (1992) *Rhythms in Fishes*. New York, Plenum Press.

Fletcher, D. J. (1984) The physiological control of appetite in fish. *Comp. Biochem. Physiol.*, **78A**, 617–628.

Jobling, M., Arnessen, A. M., Baardvick, B. M., Christiansen, J. S., and Jørgensen, E. H. (1995) Monitoring feeding behaviour and food intake: methods and applications. *Aquaculture Nutrition*, **1**, 131–143.

Talbot, C. (1993) Some aspects of the biology of feeding and growth in fish. *Proc. Nutr. Soc.*, **52**, 403–416.

Westertep-Plantenga, M. S., Steffens, A. B., and Tremblay, A. (1999) Regulation of food intake and energy expenditure. Milan: DRA.

4

Digestive physiology and nutrient digestibility in fishes

The digestive physiology of fishes is based primarily on a large number of, sometimes very old, anatomical studies. Much more recent purely physiological studies have examined areas as widely varying as regulation of digestive transit times, molecular structure of enzymes, organisation of absorptive cells and the nature and role of hormones in the digestive tract. In addition, studies of overall digestive budget, or digestibility, have provided results that complete the work of the physiologists and are of prime importance for applied nutrition and the formulation of feeds.

4.1 ANATOMY AND PHYSIOLOGY OF THE DIGESTIVE TRACT

4.1.1 Anatomy of the digestive tract

The class of fishes, like all other vertebrate classes, shows great diversity in the digestive tract, far in excess of that of the external anatomy. Zoologists try to cite their observations in relation to two main axes: the evolution of the species on the one hand and adaptation to diet on the other. Only a few basic features of the anatomy and function of the digestive tract will be given in this book.

4.1.1.1 General features

Very primitive fish (elasmobranchs including sharks and rays, and chondrosteans including sturgeons, for example) have a very specific type of digestive tract. A well-differentiated stomach is followed by a normal intestine, followed by an intestine containing a unique structure: the spiral valve. This structure is derived from the spiral folding of the intestine followed by the partial fusion of the epithelium, resulting in a structure reminiscent of a spiral staircase. It allows high absorption capacity for a short length of intestine.

As in other phyla, teleosts have mouths which are adapted to their feeding behaviour. Digestive tract anatomy is fairly homogeneous in the primitive families, but is variable in the more evolved families which show plasticity with regard to food, sometimes including types that appear entirely homogeneous from the point of view of external anatomy. There are types with or without stomachs; types with short or long oesophagi; with straight or long, contorted intestines; many intermediate types having different forms of folds or coils. The number of pyloric caeca varies from none to nearly a thousand.

From other points of view, the digestive apparatus of fish has some constant features:

- the oesophagus always has a striated musculature, allowing the animal to regurgitate later;
- the intestine resembles the small intestine of mammals, or a bird's intestine and shows a certain amount of differentiation, reminiscent of the succession of the duodenum, jejunum and ileum of mammals, although there is no true colon;
- the intestinal wall has only three layers: mucous, muscular and serous, since the absence of '*muscularis mucosae*' impedes the distinction between mucosa and sub-mucosa;
- the intestinal villi are not well differentiated and are more like pseudo-villi or folds;
- some enterocytes retain their endocytotic function, which is characteristic of newborn mammals, throughout the life of the fish;
- the provision of digestive enzymes is relatively close to that found in the higher vertebrates.

4.1.1.2 Main organs

Mouth The mouth is used both for intake of water and ingestion of food. Its function is complex owing to the non-compressible nature of the surrounding environment and by the presence of the gills. There is coordination between the jaws, the 'tongue', the roof of the mouth, the gill arches and the opercula. Teeth, when present, are rarely used for mastication. They can be used for catching and holding prey (buccal, maxillary, lingual or vomerine teeth) or filtration (pharyngeal teeth, or those associated with the gills which retain particles and divert them towards the

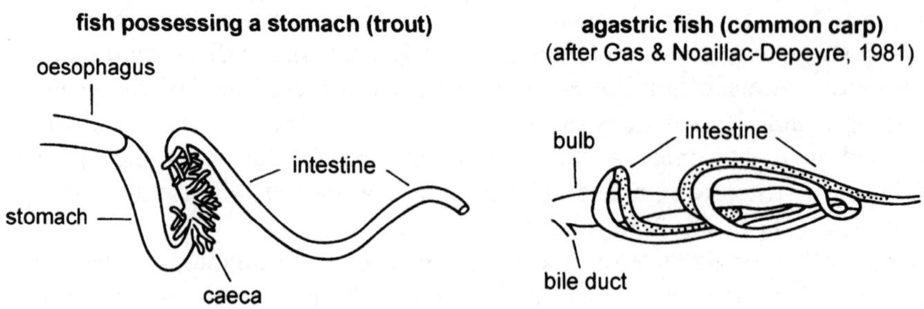

Figure 4.1. The two main types of digestive tracts found in fish.

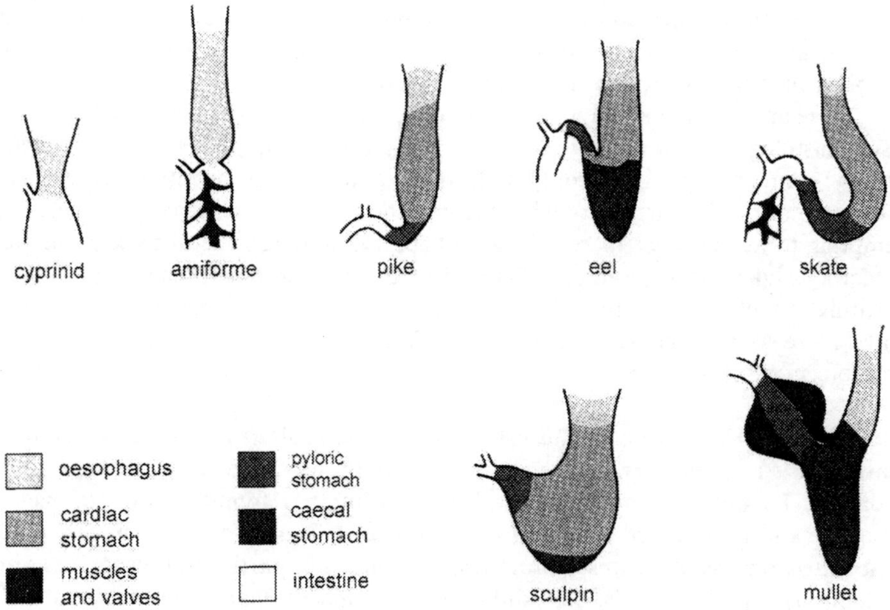

Some types of intestine
(after Bertin, 1958)

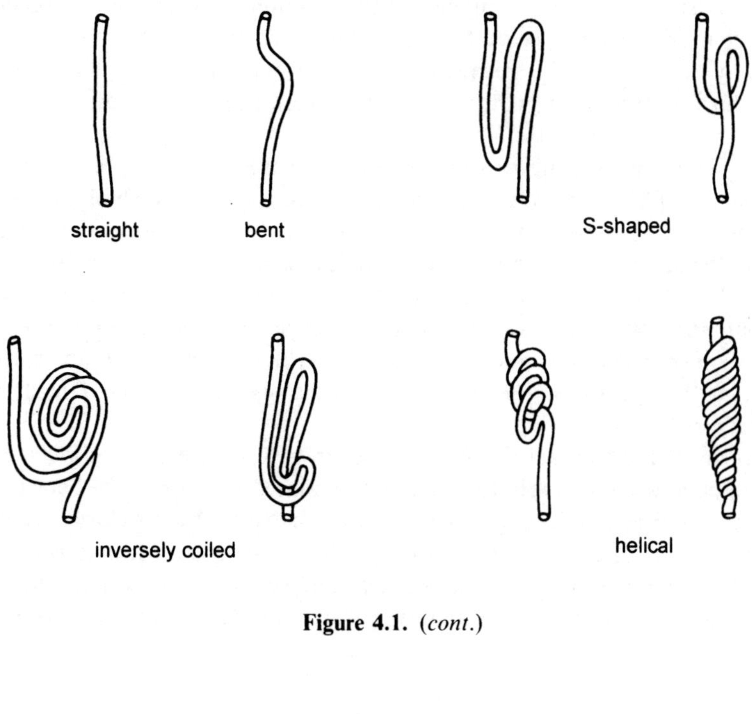

Figure 4.1. (*cont.*)

stomach). There are two main types of behaviour: capture of prey and suction; filtration is a sort of extension of suction and is used for food collection. Some fish have teeth for chewing. Cyprinids have pharyngeal teeth on the lower jaw and a masticating plate on the upper jaw, sea bream can crush shellfish and some herbivorous fish have special masticatory systems.

There are no salivary glands in the mouth. However, taste buds have been found using histology and scanning electron microscopy techniques and their function clearly demonstrated by feeding behaviour and applied nutrition experiments. There are also other organs outside the mouth that have chemoreceptive cells; it is tempting to assume that these cells are olfactory, although the distinction between olfactory and gustatory senses is blurred in the aquatic environment. The role of locating attractants is attributed to external chemoreceptors while gustatory papillae have more varied functions, ultimately linked to feeding behaviour, cessation or reinforcement of feeding.

Oesophagus The oesophagus in fish is nearly always short and wide, which in the past resulted in authors stating that the mouth was directly connected to the stomach. Its mucous membrane is distinct from that found in the stomach or intestine and is equipped with many mucous cells. The oesophagus in fish is very distinctive because of its musculature which is wholly or partly striated and contracts voluntarily. In other words, the fish has greater capacity than higher vertebrates for regurgitating food if it is not of suitable size, texture or taste. In some predatory species, the musculature has evolved to apparently allow the holding of prey which is still alive. Long oesophagi are found in some fish such as the eel, where some parts perform an osmoregulatory function during catadromous migration.

There is usually a cardiac sphincter at the junction between the oesophagus and the stomach in freshwater species, but not in saltwater species. This difference can be explained by the fact that the freshwater fish tries to minimise its water intake, while the saltwater fish must drink continuously to ensure osmoregulation.

Stomach The stomach is never present in the larval stage and appears around the time of metamorphosis, or sometimes later. Certain families, such as the cyprinids, have no stomach; in other families, it is absent only from certain genera or species. Sometimes the absence of a stomach has a logical explanation: fish feeding on corals ingest such high levels of calcium carbonate that it would be difficult to reduce pH to a level compatible with normal functioning of pepsin. In general terms, fish without stomachs are most often microphages or herbivores rather than carnivores. However, in the cyprinids, for example, agastric species are found feeding on a wide variety of foods.

The form of the stomach is very diverse, from a simple bulge converted to an elongated pouch when it is full, to a well differentiated sac. It may be bent to form a Y or J shape and is usually separated from the intestine by a pyloric sphincter or valve. It is capable of considerable distension in carnivorous fish that swallow their prey whole; microphagous species have much smaller stomachs. Longitudinal muscle fibres predominate. Despite these anatomical variations, all stomachs are relatively homogeneous in histological structure and differ little, at least in function, from those of higher vertebrates. The epithelium is endodermic and there are numerous villi covered in secretory cells producing digestive proenzymes and hydrochloric acid. Other types of cells secrete hormones or mucus. Usually mucus-secreting cells are found only in the distal part.

In some herbivorous fish such as the mullets, the pyloric region forms a true differentiated organ with toughened walls, surrounded by very thin circular musculature, following a stomach which has lost its secretory functions. By analogy with birds, this organ has been called a 'gizzard' and performs a purely grinding function. The stomach is the only region of the digestive tract that contains acid which allows the denaturation of proteins, the start of hydrolysis and kills some ingested bacteria. The pH of the stomach is generally 2–3, but can reach around 5, especially in marine fish that drink water at pH8 continually.

Intestine The terminology used by anatomists to describe the intestines of fish is extremely confused. Sections such as the anterior, middle and posterior intestine, duodenum, jejunum, ileum often have different meanings depending on author (for example, one region called the mid-gut by one author would be called the distal region by another). This confusion arises mainly from the absence of marked external anatomical differentiation and makes the interpretation of various publications very difficult. Only cytology can be used to distinguish three or four regions which have distinct roles.

The proximal intestine, often called the duodenum, is present in fish with no stomach and has a bulge which is highly developed in some families such as the cyprinids. This organ can at first glance be confused with the true stomach and its function is to store food temporarily after feeding. It is distinguished from the stomach by its other functions: connection to the bile duct is always above the bulge while it is always below a true stomach; the pH in the bulge is never acidic; no gastric type secretory cells are found and there is no pepsic activity.

In those fish with stomachs, and only those, the proximal part of the intestine often has blind-ending diverticula called caeca. Numbers of these organs vary from zero to several tens or hundreds and reach nearly a thousand in the saithe. In certain families, such as the salmonids, the number of caeca is clearly distinct between species, depending on genetic factors and can be modified by selection. It also depends on water temperature during development. However, it does not appear to be correlated in any way with nutritional characteristics. Histological studies have not revealed any marked differences between the true intestine and the caeca, whose function is still poorly understood: they are not the site of any specific secretion or absorption; in cold water species, they do not shelter a special flora. In warm water species, however, flora that can actively synthesise vitamins can be found there, or even cellulolytic flora, such as that in *Kyphosus cornelii*, an Australian herbivorous fish which is related to the sparids. However, some strictly herbivorous fish do not have these diverticula. Currently, the great majority of ichthyologists consider the caeca as an extension to the surface area of the intestine and, as such, a 'strategy' to increase the absorptive surface area of the digestive tract.

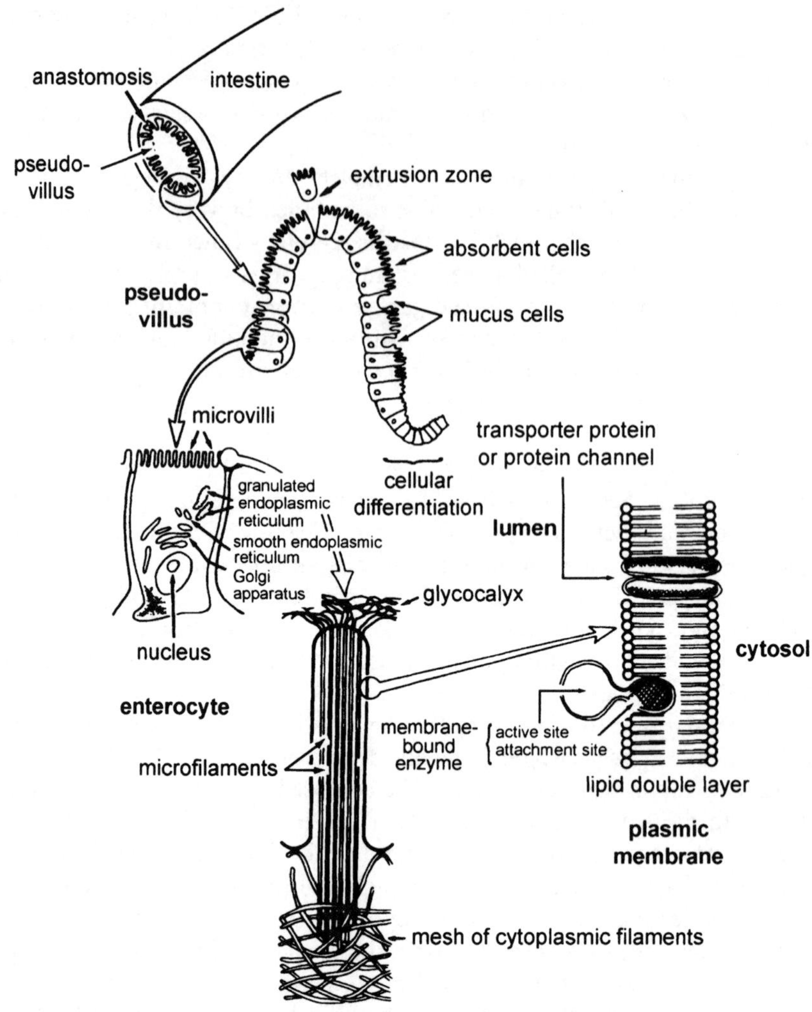

Figure 4.2. Diagram of the anatomy and histology of the intestine, from a section of the organ (macroscopic scale) to the molecular structure of the plasmic membrane of the microvilli.

Histologically, the intestinal mucosa is similar to that of the higher vertebrates, with pseudovilli increasing the surface area by approximately twentyfold (Figure 4.2.). Enterocytic cells multiply in areas equivalent to crypts, differentiating as they migrate along the pseudovillus into mucus or absorbent cells, degenerating when they finally reach the tip where they slough off. Enterocytes are clearly visible on the luminal surface which is covered in closely-packed microvilli which increase the surface area by a further factor of twenty. The microvilli have a series of polysaccharide filaments running through them which elongate in light, where they form a filtering network called the glycocalyx. Their membrane contains

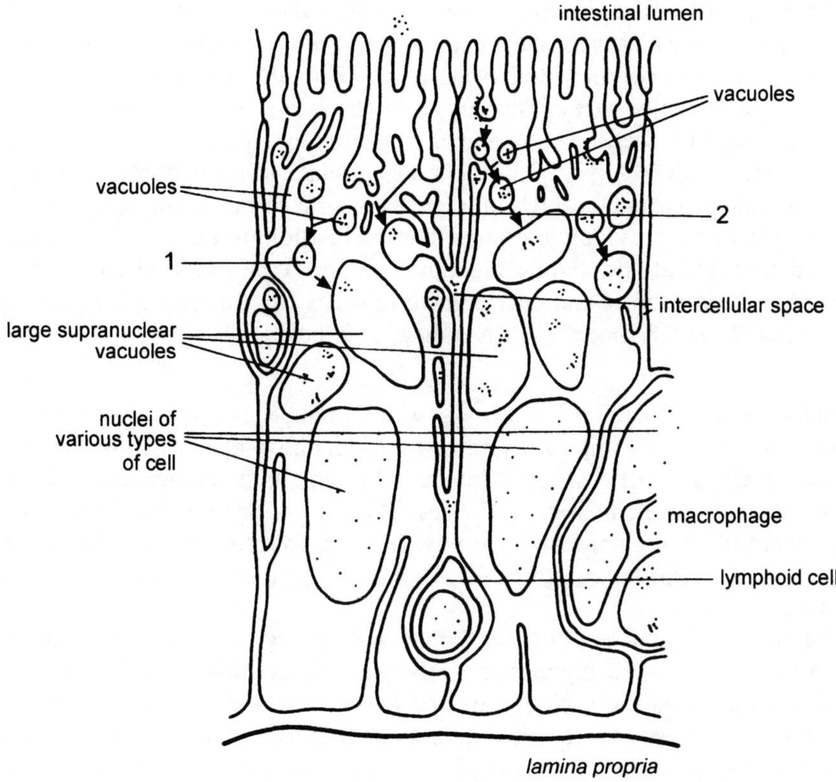

Figure 4.3. Endocytotic cells, characteristic of the posterior intestine in fish (after Vernier and Size, 1989). Protein macromolecules in the intestinal lumen penetrate via localised invaginations at the base of the microvilli. They form vacuoles which can either (1) fuse to form large supranuclear vacuoles, where the molecules are digested, or, (2) reach an intercellular space, where they come into contact with lymphoid cells and macrophages.

membrane-bound enzymes, transporter proteins and ionic channels. All these protein molecules are enclosed within the lipid double layer. Abundant formation of two types of lipid occurs in the cells of the duodenum after a meal: particulate lipids and 'spread-out' lipids. This area obviously plays an active role in the absorption of lipids.

The mid-gut, sometimes called the jejunum, is difficult to separate into anterior and posterior parts by its external appearance, although it is often more brown in colour than the proximal intestine. Histologically, it shows some notable adaptations: the enterocytes have many invaginations running through them at the bases of the microvilli. Between these and the nucleus are very large vacuoles (Figure 4.3), similar to those cells found in the new-born mammalian ileum that ensure the absorption by endocytosis of protein molecules and their intracellular digestion. In fish, these cells are present at all ages and evidently carry out absorption of proteinaceous material.

The distal part of the intestine, sometimes called the ileum, is usually very short. The enterocytes only have very short microvilli and show few of the characteristics of absorbent cells; however, they do have large numbers of mitochondria and are thus assumed to play a major part in osmoregulation by means of the selective absorption of mineral ions. This section is sometimes elongated by a preanal zone or a rectum.

Intestinal length is very variable. There is a relationship between the length of this organ and body length and with diet: in general terms, carnivores have short intestines and herbivores long intestines. This correlation (in the broadest sense) can be found throughout the animal kingdom, but must be qualified as a result of differences in transit times and development of caeca etc. Folding of long intestines takes various forms: S shape, inverse coiling, or helical.

Associated glands The liver is always well developed in fish. Its anatomy is extremely diverse, according to taxon, and it differs histologically from the liver of mammals because of the absence of lobules, although its functions are more-or-less the same as those in the higher vertebrates. There is abundant secretion of bile, the gall bladder only being absent very rarely. In some fish, the liver can become very large and fulfil a clear 'adipogenic' role, acting as a lipid reserve organ, as found in some birds.

In contrast, the pancreas does not form a discrete organ, except in some fish such as the elasmobranchs, the 'dipneustes' or the pike, for example. In almost every case there is a diffuse collection of cells around the 'duodenum' and around the caeca, when these exist; these islets or small glands can interpenetrate neighbouring organs and form a 'hepatopancreas' (as found in the common carp). Enzymatic secretions flow out through the bile duct. Endocrine cells (islets of Langerhans), which secrete insulin, exist alongside these digestive cells.

4.1.1.3 Other functions

The digestive tract, with its associated glands, is considered as the main endocrine organ in higher vertebrates. It is difficult to say if this is the case with fish, since their endocrinology has only recently been studied. In the wall of the digestive tract itself, endocrine cells secreting cholecystokinin (CCK), gastrin, cerulein and glucagon have been found, but the significance and role of these secretions remain poorly understood.

The role played by the digestive tract in active immunity is even more poorly understood. There are no true Peyers patches, which allow contact between lymphocytes and bacteria (role in active immunity), although this contact does clearly occur due to lymphocytes dispersed in the epithelial tissue of the intestine (Figure 4.3).

In some families (Cobitidae, Callichthyidae), intestinal respiration occurs, carried out by formations of capillaries which are reminiscent of pulmonary alveoli. Finally, there are some even more mysterious cells to be found in the epithelium of the digestive tract, which are described as pear or rod-shaped, depending on author, and are also found in the kidneys or gills. The hypothesis that they were some form of parasite or symbiont has now been abandoned. Their role may be secretory, but remains unknown.

4.1.2 Digestive enzymes

4.1.2.1 General

The digestive enzymes of fish, like those of all other vertebrates, fall into three main categories:

- enzymes secreted by the pancreas and, to a lesser degree, by the stomach, in the form of granules of zymogen or inactive proenzymes and mixed in a digestive juice with specific ionic composition and pH. These proenzymes are never of buccal origin in fish and are activated in the stomach and especially in the duodenum;
- membrane-bound enzymes that are found to a small extent in the chyme, but which under normal conditions only act when linked to a microvillus membrane. These enzymes are all intestinal, breaking down fragments of macromolecules filtered by the glycocalyx. They are situated close to transport systems, which ensure absorption into the enterocyte cytoplasm;
- cellular enzymes of the digestive tract, situated in areas other than the wall, in lysosomes for example. It is sometimes difficult to tell if these enzymes participate in extracellular digestion, since their activity can be detected *in vitro* after cell lysis.

The occurrence of digestive enzymes is not the same in all species. For example, pepsin is always absent in fish that have no stomach and chitinase has only been found in certain species. The organ that secretes certain enzymes can also vary from species to species; lipases and amylases have been found in the stomach. It should, however, be noted that research on enzymatic activity in segments of the digestive tract often only demonstrates qualitative characteristics. More confusion arises from peristalsis, prey enzymes and intestinal flora activity.

Enzymes produced from zymogens (gastric and pancreatic) The main enzymes of this type are listed in Table 4.1. Proenzyme activation occurs by the breaking of a peptide which masks the active site. In the case of pepsin, hydrolysis is started by the action of hydrochloric acid and is autocatalytic (hydrolysis of pepsinogen by pepsin). In the case of pancreatic enzymes, there is initial activation of trypsinogen by enterokinase, then the commencement of a chain of events by the action of trypsin on chymotrypsinogen, the following enzymes being activated successively: elastase, collagenase, carboxypeptidases A and B, phospholipase and colipase. With the exception of the lipase-colipase couple, these enzyme types then act on macromolecules through internal splitting ('endohydrolysis'), creating smaller molecules which cannot be absorbed in this form.

Proteases The proteases are represented by a gastric enzyme, pepsin, and four pancreatic enzymes: trypsin, chymotrypsin, collagenase and elastase. The latter are

Table 4.1. Main enzymes secreted in the form of proenzymes in fish.

	Enzyme	Activity	Organ	Species
Proteases	pepsin*	hydrolysis of internal bonds	stomach	species with stomachs
	trypsin	"	pancreas	all species
	chymotrypsin	"		
	elastase	"		
	collagenase	"		
Peptidases	carboxypeptidases A and B	hydrolysis of external bonds	pancreas	all species
	carboxylesterase	hydrolysis of peptides		
Glucosidases	amylase	hydrolysis of starch ($\alpha 1 \rightarrow 4$ bonds	pancreas	all species
	chitinase	hydrolysis of chitin	stomach pancreas other tissues?	species that eat mainly insects or crustaceans
Lipases	pancreatic lipase colipase	hydrolysis of triacylglycerols (especially at α position)	pancreas	all species
	esterases	hydrolysis of triacylglycerols and other lipids		
Nucleases	ribonuclease	hydrolysis of nucleic acids	pancreas	(poorly understood)

* A chymosine has been found in several species; it is an isoenzyme of pepsin which coagulates milk.

endopeptidases, belonging to the family of serine proteases, enzymes which possess the same catalytic site, containing a serine residue. They are complementary in action, each one acting on well-defined and different peptide links (Table 4.2). Depending on the amino acid (AA) sequence in the peptide chains, hydrolysis results in free amino acids or peptide residues of very variable size that will be hydrolysed, often very incompletely, by peptidases.

Lipases The structure of lipase is very similar to that of the serine proteases and, along with amylase, it is the only pancreatic enzyme to be secreted directly in its active form. It is only significantly active in the presence of another molecule, colipase, which is secreted by the same organ. The main role of colipase is as

Table 4.2. Sites of activity of the main proteolytic enzymes found in fish.

Enzyme	Bond hydrolysed
Pepsin	NH_2 of aromatic diacids and amino acids
Trypsin	COOH of arginine or lysine
Chymotrypsin	COOH of aromatic amino acids
Elastase	aliphatic amino acids (especially active on elastane)
Carboxypeptidases	amino acids with free COOH group
Aminopeptidases	amino acids with free NH_2 group
di and tri-peptidases	bonds in di- and tripeptides

follows: in contrast to those enzymes that act in an aqueous environment, lipase acts on lipid compounds which it can only attack at the interface of globules suspended in the aqueous phase. The presence of bile salts, while increasing the interface by raising the surface tension, renders the 'positioning' of the lipase on the lipid droplets very difficult. Colipase has a double affinity and thus allows the formation of 'lipid droplet–lipase–colipase' complexes, where the lipase is to some extent anchored to the lipids (Figure 4.4). Recent studies have shown that, after attachment to lipase, colipase uncovers the catalytic site of the first enzyme. The considerable activation (*sensu lato*) of the latter thus results from two distinct phenomena. Fish lipase differs from that of higher vertebrates in the sense that it shows a similar affinity for the different fatty acids of triacylglycerol (triglyceride) instead of being especially active on those in α position (external). Lastly, although bile salts play a positive role in lipid emulsification, or in aiding their dispersal in the aqueous environment, their

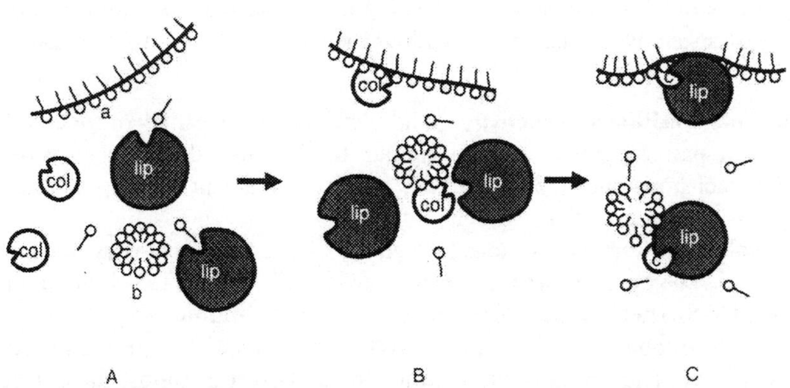

Figure 4.4. Mode of action of lipase and colipase. In A, lipase and colipase are free and bile salts (b) cover the surface of the lipid globule (a), so the lipase cannot act. In B the colipase attaches to the globule. In C, lipase is anchored due to the colipase and can hydrolyse the triacylglycerols (after Pérez *et al.*, 1986).

main role is further on, after the hydrolysis of triacylglycerols by the lipase-colipase complex.

The other lipolytic enzymes in fish are poorly understood. Esterase and phospholipase activity are frequently found. It is difficult to confirm if this type of enzyme is secreted by the stomach or even the pancreas, because in general lipases tend to be non-specific.

Glucosidases The main glucosidase is $\alpha 1 \rightarrow 4$ glucosidase, or amylase, which can hydrolyse $\alpha 1 \rightarrow 4$ bonds in amylose or linear fragments of amylopectin or glycogen, but not the $\alpha 1 \rightarrow 6$ bonds that form the branches in the latter molecules. This enzyme has been found in all fish, including the carnivorous fish of the high seas, which never encounter starch in their natural food. By externally attacking the linear chains, it hydrolyses them into maltose, leaving at the point of the branches residues of variable size and short chain dextrins.

There are several other glucosidases that are generally low in activity: cellulase ($\beta 1 \rightarrow 4$ glucosidase) is often detected, but it is usually involved with exogenous enzymes of bacterial origin. It is only recently that more convincing evidence has been obtained to support the presence of an endogenous cellulase in grass carp, even with the use of terramycin, an antibiotic which inhibits all bacterial development in the digestive tract. The presence of laminarinase ($\beta 1 \rightarrow 3$ glucosidase) has been shown in fish that consume algae. It allows the partial hydrolysis of laminarin, a structural polyholoside which, in algae, plays a similar role to that played by cellulose in terrestrial plants. The evidence in favour of secretion by the pancreas is not very convincing. However, another enzyme that hydrolyses complex β-link carbohydrates is quite abundant: it involves chitinase which splits $\beta 1 \rightarrow 4$ N-acetyl-glucosamine links in chitin to form chitobiose (a dimer of N-acetyl-glucosamine). This polymer is found in fungi and also in invertebrates, in particular in worms and arthropods. Chitinase activity (also present in some mammals) has been related to feeding behaviour by some authors; it is higher in fish that feed on crustaceans or insects and especially in those that swallow their prey without mastication.

Secretion and conditions for activity The secretion of digestive enzymes follows a postprandial pattern which is quite similar to that found in higher vertebrates. Digestive tract hormones (gastrin, CCK, etc) participate in starting off the release of grains of zymogen and bile.

The optimal physical conditions for activity of these different enzymes are shown in Figure 4.5. The optimal pH for pepsin is around 2–3, while for all other enzymes, it is around 7–8. There is thus clear adaptation to pH conditions in the stomach and to those in the duodenum. There are few data available on the optimal pH for other gastric enzymes. The optimal temperature for activity for almost all fish enzymes (not only digestive enzymes) is between 30 and 40°C (Figure 4.5). It is thus always much closer to the internal temperature of homoeotherms than to that of ectotherms. However, it should be noted that enzymes that are particularly active at low temperatures are found in cold water fish.

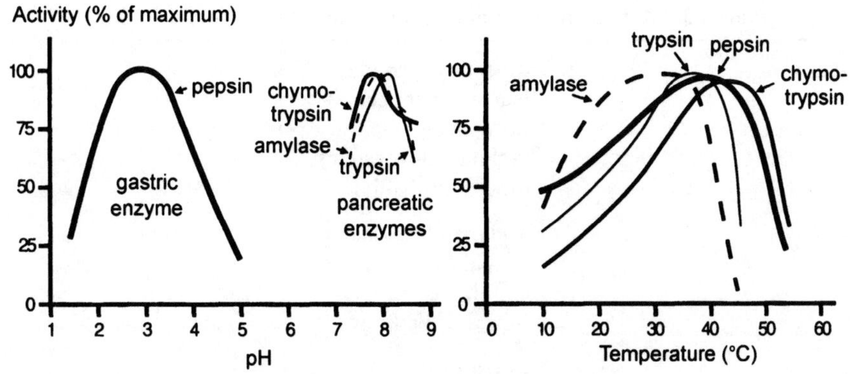

Figure 4.5. Optimal pH and temperatures for the principal gastric and pancreatic enzymes in fish (after Uys and Hecht, 1987). With kind permission of Elsevier Science.

Membrane-bound enzymes of the intestine The fish intestine, like that of other vertebrates, does not secrete enzymes. Some glands can be found in the proximal part of the intestine of gadoids, but these only secrete ions. Activity found in the intestinal fluid can originate from the food, stomach, pancreas or bacteria. It can also be produced by cells, either lysed or not, which slough off at the end of villi, or from the rupture of the peptide chain that normally attaches the membrane-bound enzymes to the microvillus wall. This activity is thus an artefact.

Membranous enzyme activity, under physiological conditions, can only be studied on the intestinal wall itself; it can be found *in vitro* in brush border preparations, but little is known about it. The principal enzymes (Table 4.3.) hydrolyse fragments of macromolecules that are formed by the action of gastric or pancreatic enzymes, i.e. peptides, disaccharides (mainly maltose) and short-chain dextrins. Maltase acts on maltose in the same way as isomaltase acts on branched dextrins, resulting in the formation of glucose that can be directly absorbed. Peptidase action does not result in complete hydrolysis into amino acids, but in a complex and variable mixture of amino acids and peptides, only the smallest of which (2 to 4 bases) are absorbable. The action of these enzymes is shown in Figure 4.2. It is now known that hydrolysis by membrane-bound enzymes always occurs very close to absorption sites. It has even been demonstrated that, in the case of γ-glutamyl transferase, the same protein carries out both hydrolysis and transport.

Other enzymes Aminopeptidases are exopeptidases that act on the amine extremities of low molecular mass peptides which are formed as a result of endopeptidase action. These are not the enzymes that are secreted in an inactive form in zymogen granules but pancreatic lysosomal enzymes. Leucine-alanine peptidase is an intestinal lysosomal enzyme. The esterases which are frequently found in histoenzymology are endocellular enzymes that play a role after particles undergo endocytosis. Activity levels of these enzymes are much higher in young larvae than in juveniles

Table 4.3. Main membrane-bound intestinal enzymes found in fish.

Peptidases	dipeptidyl peptidase IV leucine aminopeptidase phenylalanine-glycine peptidase γ-glutamyl-transferase others (probably)
Glucosidases	α-glucosidase (maltase) isomaltase chitobiase others (little known)*
Lipases	esterases lipase (little known)
Various	acid phosphatase alkaline phosphatase
Nucleases	(uncertain localisation)

* Lactase has been found in several species; like isomaltase, it may possess a second active cycle, with the presence of a site responsible for ceramidase activity (hydrolysis of the carbohydrate residues of heteroproteins).

Table 4.4. Amylase activity in some fish species in comparison with that found in goldfish (standard 100) (after Nagayama & Saito, 1968).

Species	Relative activity
Goldfish	100
Grass carp	84
Tilapia	44
Common carp	35
Silver carp	31
Trout	8
Eel	1
Yellowtail	1

(Chap. 12). Although often detected, bacterial enzymes usually have very low activity.

4.1.2.2 Total enzyme activity

While the list of digestive enzymes appears to be very close to that known for higher vertebrates, their total activity is not the same. Total proteolytic activity is always very high, even in species described as herbivorous. Lipolytic activity is very variable, ranging from considerable values (Atlantic salmon) to moderate activity (turbot). Amylolytic activity is detected in all fish from the start of their life and is clearly higher in herbivores than in omnivores (Table 4.4), thus showing good adaptation to feeding habits in the natural environment. In the sole, a predator of molluscs rich in glycogen, amylase activity is high. Secretion of amylase also seems to increase following an increase in water temperature.

4.1.3 Nutrient absorption

Absorption of nutrients from the intestinal lumen into the enterocytic cytosol and then into the blood vessels occurs via two distinct processes:

- absorption of particles or macromolecules without any of the digestive processes described above occurring;
- absorption of small molecules by processes related to simple diffusion, facilitated diffusion or active transport.

4.1.3.1 Absorption by endocytosis

The role of endocytic cells (Figure 4.3) located mainly in the distal part of the intestine has been demonstrated using immunofluorescence techniques. Invaginations situated at the base of the microvilli play the role of receptors, through which the protein molecules pass into the cytosol without losing their functional properties. Some of these molecules are subjected to intracellular digestion in vacuoles and the rest leave the cell by exocytosis through the basolateral cell wall and reach the layer called the *lamina propria*. They can thus exert an antigenic activity with regard to the animal by contact with lymphocytes. This phenomenon can be put to good use for vaccinations or the oral administration of active protein molecules.

In strictly nutritional terms, it is difficult to quantify the importance of absorption by endocytosis, but it appears that only a very small proportion, of the order of 1–6% at most, of food proteins are absorbed via this mechanism. It is therefore not taken into account as a means of amino acid input in fish. However, this phenomenon, which ensures all protein absorption in the very primitive phyla, could be the origin of many immune-type reactions that correspond to a component of the 'quality' of food proteins, which have been little studied.

4.1.3.2 Absorption by diffusion or transport

Absorption by simple diffusion across membranes is directly proportional to the concentration gradient (Figure 4.6) and is described by non-asymptotic kinetics. It only occurs in the presence of a gradient and is only rarely observed for hydrosoluble molecules. It appears, however, to be the rule for products of lipid degradation.

Absorption by facilitated diffusion and active transport can be recognised from their asymptotic curve kinetics and correspond to the intervention of protein molecules inserted into the lipid bi-layer in membranes (Figure 4.2). These proteins function in an almost entirely analagous way to those of enzymes and the same equations are applicable to them, if concentrations are replaced with figures for flux (Michaelis-Menten kinetics, Figure 4.6). The classical example is that of glucose or amino acid transport, which involves a sodium-dependent process. Transport

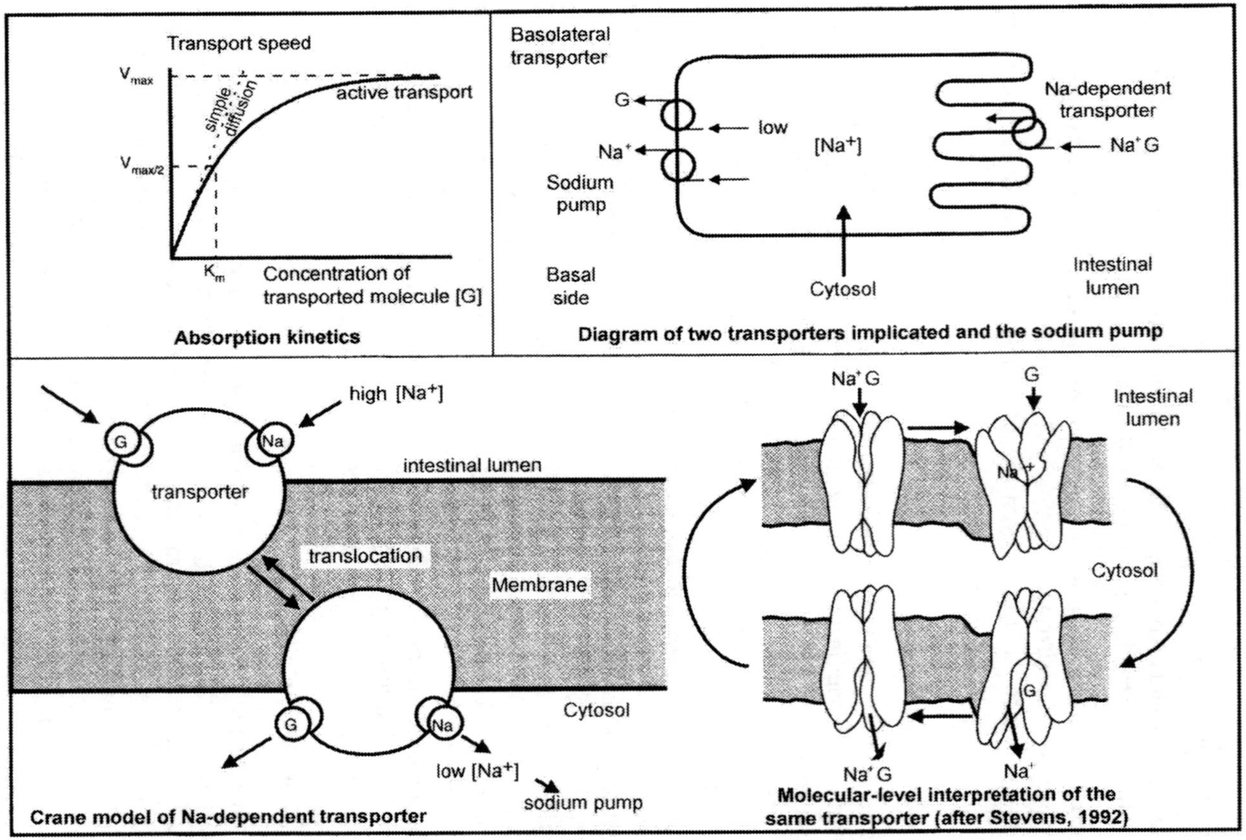

Figure 4.6. Mechanism for glucose (G) absorption by active transport. The principle is the same for other substances such as amino acids and peptides.

from the lumen to the cytosol of enterocytes is in fact co-transport of the organic molecule and a sodium ion (Figure 4.6). Sodium ions follow a decreasing gradient, with the low intracellular sodium concentration being maintained by the sodium pump located on the basement membrane. This 'pump' extrudes ions towards the internal compartment, using energy provided by ATP hydrolysis. Glucose or amino acids are thus accumulated in the cytosol, despite the increasing gradient. They then diffuse across a decreasing gradient across the baso-lateral cell wall by facilitated diffusion (Figure 4.6). The assemblage of these mechanisms equates to active transport which also ensures the absorption of a large quantity of peptides (usually mono-, di- or tripeptides).

Facilitated diffusion also relies on the existence of carriers. However, in contrast to active transport, it only operates in the direction of decreasing gradients and does not require energy. While the molecules that cross the brush border are transformed in the cytosol (for example, isomerisation, or reincorporation of fatty acids into triacylglycerols), the decreasing light gradient towards the cytosol is maintained. Facilitated diffusion is possible. Although this is quite rare, like simple diffusion it can play a quantitatively significant role.

As in other animals, lipid absorption occurs via particles called micelles which are formed from fatty acids, monoacylglycerols (residues from the hydrolysis of triacylglycerols by pancreatic lipase) and also phospholipids (alimentary or biliary), cholesterol and bile acids. The various lipid compounds, as well as the lipid soluble components found in the micelles in small quantities (group A vitamins and carotenoids), diffuse across the lipid bi-layer after disaggregation of the micelle. There is resynthesis of triacylglycerols, forming 'particular lipids', lipoproteins that are equivalent to the chylomicrons found in the endoplasmic reticulum and Golgi apparatus in mammals. These particles are a form of transport. They leave the enterocyte by exocytosis and reach various organs via the blood or lymphatic vessels. The 'spread-out lipids' in the enterocytes are a form of storage unique to fish.

Liposoluble vitamins are absorbed via micelles while hydrosoluble vitamins diffuse across membranes either by means of specific carriers or via passive diffusion. Minerals are absorbed either with the aid of carriers or via ionic ducts. Oligo-elements usually first attach to an amino acid which acts as a ligand.

4.1.4 Digestive transit time

Food taken in by the fish can remain in the stomach (when there is one) for relatively long periods (often several tens of hours when the temperature is less than 10°C). The contribution made by this organ is spectacular in the case of those fish that swallow their prey whole. This digestion commences on the exterior teguments and occurs as a result of intense proteolytic activity (in particular, elastase activity) along with chitinase activity. In other cases, and in particular with compound foods, evacuation mechanisms appear to be multifactorial; they are a function of the species and size of fish, the type and amount of food, as well as temperature. It is generally agreed that the kinetics of evacuation can be described by a differential

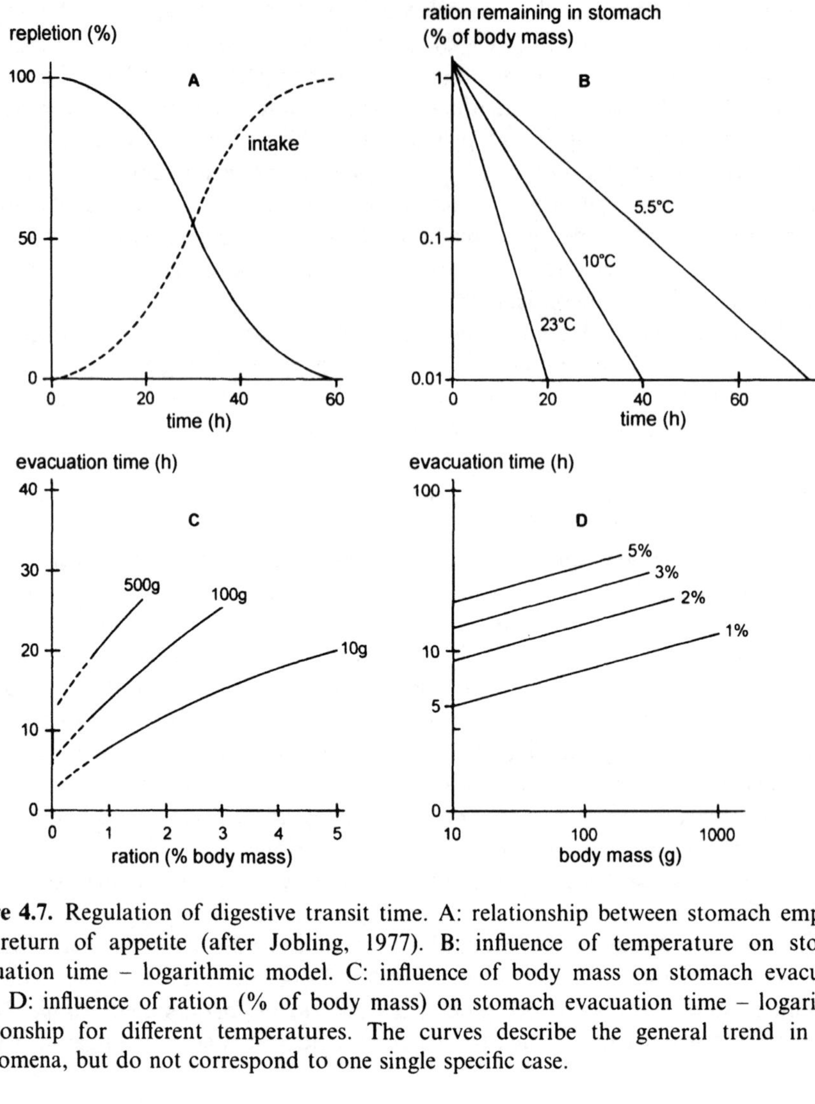

Figure 4.7. Regulation of digestive transit time. A: relationship between stomach emptying and return of appetite (after Jobling, 1977). B: influence of temperature on stomach evacuation time – logarithmic model. C: influence of body mass on stomach evacuation time. D: influence of ration (% of body mass) on stomach evacuation time – logarithmic relationship for different temperatures. The curves describe the general trend in these phenomena, but do not correspond to one single specific case.

equation affected by the filling of the stomach: the fuller the stomach, the quicker it will empty and, at a certain level of emptiness, the organism no longer has the sensation of being replete and appetite returns.

For a given species and food, gastric emptying time is a direct function of the level of consumption; it is also almost proportional to the logarithm of body mass (Figure 4.7). For the same level of consumption, evacuation will be more rapid in a small fish; however, for a given quantity of food taken in, the inverse will apparently be true, the same amount corresponding to a higher intake rate in a smaller fish. Transit time accelerates with increasing temperature (Figure 4.7) with a Q_{10} very close to 2. In agastric species, evacuation of the entire digestive tract appears to follow kinetics similar to that for stomach emptying, the intestinal bulge playing a similar role to the stomach. However, fish tending to be herbivorous have much faster transit times than those tending to be carnivorous.

Knowledge about the amount of food voluntarily taken in at one meal and the kinetics of gastric emptying thus allow, in the first instance, the calculation of the number of meals to distribute and the total daily ration.

4.1.5 Overall view of digestion-absorption phenomena

Digestive efficiency—in its broadest sense—results from digestive processes (in the strict sense) and thus from enzyme secretion, absorptive and transit processes that affect the period during which any of the processes may occur.

The interaction of these three phenomena has not been studied in detail, except in relation to some factors that may strongly influence them, such as fish age (or size), nutrient balance or temperature. With the exception of pathological cases (for example the almost complete disappearance of pancreatic enzymes following pancreatic disease in salmonids or diarrhoea accelerating transit time), the interrelationships between these processes appear to result in remarkable regulation. For example, a fall in temperature decreases enzyme activity and rates of absorption processes, slows transit and decreases ingestion. However, in eurythermic fish at least, it hardly affects the balance. An increase in protein, carbohydrate and lipid content of the food leads to rapid stimulation of trypsin, amylase and lipase secretion, respectively. This regulation is neither instantaneous nor perfect, but regulation of absorption mechanisms (as found in higher vertebrates) is generally sufficient to maintain the overall efficiency of processes at practically unchanged levels after an adaptation period which is almost certainly of the order of a day. Similarly, during growth (except during metamorphosis), requirements vary in relation to body mass and, if the food is taken readily, ingestion increases in parallel with requirements. Despite this, digestive balance almost always remains unchanged, proving that there is an equilibrium between ingestion, enzyme secretion, absorption and transit time.

The role played by salinity has not been extensively studied. Transferring an euryhaline fish from freshwater into saltwater speeds up transit time and can decrease digestibility. By itself the functioning of sodium-dependent absorption mechanisms appears to depend on mechanisms which are different in marine fish and freshwater or terrestrial animals (Ferraris and Ahearn, 1984).

4.2 DIGESTIBILITY OF NUTRIENTS

Digestibility constitutes a global measure of a group of phenomena summarised in the first section and which result in the intestinal absorption of nutrients from the food bolus. For various reasons, part of this cannot cross the intestinal mucosa and is eliminated through the anus as faeces, in the same way as wastes of metabolic origin (mucus, desquamated cells, secreted enzymes) or those from the bacterial flora. As a result, the nutritive value of a food depends not only on its nutrient content but also on the capacity of the animal to digest and absorb it. Digestibility is the digestive balance of ingested nutrients absorbed by the animal.

In aquaculture, studies of digestibility have three objectives: a better understanding of the potential utilisation of nutrients; an improvement in quality of food for the fish; and, lastly, a decrease in wastes of food origin, allowing the preservation of environmental quality in general, and water quality in particular.

4.2.1 Digestibility coefficients

Digestibility coefficients (DC) allow the quantification of digestibility. They can be defined in two ways, depending on whether or not the possible presence of a fraction of endogenous origin (mainly for lipids, amino acids and minerals) is taken into account in the faecal waste.

Thus, for a given food, the apparent digestibility coefficient (ADU) and the true coefficient of digestibility (TDC) can be defined as:

$$\mathrm{ADC} = \frac{\text{ingested} - \text{faecal}}{\text{ingested}} \times 100$$

$$\mathrm{TDC} = \frac{\text{ingested} - (\text{faecal} - \text{endogenous faecal})}{\text{ingested}} \times 100$$

In theory, ADC depends on the physiological state of the fish and ingestion rate and allows the aptitude of the animal to retain or utilise a food ration to be evaluated. TDC depends mainly on the type of diet and digestive capacity of the species and allows evaluation of the suitability of a food to provide the animal with utilisable nutrients. It is usually ADC that is estimated since determination of the endogenous fraction, which is low in fish, is difficult.

4.2.2 Methods for studying digestibility

The majority of studies concern salmonids because they are relatively easy to rear. However, nutrient digestibility is now also being studied in other species, such as eel, turbot, carp, Channel catfish, etc. The measurement principles require data on ingestion and faecal emission. Measurements can be made either directly or indirectly.

4.2.2.1 Direct method

This method requires the measurement of all food ingested and all faeces produced, corresponding to one or more identifiable meals. A metabolism chamber, the principle of which is the same as for terrestrial animals, can be used for this purpose. All other methods for collecting faeces are also suitable, as long as they allow quantitative collection of faecal matter derived from food rations distributed to the fish.

4.2.2.2 Indirect method

In contrast to the method described above, this method does not require measurement either of the entire food consumption or of faeces production. Its originality resides in the utilisation of an inert, indigestible, non-absorbable marker, which is included in the food ingredients or added to the diet. Because of its unique properties, after passing through the gut in the food bolus, this substance is found in its entirety in the faeces (food ingested × marker content of food = faeces excreted × marker content of faeces). The increase in concentration of the marker, in relation to that of the nutrients, allows quantification of the disappearance of these nutrients and this disappearance is considered as absorption. There are generally two main groups of markers: internal markers or substances already present in the food (for example lignin or silica) and external markers or substances that are added to the food (for example chromium oxide, rare earths or coloured particles). Within this wide range of substances, there is none known that unites all the qualities of an ideal marker:

- being absolutely inert, with no behavioural or physiological effects on the animal;
- is neither absorbed nor metabolised;
- has the same transit time as the food bolus or the food being studied;
- does not influence the phenomena of absorption, secretion, digestion or gastro-intestinal motility;
- can be easily and rapidly measured out.

A choice must therefore be made in relation to the objectives pursued or the material requirements. Currently, the marker classically used for digestibility studies in the fish, using the indirect method, is chromium oxide, Cr_2O_3, at an incorporation rate which varies from 1–2%. However, interest has turned more recently to acid-resistant ash.

Under these conditions, digestibility of dry matter (expressed as a percentage) is calculated using the formula:

$$\text{ADC} = 100 - \left(100 \times \frac{\%\ \text{marker in food}}{\%\ \text{marker in faeces}}\right)$$

Similarly, the digestibility of a nutrient is obtained from the formula:

$$\text{ADC} = 100 - \left(100 \times \frac{\%\ \text{marker in food}}{\%\ \text{marker in faeces}} \times \frac{\%\ \text{nutrient in faeces}}{\%\ \text{nutrient in food}}\right)$$

The digestibility of a nutrient in a particular raw material can be measured in different ways.

- It can be distributed alone, with no other feed. Thus the gross digestibility of the raw material is obtained. However, save in a few cases, the raw material does not constitute a balanced diet and the digestibility of these nutrients could differ when incorporated into particular feeds.

- It can be added to a basic diet, the digestibility of which is known, ration increasing in proportion. This method is rarely used.
- It can be incorporated into a basic diet, the digestibility of which is known, ration remaining constant. This incorporation is carried out either at one single rate (30% of the diet, most of the time) or, if greater precision is sought, at several levels (for example, 15, 30 and 45% of the diet). The estimation of the ADC of the tested raw material is therefore made either from the difference from the basic diet or by extrapolation from the latter. The latter method has been used in particular to measure the digestibility of starch in cod, trout and carp.

Digestibility is generally additive in nature, so that the digestibility of a food can be predicted from digestibility values of its main constituents. However, cases where digestibility is not additive have also been reported in fish. Thus, an excess of starch can have a depressive effect on protein digestibility but could also influence that of starch itself. In practice, it is recognised that nutrient digestibility is not affected by the balance of the feeding regime. It involves a hypothesis which is generally close to reality but which has not been conclusively proven.

4.2.3 Faeces

In a rationally managed intensive rearing system – where food is not refused by the animal – faeces constitute the part of the food which is not broken down following incomplete digestion, in addition to metabolic waste (mucus from desquamated cells, the remnants of enzymatic secretions and bacterial flora). However, the latter only represent a very small fraction, in quantitative terms, in relation to the remaining part of undigested food. Thus, in the trout, when 100 g of dry food contains about 6% nitrogen, faecal metabolic losses are only 0.1 g of nitrogen while true digestive losses are as high as 0.4–0.6 g nitrogen. The amount of solid waste varies in relation to the different basic materials used in the manufacturing of the compound feed and also in relation to the different proportions of nutrients (proteins, lipids and carbohydrates).

4.2.3.1 Composition

Between 3 and 6 g of faeces are produced per day by a 200 g trout, containing 80–85% water. The average composition of the undigested fraction of food consists of 15–20% 'proteins' ($N \times 6.25$), less than 1% lipid and 15–20% minerals, with the remainder consisting of various carbohydrates and fibres. However, this composition varies markedly depending on the food type. In marine fish, faeces are overloaded with minerals from the water and this overload depends on the sampling method used.

4.2.3.2 Factors of variation

In the trout, emission of faeces is continuous throughout the 24-hour period, with a marked, but not systematic, increase during the night. This production follows a rhythm reminiscent of a sinusoidal function (Figure 4.8). The variations in the

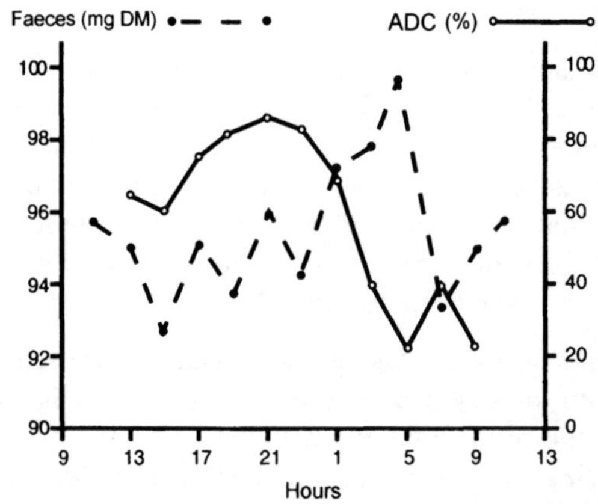

Figure 4.8. Example of the change in faeces emissions and 'instantaneous ADC' during the day–night cycle (experiment on trout).

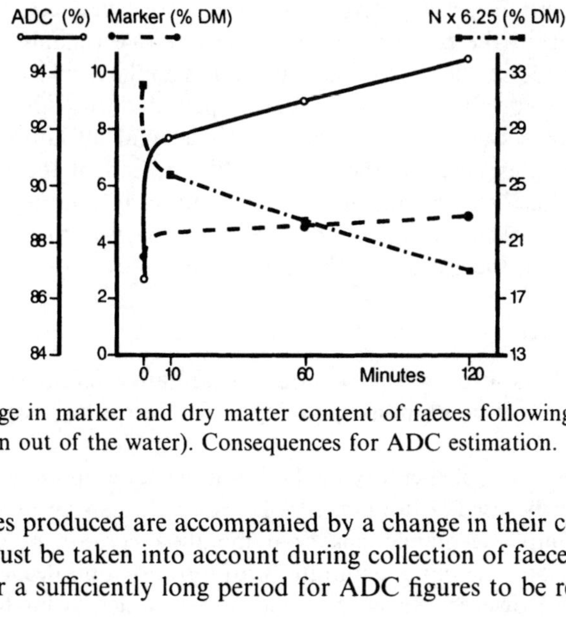

Figure 4.9. Change in marker and dry matter content of faeces following their emission (if they are not taken out of the water). Consequences for ADC estimation.

amount of faeces produced are accompanied by a change in their composition. This phenomenon must be taken into account during collection of faeces, which must be carried out over a sufficiently long period for ADC figures to be representative.

4.2.4 Faeces collection

Collection of faeces from fish is more complex than from terrestrial animals because of the nature of the environment – water – with the associated phenomena of leaching, 'dissolution' and crumbling. These factors begin in the first five minutes of faeces immersion and continue for several hours (Figure 4.9). They are at the origin of the overestimation of the amount excreted (for example, of the order of 10–15% in one hour for nitrogen) and thus the overestimation of ADC. Faeces collection techniques are thus of particular importance. They can be classified in two groups depending on whether or not the fish is removed from the water.

4.2.4.1 Fish out of water

Collection using abdominal pressure This method consists of anaesthetising the fish and expelling the faeces by force by applying gentle pressure to the abdomen. In the trout, this is possible as a result of its straight intestine, but this method cannot be used for all fish (for example carp). In addition, by pressing in this way on the fish's abdomen, urine, genital secretions and sometimes even blood can also be expelled. This technique leads to widespread underestimations of ADC. For this reason, the following two methods have been developed:

- Collection by anal suction: Using a glass canula connected to a vacuum flask, samples are taken directly from the anus. The vacuum is regulated so that only faeces found in the rectal bulb are sampled.
- Dissection: Using this method, the contents of the rectum are sampled correctly, but the fish is sacrificed.

These three techniques are simple to use but have been subject to criticism. Samples taken in this way are in fact digestive contents and not true faeces. This is because there is absorption in the posterior part of the intestine (mainly of proteins). In addition, these methods only allow instantaneous sampling, which is not representative of the faeces produced over a 24-hour period during the entire digestive process. The average transit time for a trout is around 30 hours. It is obvious, therefore, that samples taken at the start, middle or end of the transit time will not be the same. Lastly, handling the fish, in addition to limiting the numbers of fish which can be worked on, has the drawback of stressing them so that samples cannot be taken from them again in the near future.

4.2.4.2 Fish in water

In order to avoid these problems of non-representative faeces sampling, several sampling techniques for continuous sampling of fish faeces have been developed.

Siphoning of faeces The elementary method of immediate siphoning of faeces after their emission can be used. Although it is laborious, it can be used as a point of reference. This simple technique has been modified because of its constraining character and has become more elaborate, with filtration and decantation columns (Figure 4.10, top). However, this method should be avoided owing to the dissolving and crumbling phenomena described above, especially at the point when the faeces are recovered. The faeces are also exposed to bacterial action before they are collected. Lastly, for marine fish, this method leads to the collection of a large amount of salt contained in the water, which must be evaporated. This excess salt can impair the combustion of faeces samples and thus the measurement of the ADC of energy.

Continuous water filtration The basic method rests on the following principle: water exiting the tanks is filtered continuously by moving metal screens (Figure 4.10, bottom). Faeces suspended in the water are thus quickly removed by this action.

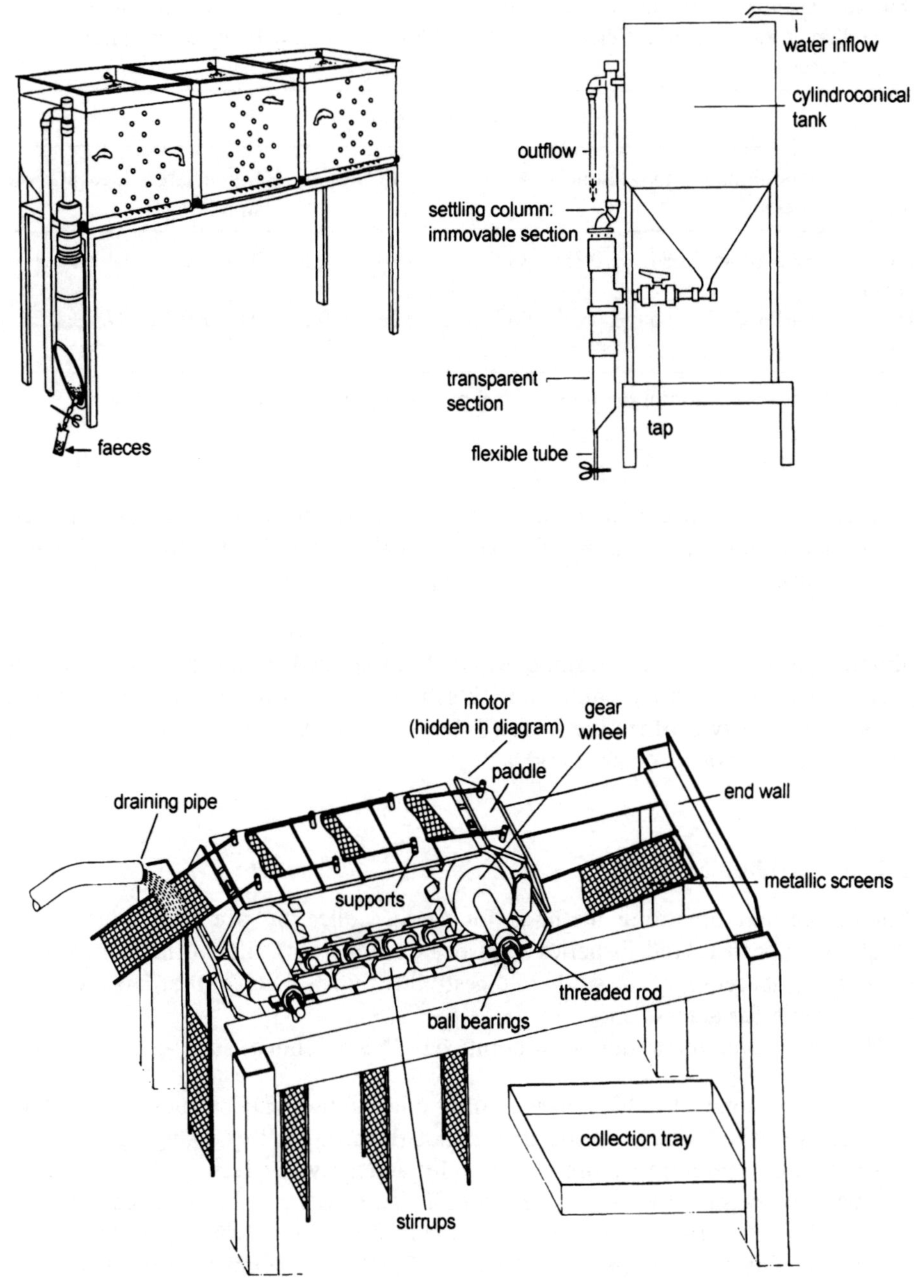

Figure 4.10. Apparatus for collecting fish faeces. Top: apparatus with settling column (from Guelph); bottom: apparatus with continuous filtration (from INRA-Saint-Pée). After Cho *et al.*, 1985; Spyridakis *et al.*, 1989; Choubert *et al.*, 1983.

Table 4.5. Apparent digestibility of proteins and lipids in the sea-bass using the faeces collection method (mean ± standard error) (after Spyridakis *et al.*, 1989). By kind permission of Elsevier Science.

	Collection method					
	Abdominal pressure	Dissection	Anal suction	Filtration	Immediate siphoning	Decantation
ADC of proteins	$82.5 \pm 1.4^{a*}$	84.4 ± 0.8^{ab}	86.6 ± 0.3^{b}	90.4 ± 0.6^{c}	90.6 ± 0.3^{c}	94.2 ± 0.1^{d}
ADC of lipids	94.1 ± 0.8^{a}	95.0 ± 0.4^{ab}	96.3 ± 0.4^{b}	96.0 ± 0.2^{b}	97.3 ± 0.2^{b}	97.1 ± 0.3^{b}

* Values with the same exponent (on the same line) do not differ significantly.

However, this method cannot be used for species which produce very soft faeces, such as the turbot; the addition of binder to make faecal waste more solid is not always sufficient.

Validation of these different techniques has been studied by measuring the digestibility of proteins in sea bass and chinook salmon. Results from filtration methods, compared to the two other methods (Table 4.5.) are very close to the reference method (extemporaneous siphoning).

4.2.5 Standardisation

Although certain operating methods for faeces collection are recommended by EIFAC (European Inland Fisheries Advisory Commission), there is no standardisation, in the strict sense, of the study of digestibility in fish. Each author has their own unique criteria for assessment.

However, some methodological points have been defined, such as:

- adaptation time of fish to new feeding conditions. ADC values usually take definitive values characteristic of the diet three days after changing the food. Adaptation therefore occurs very rapidly and only the latent effect of transit time must be taken into consideration. Thus, faeces can be collected from the third day after distribution of the diet, the digestibility of which is being studied;
- frequency of distribution of daily ration of food. The number of meals per day does not seem to have any effect on food digestibility. However, it does affect transit time and the kinetics of food absorption. In practice, a single daily meal is sufficient for the trout;
- the number of replicates to be carried out and the duration of collection with respect to obtaining a given precision for ADCs. The combinations 'number of

Table 4.6. Optimisation of the relationship: number of tanks (T) × number of days (D) required to obtain a given relative precision in ADC for various components (after de la Noüe *et al.*, 1980).

Component	Precision		
	1%	2%	5%
Dry matter	> 5T 14D	> 5T 14D	3T 10D
Nitrogen	2T 6D	2T 4D	2T 3D
	3T 5D	3T 3D	3T 2D
	4T 4D		
Energy	> 5T 14D	> 5T 14D	3T 5D
			4T 4D

repetitions × number of days' to be used for a given precision have been established *a posteriori* (Table 4.6).

4.2.6 Nutrient digestibility: results

4.2.6.1 General properties

If it is assumed that emission of endogenous faeces as well as digestive capacity in its widest sense are constant no matter what amount of feed is ingested, TDC is independent of the latter, while ADC is not. For a very low rate of ingestion, this coefficient can have a very high negative value, to approach TDC asymptotically, in direct proportion to increasing ingestion. In practice, ADC is measured in the area where it is close to TDC and thus becomes almost constant. However, for high intake levels, there is a decrease in CDA which can thus be attributed to 'saturation' of the digestive capacity.

The ADC of a nutrient can be measured very precisely (coefficient of variation 2–3% for a diet, 4–5% for an ingredient contained within it), but variability due to methodology is great. As a result, it is very difficult to compare ADC values obtained in different laboratories, using distinct species and various raw materials. Reliable figures are therefore still no more than a relatively poor data bank. However, some results have been obtained (Table 4.7).

4.2.6.2 Proteins

Fish generally digest proteins with an ADC exceeding 90%, a level equal or superior to those observed in terrestrial vertebrates. Digestibility of proteins from a given source varies relatively little between fish species. For a given species, it is very constant, although it sometimes increases slightly with fish size. It is practically independent of ingestion rate and temperature. It is not affected by the presence of lipids in the diet, even at high concentrations.

Digestibility is, however, a function of the nature of the proteins themselves, or their source and the treatments they may be subjected to, even the size of the food

Table 4.7. Coefficient of digestive utilisation of certain primary ingredients in trout (%) (after Cho *et al.*, 1982).

Primary ingredients	Dry matter	Proteins	Lipids	Energy
Lucerne flour	39	87	71	43
Blood meal	91	99	–	89
Maize	–	95	–	39
Cornflour gluten	–	96	–	83
Hydrolysed feather meal	75	58	–	70
Fishmeal	85	92	97	91
Bonemeal	78	85	73	85
Rape oilcake	35	77	–	45
Cooked soya	78	96	94	85
Soya oilcake	74	96	–	75
Fish protein concentrate	90	95	–	94

particles. Proteins of animal origin are generally more digestible than those of plant origin. Some technological treatments applied to plant proteins bring about a marked improvement in ADC by destroying antinutritional factors. Thus, cooking whole soya grain leads to an increase in ADC from 70 to 85% (Chapter 19).

4.2.6.3 Lipids

Lipids with low melting points are well utilised by fish (ADC > 95%), whatever their origin. Their ADC appears to be better when the incorporation rate is high, provided the fatty acids are protected from oxidation phenomena. However, digestibility of saturated fatty acids decreases as chain length increases and, for identical chain lengths, it increases with the degree of unsaturation. Thus lipids containing high levels of saturated fatty acids with medium chain lengths (for example tallow), have a relatively low ADC, especially when water temperature is low. This phenomenon results from the solid state which renders the emulsification of lipids more difficult. Thus, in salmonids, the ADC of lard, which has a melting point of 28–48°C, changes from 70 to 78% when the temperature increases from 5 to 15°C, while under the same conditions, the ADC of oils with melting points less than 0°C, remains almost unchanged (90–93%).

4.2.6.4 Carbohydrates

Loss of carbohydrates through faecal waste is the most significant, although digestibility of simple sugars (glucose and sucrose) is close to 100%. In fact, the digestibility of starch, which is the only source of carbohydrates likely to be incorporated into fish feeds for economic reasons, is often of the order of 70–80% and can be less than 50%. The ADC of starch varies with amylase activity, which is species specific (see p. 40 and Table 4.4). It is also a function of water temperature: in general terms, it is higher in warm water fish than in those from temperate zones. In the latter, a rise in water temperature can also improve starch digestibility; this phenomenon has been very

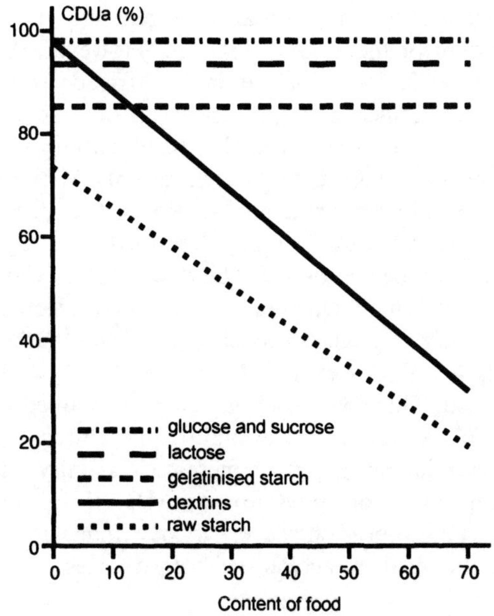

Figure 4.11. Diagramatic variations in CDU of different carbohydrates in relation to level of incorporation.

clearly observed, both in trout between 8°C and 18°C and in turbot between 13°C and 18°C, the ADC of pure starch changing in the latter from 25% to 48%.

Above all, starch digestibility depends on its nature, i.e. the relative proportions of amylose and amylopectin (Chapter 8) as well as the size and integrity of the starch grain. Although amylose in its pure form is more easily attacked by amylase than amylopectin, it confers a crystalline structure on the grain which renders it less sensitive to amylase. As a result, the ADC of starch decreases when amylose content is increased and also when grain size is increased. Thus starch grains from tubers and leguminous plants are larger and less digestible than those from cereals. All thermal or hydrothermal treatment that alters the grain structure increases the CDU of starch; this is the beneficial role of gelatinisation (Chapter 8). Dry heat treatment of pure starch, which results in very partial hydrolysis of the initial molecules into dextrins (commercial product), has similar effects.

When a starch is not very digestible, the ADC generally decreases with the incorporation level, while it can be considered constant for simple sugars or completely gelatinised starches (Figure 4.11).

In monogastric animals, the digestibility of starch is often negatively correlated with the size of the particles of feed ingredients. Data on this topic are scarce in fish.

4.2.7 Other variables

In fish, as in other animals, the digestibility of different nutrients, and in particular complex carbohydrates, varies from one species to another. The significance of intake levels of the diet is still controversial. It is generally assumed that nutrient digestibility is independent of the amount ingested. However, digestibility is negatively correlated with intake levels for protein in the African catfish and for starch in the trout. In addition, in all species, slight restriction of food intake allows better utilisation of nutrients and can be partly explained by an improvement in ADC.

Other factors can also affect ADC to different extents. Without focussing on the methodology, which can induce considerable artefacts, the digestive flora and two environmental factors – temperature (already described with regard to starch) and salinity of the water – should be mentioned. The role of the digestive flora appears to be negligible, at least for fish in temperate zones. The effect of temperature on digestibility is more controversial. According to some authors, the ADC of proteins and energy is a little lower at 8°C than at 18°C, while for others, it is only very slightly affected. This contradiction can be explained by the variation in the ADC of starch which is often raised at high temperatures.

It appears that, in euryhaline species, an increase in salinity leads to a decrease in protein digestibility. The effects of salinity on the ADC of lipids and carbohydrates are less well known. They appear, however, to be similar to those observed for proteins and could be the result of accelerated transit times.

4.3 CONCLUSION

The organisation of the digestive tract in fish is both very varied and very different from that found in the higher vertebrates. In contrast, the enzymes involved are similar – sometimes very similar – to those in terrestrial vertebrates, at least in qualitative terms. This is not the case in quantitative terms, where high protease activity contrasts with the often low activity of glucosidases. The main mechanisms of digestion and absorption are now well known. They are strongly influenced by ectothermy, but there is remarkable regulation of digestion, absorption and transit, so much so that the digestive balance is influenced very little by dietary or environmental factors. However, many aspects remain unstudied. Interspecific variations are considerable.

In overall terms, from digestive physiology to nutrient digestibility, which is more difficult to measure than in terrestrial animals, fish always present unique features. Their remarkable aptitude for digesting proteins contrasts with their variable and lower aptitude for digesting carbohydrates. Data are already available that can be applied to food formulation. However, these only concern a few species and remain very incomplete. Progress in the areas of nutrition and feeding will occur necessarily from a better knowledge of the physiology of digestion and food digestibility.

4.4 BIBLIOGRAPHY

Bergot, P. (1981) Structure de l'appareil digestif. II. Structure et fonction des caeca pyloriques. In: Fontaine, M. (ed) *Nutrition des Poissons.* CNRS, Paris, 45–53.

Bertin, L. (1958) Appareil digestif. In: Grassé, P.-P. (ed.) *Traité de Zoologie. Anatomie, Systématique, Biologie. VIII. Agnathes et Poissons, Fascicule No. 2: Anatomie, Ethologie, Systématique.* Masson et Cie, Paris, 1248–1302.

Cho, C. Y., Kaushik, S. J. (1985) Effect of protein intake on metabolizable and net energy values of fish diets. In: Cowey, C. B., Mackie, A. M. (eds) *Nutrition and Feeding in Fish.* Academy Press, London.

Choubert, G., de la Noue, J., Luquet, P. (1983) Un nouveau collecteur automatique quantitatif de fèces de poissons. *Bull. Fr. Piscic.*, **288**, 68–72.

De la Noue, J., Choubert, G., Pagniez, B., Blanc, J. M., Luquet, P. (1979) Digestibilité chez la truite arc-en-ciel (*Salmo gairdneri*) lors de l'adaptation à un nouveau régime alimentaire. *Can. J. Fish Aquat. Sci.*, **37**, 2218–2224.

Ferraris, R. P., Ahearn, G. A. (1984) Sugar and amino acid transport in fish intestine. *Comp. Biochem, Physiol.*, **77A**, 397–413.

Gas, N., Noaillac-Depeyre, J. (1981) Structure de l'appareil digestif. I. Organisation, ultrastructure et fonction du tube digestif des téléostéens d'eau douce. In: Fontaine, M. (ed.) *Nutrition des Poissons.* CNRS, Paris, 19–44.

Gropp, J. M., Tacon, A. G. J. (1994) *Report of the EIFAC workshop on methodology for determination of nutrient requirement in fish.* EIFAC Occasional Paper 29, FAO, Rome.

Jobling, M., Gwyther, O., Grove, O. J. (1977) Some effects of temperature, meal size and body weight on gastric evacuation time in the dab *Limanda limanda* (L.). J. Fish Biol., **10**, 291–298.

Laplace, J. P., Corring, T, Rérat, A., Demarne, Y. (1986) Digestion. In: Pérez, J. M., Mornet, P., Rérat, A (eds) *Le Porc et son Elevage. Bases Scientifiques et Techniques.* Maloine, Paris, 1–120.

Nagayama, F., Saito, Y. (1968) Distribution of amylase – and ß-glucosidase and ß-galactosidase in fish. *Bull. Jpn. Soc. Sci. Fish*, **34**, 944–949.

Pérez, J. M., Mornet, P., Rérat, A. (1986) *Le Porc et son Elevage. Bases Scientifiques et Techniques.* Maloine, Paris.

Smith, L. S. (1989) Digestive functions in teleost fishes. In: Halver, J. E. (ed.) *Fish Nutrition.* Academic Press, San Diego, 331–421.

Spyridakis, P., Métailler, R., Gabaudan, J., Riaza, A. (1989) Studies on nutrient digestibility in European sea bas (*Dicentrarchus labrax*). I. Methodological aspects concerning faeces collection. *Aquaculture*, **77**, 61–70.

Stevens, B. R. (1992) Vertebrate intestinal apical membrane mechanisms of organic nutrient transport. *Ann. J. Physiol.*, **263**, 458–463.

Uys, W., Hecht, T. (1987) Assays on the digestive enzymes of sharptooth catfish, Clarias gariepinus (*Pisces: Claridae*). *Aquaculture*, **63**, 301–313.

Verigina, I. A. (1991). Basic adaptation of the digestive system in bony fishes as a function of diet. *J. Ichtyol.*, **31**, 8–20.

Vernier, J. M., Sire, M. F. (1989) L'absorption intestinale des protéines sous forme macromoléculaire chez les vertébrés. Implications physiologiques. *Année Biologique*, **28**, 255–288.

5

Nutritional energetics

Like all animals, fish require energy to carry out their vital functions. Under aerobic conditions, the only energy that the organism can utilise comes from the oxidation of organic compounds (carbohydrates, lipids and proteins) which are provided by the digestion of food and the turnover of cells and tissues. The energy requirements of fish depend on the species and physiological stage of the animal itself. They also vary in relation to environmental factors and, in particular, water temperature, their ectothermy having an acute effect on energy metabolism. Nutritionists have tried hard to analyse the partition of energy from food towards anabolism or catabolism. The study of the metabolic utilisation of nutrients, i.e. the various stages of intermediate metabolism and energy production at the cellular level, has constituted a complementary approach, allowing a better interpretation of nutritional data.

5.1 ENERGY PRODUCTION AT THE CELLULAR LEVEL

5.1.1 General

During intra- or extra-luminal digestion, macronutrients provided by food are broken down into amino acids (AA) from proteins, fatty acids (FA) and glycerol from triacylglycerols (TAG) as well as hexoses, mainly glucose, from polyholosides (Chapter 4). Once absorbed nutrients are either incorporated into cellular structures after transformation, through anabolic reactions that allow energy accretion for the organism, or catabolised during exergonic, i.e., energy-generating, reactions. Exergonic reactions are coupled to the synthesis of molecules with at least one energy rich bond the main one being ATP.

The energy required for processes such as syntheses, transmembrane transport and muscular contractions is released from the hydrolics of such molecules. During catabolic processes, energy not stored in the form of ATP or its equivalents is dissipated as heat, transformations of chemical energy into another form of

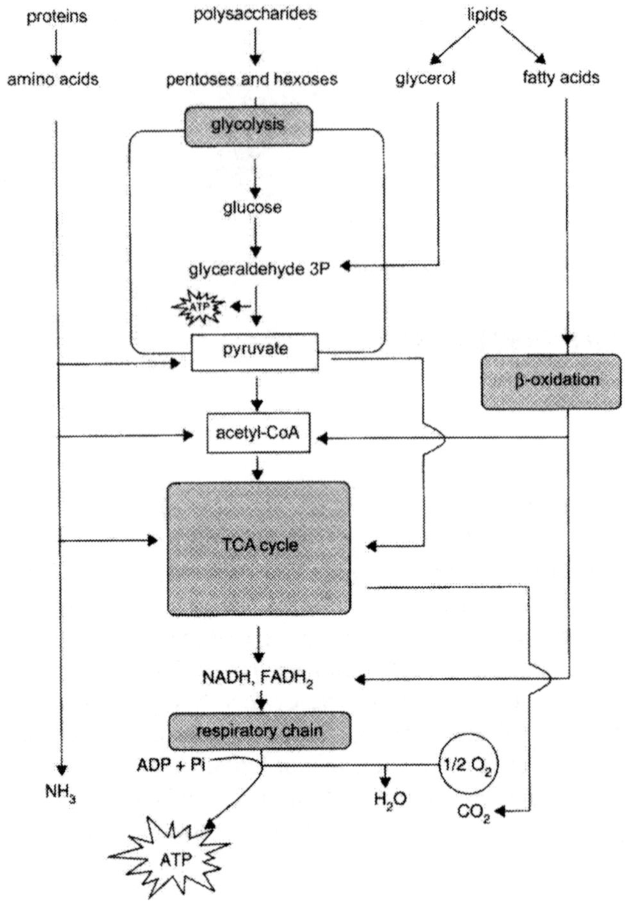

Figure 5.1. Main routes of energy production.

energy always occurring with a yield of less than 100%, according to the 2nd principle of thermodynamics.

Energy production occurs aerobically in cells in three stages (Figure 5.1):

- formation of acetyl-coenzyme A (CoA) by oxidation of FA, AA or pyruvate provided mainly by glycolysis;
- degradation of acetyl groups through the tricarboxylic acid (TCA) cycle, also called the citric acid or Krebs cycle, with the liberation of CO_2, reduced nucleotides and hydrogen atoms;
- transport of hydrogen electrons towards molecular oxygen, accompanied by phosphorylation, coupled with ADP and generating ATP. At any time, energy

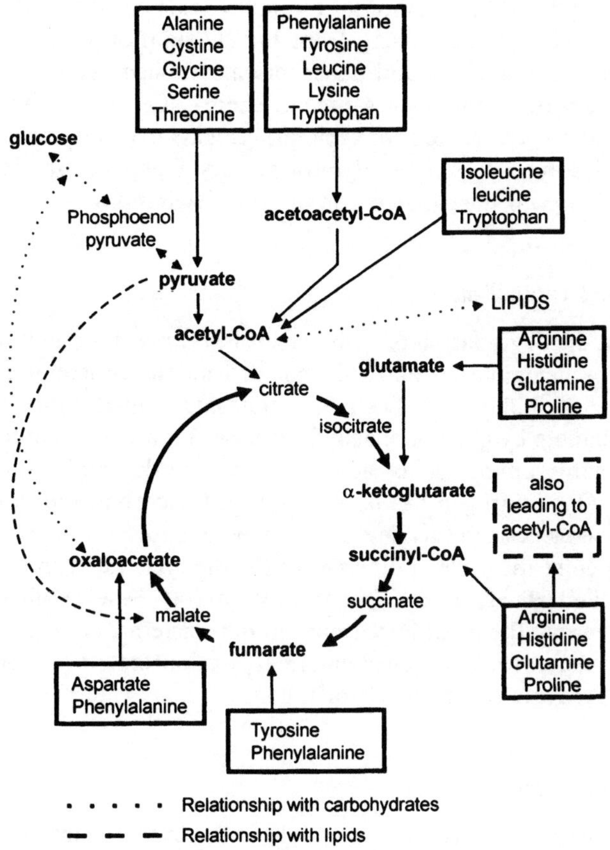

Figure 5.2. Potential entry sites for amino acids into the TCA cycle in fish liver.

can be released or stored, depending on cellular demands, as a result of ADP/ATP coupling, which allows energy distribution within the cell.

In fish, as in other vertebrates, the TCA cycle is the common site for utilisation of energy substrates under aerobic conditions. In contrast to birds and mammals, fish utilise small amounts of glucose as a substrate for cellular oxidation. One of the main characteristics of fish is their use of amino acids as energy substrates. Amino acids can enter the cycle at various points (Figure 5.2) and be responsible for an increased portion of its activity. Thus, in fish, amino acids, rather than glucose, are used preferentially as energy sources. Fatty acids also contribute to ATP production. When the diet contains high levels of lipids, a decrease in oxidation of amino acids is observed; this results in a protein sparing effect.

There are many differences between tissues; the highest levels of oxidation are observed in tissues that are richest in mitochondria: the liver, heart and gills as well as the red muscle which permits slow, continuous contractions. In the liver, keto-acids resulting from the catabolism of amino acids constitute the main energy substrates, with acetyl groups arising from the β-oxidation of fatty acids. In the white muscle, with rapid fibres and anaerobic metabolism, conversion of glucose into lactate appears to be the main source of energy. This tissue, hypertrophied in comparison to red muscle, is used for rapid movements over very short periods. The oxidative and glycolytic capacities of muscle, vary from one species to another, apparently in relation to the capacity for sustained swimming.

5.1.2 Amino acid catabolism

The first stage in the degradation of amino acids is a transamination: the amino acid is transformed by deamination into α-ketoacid, with the concomitant synthesis of glutamate from α-ketoglutarate. The α-ketoglutarate is then reformed by the deamination of glutamate by glutamate dehydrogenase (Figure 5.2). Glutamate is transaminated by many amino acids and the liver is the main site for such a transformation. Once nitrogen has been removed, the carbon skeleton can be used either for gluconeogenesis via oxaloacetate, or be oxidised in the TCA cycle. Since the amino acids enter the cycle at different levels, the cycle rate is not only regulated by the activity of citrate synthetase, but also by entry at the level of α-ketoglutarate. In quantitative terms, the most important aminotransferases are the aspartate and alanine aminotransferases (also called glutamate oxaloacetate transaminase or GOT and glutamate pyruvate transaminase or GPT).

5.1.3 Lipid catabolism

Although lipid synthesis in fish occurs mainly in the liver, its degradation occurs in all tissues almost. Mechanisms for transport into the mitochondria and for fatty acid oxidation are similar to those found in other vertebrates. The fatty acids liberated from the hydrolysis of TAGs by lipase are transported into the mitochondria by acyl-carnitine transferase system. They are then broken down by successive cycles of β-oxidation, the result of which is the release of 2-carbon fragments in the form of acetyl-CoA (Chapter 7) which enters the TCA cycle at the citric acid level.

In mammals, the acetyl-CoA produced by the β-oxidation of fatty acids can lead to the formation of ketone bodies, when entry into the TCA cycle becomes saturated. In fish, the level of ketone bodies is kept at a low level due to the existence of a 'futile cycle', very active process which does not generate substrate, but leads to energy losses in the form of heat, aceto-acetate $\longleftrightarrow$ aceto-acetyl CoA.

5.1.4 Carbohydrate catabolism

Glycolysis is the main route for glucose catabolism in fish tissues. It transforms glucose into pyruvate, with a net gain of 2 ATP/molecule of glucose, also generating a small amount of energy. However, this represents only about 2% of the energy corresponding to complete combustion (enthalpy) of glucose.

As in other animals, glycolytic reactions are all reversible and can thus function in the direction of gluconeogenesis, with the exception of three stages, catalysed by hexokinase, phosphofructokinase and pyruvate kinase respectively. All three of these are limiting stages of glycolysis, hexokinase having especially low activity. All the glycolytic enzymes have been detected in the tissues of the different species of fish that have been studied (Chapter 8). In fish liver, glycolysis provides precursors for the biosynthesis of various molecules rather than producing pyruvate for oxidation. In aerobic cells, pyruvate leads, by a decarboxylation, to the formation of acetyl-CoA, which can then be oxidised in the TCA cycle. In white muscle, under anaerobic conditions, the conversion of pyruvate into lactate is catalysed by a lactate dehydrogenase.

Glycolysis is not the only route for breakdown of carbohydrates. The pentose cycle, or shunt, also functions in fish. It generates the NADP required for fatty acid synthesis (Chapter 8). Its relative activity depends on temperature.

5.1.5 Tricarboxylic acid (TCA) cycle

The eight reactions within this cycle that occur in the mitochondria function in a similar way in mammals and fish. The enzyme that controls the entry of acetyl-CoA into the cycle (at citrate synthetase level) has the same characteristics in trout liver as in mammals. However, there is only one active form of isocitrate dehydrogenase (the $NADP^+$ dependent form) in fish, whereas there are two in mammals. The subsequent stages leading to the formation of oxalo-acetate are not very different in fish.

5.1.6 Respiratory chain

Oxidation of the reduced forms of NAD, NADP and FAD produced during the TCA cycle, as well as during glycolysis, occurs eventually, aerobically, by a series of oxido-reductions in the respiratory chain. Pairs of electrons from intermediaries in the TCA cycle are passed along a chain of electron transport enzymes, with progressively lower energy levels. At the last stage, they reduce molecular oxygen which is the final acceptor of electrons in respiration; at this stage they actually combine oxygen and hydrogen to form water. The entire process is called oxidative phosphorylation. It creates 36 molecules of ATP per molecule of glucose, the energy equivalent of 38% of the enthalpy of this compound.

Carbon flux through the TCA cycle is regulated by the $NADH/NAD^+$ and ATP/ADP ratios; these molecules act as modulators of catabolic enzymes since their presence is essential to the functioning of these enzymes. The relative proportions of ATP and ADP in the cell give an indication of the equilibrium between energy-consuming and energy-generating reactions. The energetic state of a cell is described by the energy charge defined as:

$$EC = \frac{[ATP] + \frac{1}{2}[ADP]}{[ATP] + [ADP] + [AMP]}$$

The energy charge of liver tissue in different species of fish is lower than that found in mammals, which suggests that there is a difference in the regulation of energy metabolism. The consequences of this phenomenon remain unknown.

5.2 PARTITION OF DIETARY ENERGY

Figure 5.3 shows the partition of energy from food in the fish. Gross energy (GE), digestible energy (DE), metabolisable energy (ME) and net energy (NE) can all be used to measure the energy content of food. In all cases, energy is expressed in kilojoules (kJ) or megajoules (MJ) per gram or kilogram of dry matter (the kcalorie is still used: 1 kJ = 4.184 kcal).

5.2.1 Gross energy (GE)

GE is the same as the term enthalpy used by physicists (heat of combustion at a constant volume), conventionally designated with a + sign rather than a − sign. It can be determined by combustion of the food in a calorimeter. The GE of a food

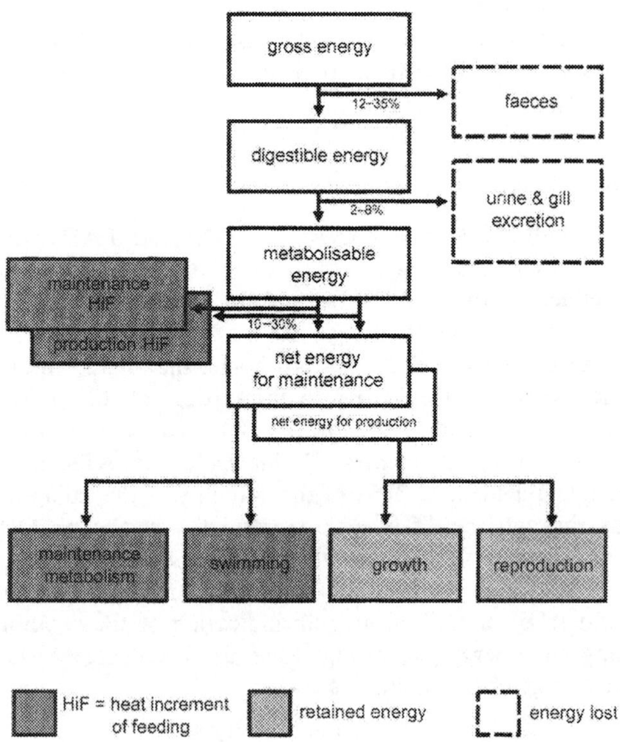

Figure 5.3. Partition of dietary energy in fish.

depends solely on its chemical composition and can be calculated by addition of the GEs provided by each organic component of the diet, mean values being 23.7 kJ.g^{-1} for proteins, 39.5 kJ.g^{-1} for lipids and 17.2 kJ.g^{-1} for carbohydrates (including fibre). It can also be calculated from the GEs of each ingredient of the diet, if they are known.

5.2.2 Digestible energy (DE)

DE corresponds to gross energy (GE) ingested, less the GE of faeces, which generally varies between 10 and 30% of the GE of the diet. The digestibility of dietary energy results from that of its ingredients. In fish, the digestibility of energy of a given nutrient is hardly influenced by the other dietary ingredients (Chapter 4), with the exception of raw starch (Chapters 4 and 8). The DE of a diet can also be determined either by measuring the GE of the food and faeces, or estimated by calculating the sum of the DEs of the proteins, lipids and carbohydrates or even by calculating the sum of the DEs of the ingredients. Thus, if the interactions between ingredients and digestibility are negligible, DE can be considered to be additive.

5.2.3 Metabolisable energy (ME)

ME is DE less the energy lost by the fish as gill and urinary excreta. Nitrogenous catabolites are, as mentioned above, generated by the deamination which precedes amino acids oxidation. In practical terms, their energy value (in kJ) is considered as the product of the amount of nitrogen excreted times the coefficient 24.7. Thus it is estimated that nitrogenous excretory products are made of (or contain) 85% nitrogen in the form of ammonia (25 kJ.g^{-1}) and 15% nitrogen in the form of urea (23 kJ.g^{-1}), although this is only an approximation.

Some researchers express the dietary energy of fish feeds in ME by analogy with systems utilised for other species, especially poultry, in which it is difficult to separate faeces and urine. However, although the DE of a diet can be calculated by adding those of each component of the diet, the metabolic losses required for the calculation of ME must be measured. Urinary and gill wastes vary, not only with the amount of ingested food and function of the protein source (according to its balance in essential AAs), but also as the equilibrium between protein energy/non-protein energy. This equilibrium can be expressed as the relationship between the quantity of digestible proteins (DP in mg) and the quantity of DE (in kJ) provided by the diet. Figure 5.4 shows that the metabolic energy losses for a given nitrogenous intake are lower when the ratio DP/DE of the food is lowered from 23 to 18 mg.kJ^{-1}. Nitrogenous excretion can be reduced by an increase in input of non-protein DE: an increased amount of non-protein nutrients is thus catabolised for energy purposes, hence decreasing protein catabolism and increasing its retention. The protein-sparing effect of lipids is clearly demonstrated in most farmed fish. Protein sparing by digestible carbohydrates such as gelatinised maize or wheat starch is more controversial. It has, however, been demonstrated in trout, Nile tilapia and the European eel.

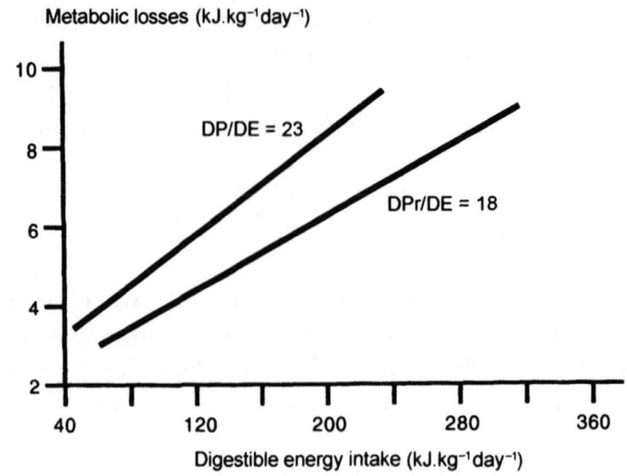

Figure 5.4. Variations in metabolic losses as a function of the DP/DE ratio of the diet.

Metabolic energy losses represent between 2 and 8% of the GE, depending on the protein source and the ratio DP/DE. Thus, ME does not relate to raw materials but only to a complete diet. In the absence of material required to measure nitrogen excretion, it is preferable to formulate fish feed on the basis of DE.

5.2.4 Net energy (NE)

The NE of a diet is its ME less energy expenditure linked to food consumption and utilisation. These losses, which correspond to heat energy, result from both mechanical work (ingestion, mastication, digestive tract movements) and biochemical work (digestion, absorption, transport and transformation of nutrients), the latter being widely preponderant. These result in an increase in oxygen consumption after food intake. This is what Rubner, in 1902, called the 'specific dynamic effect' of food, also called, 'specific dynamic action' or SDA. Other terms (calorigenic effect, thermogenic effect, food thermogenesis, etc.) have been used to describe this expenditure but the term 'heat increment of feeding' (HiF) is the most commonly accepted. HiF is mainly linked to the quantity of proteins ingested but also varies dietary carbohydrate and lipid content.

HiF expenditure increases with ration and is approximately proportional to the quantity of energy ingested. In fish, as in monogastric mammals and birds, it is mainly due to protein utilisation: it increases with the protein content of the diet (Figure 5.5). A value of 30 $kJ.g^{-1}$ digestible nitrogen intake has been proposed for trout. Figure 5.5 shows, however, that in trout HiF decreases when the lipid content of the food exceeds 15%. The same phenomenon has been demonstrated in the Channel catfish. The role played by carbohydrates in HiF is less clear. The results of a recent study on trout (Table 5.1), indicate that a diet rich in digestible carbohydrates results in a higher HiF than a lipid-rich diet, at 18, but not 8°C.

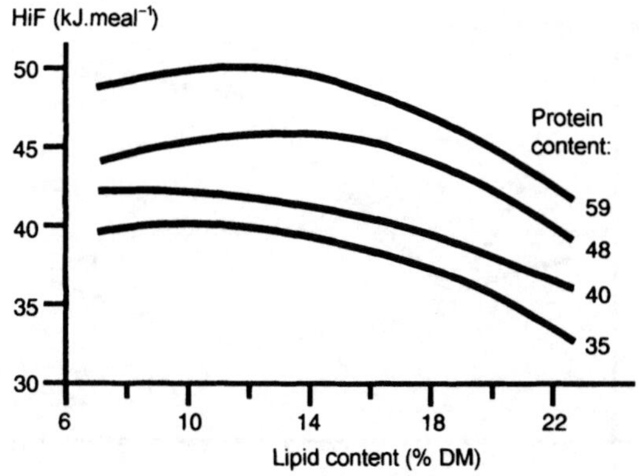

Figure 5.5. Effect of protein and lipid content of food on HiF in trout (after Legrow and Beamish, 1986).

Table 5.1. Examples of energy budgets for two diets (DP: 38%, DE: 20 kJ.g^{-1} DM) in trout. Values are expressed as a percentage of digestible energy intake.

Main non-protein energy source in the diet	Lipids		Carbohydrates	
Temperature (°C)	8	18	8	18
Metabolic losses	5.2	4.3	5.4	5.0
Expenditure	39.9	32.3	41.8	43.1
fasting metabolism	28.1	21.3	29.6	23.7
increment of feeding (HiF)	11.8	11.0	12.2	19.4
Retention	54.9	63.3	55.4	49.3
in the form of proteins	30.8	25.9	30.0	22.8
in the form of lipids	24.1	37.4	25.4	26.5

At the start of the twentieth century, researchers sought to link HiF to protein catabolism. More recently, it has been shown that in growing animals this expenditure mainly reflects the cost of protein accretion. In general terms, it results from all heat energy losses at the time of energy transformations from one state to another during metabolism. HiF thus does not have a constant value for a given food; its value is related to food utilisation and interactions between nutrients. HiF is also influenced by water temperature. For each species, there is a thermal zone in which it has a minimal value. As shown in Figure 5.6, this temperature is around 10°C for cod and 15°C for trout.

The NE of a diet (ME – SDA) is available for maintenance, muscle activity and production (somatic + gonadal growth). Although NE appears to be an interesting way of expressing food energy, no system for measuring it has yet been developed for

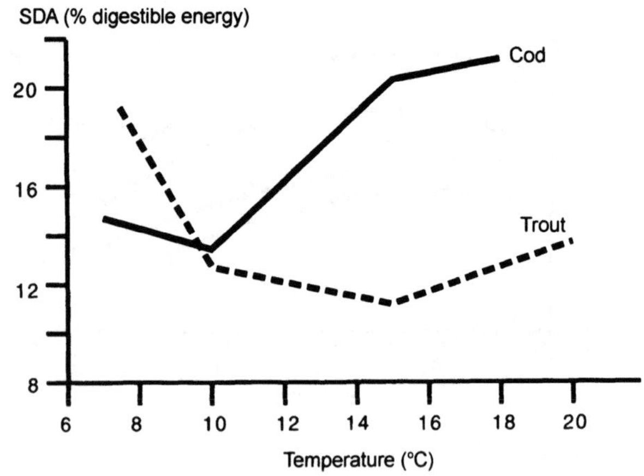

Figure 5.6. Effect of water temperature (°C) on HiF expenditure (% of digestible energy intake) in trout and cod.

fish. Methods used for higher animals are not suitable for fish which, being ammonotelic, take more of their NE from proteins than other animals.

5.3 ENERGY REQUIREMENTS OF FISH

5.3.1 Measurement principles

Energy expenditure in the form of heat can be measured using two methods: direct or indirect calorimetry. Direct calorimetry measures the heat dissipated by the organism (conductive, convective and radiant energy) while indirect calorimetry allows estimation of the heat produced by oxidative processes. The energy used by the organism comes principally from oxidative catabolism. Over a sufficiently long time period and for a given energy substrate, the quantity of oxygen consumed by an animal is proportional to its energy expenditure. It is for this reason that heat loss is generally quantified by measuring oxygen consumption.

Measurements using direct calorimetry are carried out in adiabatic calorimeters or calorimetry chambers. This method is very difficult to put into practice for fish, since basal metabolism is low and the calorific capacity of water is very high, requiring very precise heat-measuring instruments. In addition, the chambers must be small for accurate measurement precision, which means that it is not possible to work with groups of fish.

The indirect calorimeter most often relies on the measurement of oxygen consumption only. For every gram of oxygen consumed, the amount of heat produced (QO_2) is on average 13.6 kJ, but it depends on the nature of the oxidised substrates; the oxidation of a given food requires a certain amount of oxygen and generates a

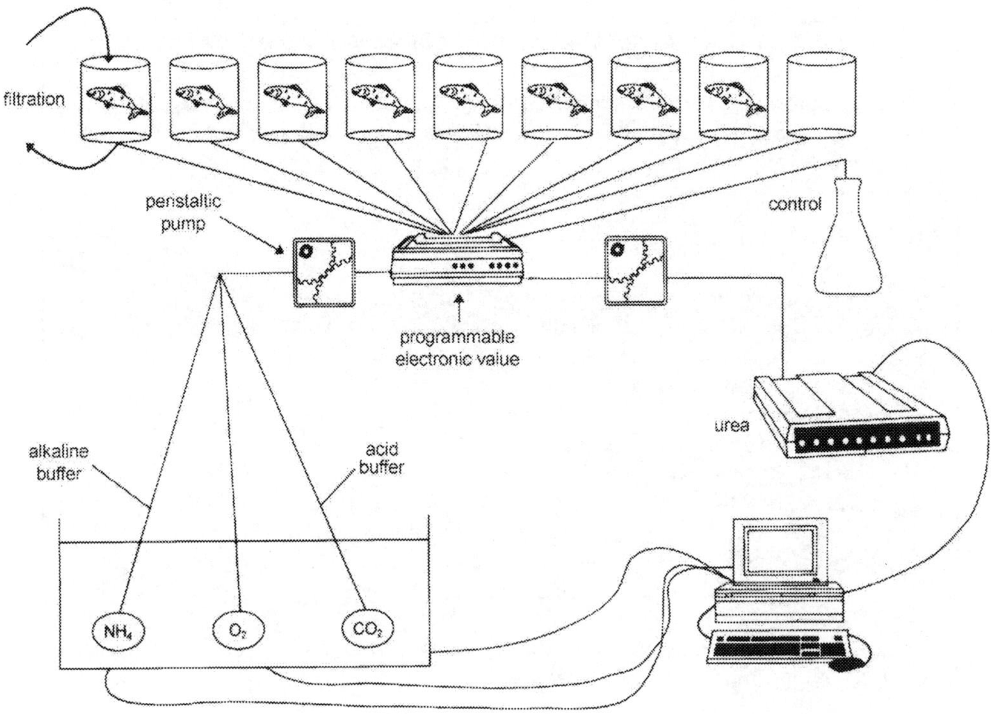

Figure 5.7. Diagram of experimental set-up for indirect calorimetry measurements in fish.

given amount of energy (Box 5.1). The evaluation of energy expenditure is more precise when the measurements of oxygen consumption are complemented by measurements of carbon dioxide and nitrogen excretion. These data allow effective estimation of the quantity of each oxidised nutrient. HiF energy expenditure is evaluated from the differences in heat production by fed and fasting fish. Figure 5.7. shows an experimental set-up for the analysis of water used for indirect calorimetry measurements in fish. Details of the method as well as data handling methods are given in Box 5.1.

5.3.2 Energy expenditure of fasted fish

Basal metabolism of terrestrial animals is measured using fasting individuals at rest. It is difficult to obtain complete rest in pelagic fish, since retaining their position in the water column requires movements and immobilisation in metabolic chambers subjects the fish to stress. Various authors have attempted to define 'standard' states for fasting, immobilised fish, 'routine' for fasting, but free-moving fish, or even 'active' for fed, free-moving fish. It seems more accurate to use the term 'minimal energetic expenditure', relative to fasting fish and including residual physical activity

Box 5.1 Principles of the indirect calorimetry method in fish

Water samples are taken from tanks containing fish and from a tank containing no fish. They are analysed, usually using specific electrodes. Gas exchanges R are calculated using the formula:

$$R = \frac{(S_1 - S_2)}{M \times (t_2 - t_1)} \times V + \frac{(E_1 - S_1) + (E_2 - S_2)}{2 \times M} \times F$$

E_1 = gas concentration in tank containing no fish at time t_1;
E_2 = gas concentration in tank containing no fish at time t_2;
S_1 = gas concentration in tank containing fish at time t_1;
S_2 = gas concentration in tank containing fish at time t_2;
M = fish biomass, in kg;
V = volume of water in the tank, in l;
F = flow of water in tanks, in $l.h^{-1}$;
t_1 and t_2 are expressed in hours and concentrations in $mg.l^{-1}$.

Energy expenditure EE in kJ is calculated using Brafield's (1985) formula:

$$EE = 11.18O_2 + 2.64CO_2 - 9.55N$$

The respiratory (RQ) and nitrogen (NQ) quotients are calculated from O_2 consumption, CO_2 output and nitrogen excretion:

$$RQ = \frac{(CO_2\text{mmoles})}{(O_2\text{mmoles})} \qquad NQ = \frac{(N\text{mmoles})}{(O_2\text{mmoles})}$$

The fraction P of proteins catabolised can be calculated from NQ. If proteins are the only substrate which is oxidised, RQ will be close to 0.97 and NQ to 0.27 (NB: these values correspond to a 'standard' protein). The fraction of proteins in total catabolism is therefore:

$$P = NQ \times 100/0.27$$

This value allows the calculation of non-protein RQ (RQ_{np}) and thus, the respective fractions of carbohydrates (C) and lipids (L) in catabolism. In practice, the RQ of carbohydrates is about 1 and that of lipids close to 0.7 (value for trioleine).

$$C = 333.33 \times RQ_{np} - 233.33$$

$$L = 100 - C$$

The quantities of oxygen consumed ($mmoles/kg^{-1}day^{-1}$) to oxidise the different nutrients, either O_2p, $O_2np \times C$ or $O_2np \times L$ for proteins, carbohydrates or lipids respectively, can thus be calculated. The oxycalorific coefficients of proteins, carbohydrates and lipids are 19.1, 21.1 and 19.6 $kJ.l^{-1}$ of oxygen consumed respectively, in the fish (Brafield, 1985); estimates of energy production from the catabolism of each different type of nutrient can be made and eventually the total energy expenditure.

When lipogenesis from carbohydrates is greater than lipid oxidation, the respiratory quotient is greater than 1. In this case, the 'oxidation' values for lipids are negative, indicating net lipogenesis (the difference between lipogenesis and lipolysis). The 'oxidation' values for carbohydrates actually represent the disappearance of carbohydrates (oxidation + conversion into lipids).

When only oxygen consumption can be measured, energy expenditure in kJ can be estimated less precisely.

than to use the expression 'standard expenditure'. The expenditure must be measured for fish that are well acclimated to their environment and starved for a sufficiently long period for the digestive tract to be empty (Chapter 4). In order to limit stress, it is also preferable to carry out measurements on groups of fish rather than on isolated individuals.

The expenditure is 10–30 times lower than in mammals, mainly because of ectothermy, ammonotelism and buoyancy:

- ectothermy: fish (with the exception of the tunas and a very small number of other taxa) have an internal temperature very close to that of the water in which they live; thus, unlike homeotherms, they do not expend energy to maintain a constant body temperature (thermoregulation). When the water temperature decreases, that of the organism decreases as a consequence and basic energy requirements are greatly reduced;
- ammonotely: in teleosts, the main nitrogenous catabolite, ammonia, is excreted directly into the water across the gills and in the urine. This involves a passive process that is very low in energetic cost (in contrast to the synthesis and excretion of urea and uric acid);
- buoyancy: this allows energy sparing in comparison to terrestrial animals which have large antigravitational muscles. However, while rapid swimming is much more costly than moving at the same speed on land, slow swimming is less costly.

5.3.3 Factors affecting minimal energy expenditure

The energy requirements of fish vary mainly as a function of their size, as well as with water temperature (Figure 5.8). For freshwater fish (10–250 g), average daily expenditure is between 25 and 45 $kJ.kg^{-1}$, depending on temperature. For one species at a given temperature, the values can vary with length of fasting period and nutritional history. Values reported in the literature for marine fish vary from one species to another more than for freshwater species, flatfish having a particularly low metabolism. The influence of the water salinity on basal metabolism of fish is again controversial, but the energy costs of osmoregulation appear to be low.

In ectotherms, as in homeotherms, the quantity of energy expended increases, in absolute terms, with the mass of the animal. The logarithm of metabolic expenditure increases linearly with the logarithm of body mass. However, the slope of this relationship is lower than 1: in all species, small animals expend more energy per unit mass than large animals. As a first approximation, the logarithmic relationship between size and metabolism is valuable, both for adults of different species and for

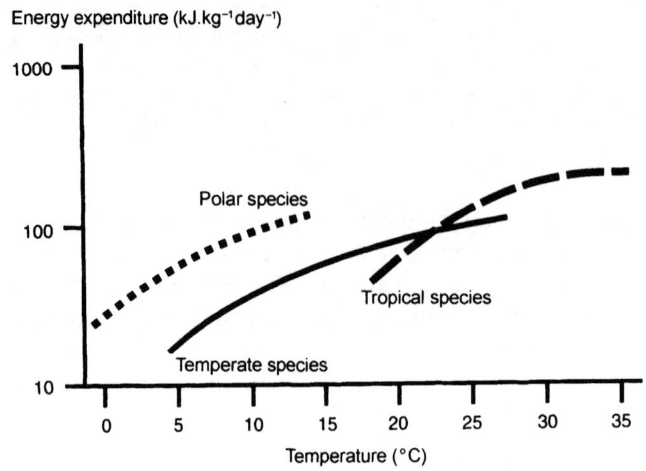

Figure 5.8. Effect of water temperature (°C) on energy expenditures of different fish species (after Brett, 1970).

different-sized members of the same species. Thus, fish larvae have considerably higher energy requirements per unit body mass than large fish. The relationship between body metabolism and body mass is described by the concept of metabolic mass (sometimes wrongly called 'metabolic weight'), i.e. the body mass of an individual raised to a power corresponding to the slope of the logarithmic relationship. For higher vertebrates, the accepted exponent is usually 0.75. For fish, cited values vary from 0.63 to greater than 1; the coefficient most often used being 0.82. However, in studies on fish, energy expenditure is still often expressed not as a function of metabolic mass, but as a function of body mass, which varies from a few mg in larvae to several kg.

The increase in minimal energy expenditure in relation to water temperature is generally expressed by Q_{10}. This criterion is independent of fish size. For a given species, it is not constant over the entire temperature range within which it lives (Figure 5.8). It is often lower in the preferred thermal zone for the species. For example, Q_{10} is about 3.5 below 8°C, for oxygen consumption in trout, and about 1.7 between 8 and 18°C. In general terms, a Q_{10} lower than 2 indicates regulation by thermoenzymes. These allow the fish to conserve a higher metabolism which could not be possible by simple physical dependence on temperature. It partially compensates for the consequences of ectothermy. In other cases, however, and in particular in the case of a Q_{10} of 3.5 as cited above, it must be assumed that metabolic regulation corresponds to different mechanisms.

When an animal is starved, energy that is vital for survival is provided by the breakdown of body reserves. During the first three days of fasting in trout, proteins are the principal energy sources. Their participation in covering basal energy requirements is about 10 times greater than in higher vertebrates. Energy expenditure is on average 20 times less than in terrestrial vertebrates, while nitrogen losses are only two times lower (100–200 mg N.kg$^{-1\,0.8}$day^{-1} in fish, in contrast to 220–320 mg N.kg$^{-1\,0.8}$day^{-1} in terrestrial mammals, representing 2–5 mg endogenous nitrogen.kJ^{-1} in fish in contrast to 0.2 mg.kJ^{-1} in terrestrial mammals). When fasting is prolonged, lipids play an increased role in the supply of energy. In trout at 18°C, for example, from the twelfth day of fasting it is equal to that of proteins.

5.3.4 Maintenance energy requirements

The energy requirements for maintenance are defined as the quantity of energy necessary to reach a level of equilibrium between energy input and expenditure, such that the energy balance is nil. It has only rarely been evaluated in fish, sometimes being equated to the energy requirements of fasted fish, although this only represents 50–60% of maintenance requirements (Figure 5.9). A feeding ration providing only the energy necessary to cover the energy requirements of starvation is not sufficient for the animal to maintain constant body mass. As shown above, food intake, digestion, absorption and metabolic utilisation of nutrients lead to particular energy expenditure or HiF. Thus it is the HiF of maintenance as opposed to the SDA of production that reflects the cost of energy accretion.

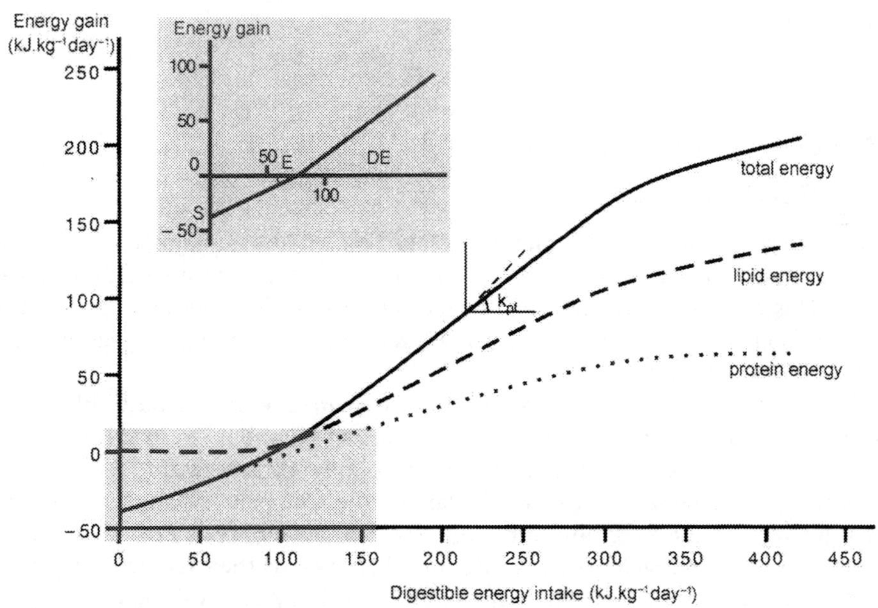

Figure 5.9. Variation in energy gain in trout as a function of digestible energy intake. The slope k_{pf} shows the efficiency of DE utilisation for energy accretion; $1 - k_{pf}$ shows HiF of production. In shaded area: change in energy budget around maintenance E. OS: budget of starved animal. km = OS/OE shows the efficiency of DE utilisation for maintenance; this ratio is less than 1 as a result of HiF of maintenance.

Table 5.2. Energy requirements for maintenance in some fish species.

Species	Body mass (g)	Temperature (°C)	Energy expenditure ($kJ.kg^{-1}day^{-1}$)
Common carp	80	10	28
	80	20	67
Catfish	10–20	25	84
	100	25	72
Trout	150	18	85–100
	300	15	60

The most reliable method for evaluating maintenance requirements consists of measuring the variations in energy content of groups of fish fed increasing amounts of energy and extrapolating or interpolating to obtain an energy gain equal to zero (Figure 5.9). Table 5.2 shows the values for maintenance requirements for three species. Variations in requirements in relation to body mass and water temperature result from the effect of these factors on minimal metabolism, which represents the principal component of maintenance expenditure. For the catfish at 25°C, as for the trout at 18°C, daily maintenance ration corresponds to approximately 0.4–0.6% of body mass.

5.3.5 Energy requirements for production

5.3.5.1 Growth requirements

When fed to satiation, or by demand feeders, fish are capable of adjusting their food consumption depending on the energy density of the food in order to attain maximal growth. This adaptation is not always exact and stomach filling may be limiting when the food is low in energy content. In general, diets should provide at least 15–18 MJ DE per kg dry matter.

The energy requirements for growth can be defined as the quantity of energy necessary to produce one kilogram of fish. In trout, this is about 15–16 $MJ.kg^{-1}$ gain in body mass at 8°C and 17–19 $MJ.kg^{-1}$ gain in body mass between 15 and 18°C. Values for growth requirements are similar for the Channel catfish, but they are clearly higher for the common carp and the sea bream, both reared at 25°C. Therefore, it is not possible to generalise for all species, as their requirements have not been measured at the same temperature and their body composition may be very different. It is supposed that the composition of the gain in body mass (protein and lipid content) influences the energetic requirements for growth (expressed in $MJ.kg^{-1}$ gain in body mass), as this is the case for terrestrial animals.

The requirements thus defined are also related to the composition of the diet: the orientation of nutrients towards anabolism or catabolism depends on the energy

Table 5.3. Relationship between digestible protein and digestible energy currently recommended for growth in different fish species.

Species	DPr (mg.g^{-1} DM)	DE (kJ.g^{-1} DM)	DP/DE (mg.kJ^{-1})
Catfish	270–244	13.1–12.8	19–21
Red drum	315	13.4	24
Nile tilapia	300	12.1	26
Common carp	315	12.1	26
Trout	330–420	15.1–17.2	22–25

Source: NRC, 1993.

balance between protein and non-protein energy. Consequently, the recommendations concerning the composition of diets are given in terms of the ratio DP/DE. These values (Table 5.3), are much higher than for pigs or chickens, where they are about 10–14 mg.kJ^{-1}, but they appear to be overestimated for some species. Several recent studies have shown that in trout the DPr/DE ratio can be lowered to 18 mg.kJ^{-1} by increasing DE to 18 MJ.kg^{-1}, without growth being adversely affected. The situation is similar for carp and sea bream.

5.3.5.2 Requirements for reproduction

The cost of reproduction exceeds that of simple gamete production. The development of secondary sexual characteristics (morphological or coloration changes), mucus production, reproductive behaviours, nest construction for certain species or migration for others lead to supplementary energy expenditure. The cost of all the processes of reproduction can only be evaluated in the long term by reproductive effort, which corresponds to the quantity of energy stored in the gametes, in relation to the quantity of energy ingested. Values for aquarium species, reported by Wooton (in Tytler and Calow, 1985), varied from 10–18% for the medaka and 16–30% for the stickleback.

In most species, male gonads at maturity represent no more than 10% of body mass. In contrast, mature ovaries can represent up to 30% of body mass in some species. The mean energy content of the eggs, measured for around 50 teleosts, is about 23.5 kJ.g^{-1} DM, whatever the egg size. In the pike, just before gamete emission, the energy content of the male gonads is about 75 kJ while that of the ovaries exceeds 1000 kJ. In salmonids, the amount of body energy stored in the eggs represents 8–15% of the energy of the whole body, corresponding to the gonadosomatic index.

There do not seem to be specific requirements for reproduction in terms of the protein/energy ratio of the diet. A DE level higher than 17 MJ.kg^{-1} DM associated with a protein content of 35% produces good reproductive performance in salmonids. When the energy supply from food is not sufficient, most species favour gonad production at the expense of somatic growth; thus females mobilise their body reserves for gonad development.

5.4 ENERGY BALANCE AND ENERGY RETENTION

The energy budget relates the different stages of utilisation of GE intake by an animal under given environmental conditions. The relationship between DE or ME intake and energy retained (R) describes the efficiency of food utilisation. R corresponds to the energy intake (I) from which are subtracted faecal energy losses (F), metabolic losses (BU) and losses in the form of heat (C), including expenditure linked to the animal itself and those linked to food utilisation (HiF). Based on the law of energy conservation, the energy budget is:

$$R = I - (F + BU) - C$$

5.4.1 Methods of measuring energy budgets

5.4.1.1 Energy budget based on measurement of losses

Digestive losses are calculated from intake and energy ADC. Metabolic losses ($B + U$ in $kJ.kg^{-1}h^{-1}$) are obtained by multiplying the amount of nitrogenous excretion (in $g.kg^{-1}h^{-1}$) by 24.7 (see Section 5.2.3). Energy expenditure is measured for fasted and fed fish (Box 5.1); the difference between the two represents HiF. It is thus possible to calculate the parts of DE found in gill and urinary losses that are utilised for HiF, maintenance and activity. The remaining part is production energy (somatic growth and gamete production).

This method allows evaluation of the different components of the energy budget, i.e. the different stages of energy transformation of the food. However, it often leads to overestimation of energy retained which can be as high as 10% as a result of underestimation of the various losses.

5.4.1.2 Energy budget based on measurement of body energy gain

The most direct evaluation of the amount of energy retained by a group of animals is the measurement of body energy (by combustion of a known quantity of dry body matter in a calorimeter) at the start and end of a given period of time. Measurements are carried out on samples of animals representative of the initial and final stages. This method has the drawback of not providing any indication of the other components of the energy budget. It must be complemented by measurements of metabolic losses, HiF and other heat expenditure. For such experiments, when the laboratory apparatus necessary to measure excretions and heat production is not available, Cho and Kaushik (1990) recommended another method for evaluating the components of the energy budget: metabolic losses are obtained from the difference between digestible nitrogen intake and nitrogen retained. Their energy value is calculated using a coefficient of 25 (which can beneficially be replaced by the coefficient 24.7, as explained above). ME is obtained by subtracting this product from DE. Total heat energy expenditure corresponds to the difference between ME intake and energy retained.

5.4.2 Energy retention

Energy retention (accretion) varies in relation to fish size and rearing conditions (Table 5.4), as well as ration size and diet composition. Figure 5.9 shows the relationship between daily gain and DE intake in two groups of trout fed over eight weeks with increasing amounts of food, body energy content having been evaluated using the method described above. The results are expressed as gain or loss in total energy, or in the form of protein per kg and per day. Energy gain increases linearly at first, in relation to DE intake, the slope k depending on the composition of the diet. However, above a certain threshold, the relationship becomes curvilinear: the increase is less significant when the amount of digestible energy intake increases, indicating that increasing HiF production is more than proportional to intake. The same trend is even more marked for energy retention in the form of protein. The slope (marked as k_p) of the relationship between daily energy gain in the form of protein and the quantity of DE intake per kg and per day shifts when the latter increases: this phenomenon results from the existence of maximal protein anabolism.

In addition, whatever the feeding regime, the form – whether protein or lipid – of energy retained varies with age (and thus size) of the fish. When animals are young, energy is stored mainly in the form of protein and then, at around two years of age in

Table 5.4. Basal energy expenditure of some fish species during fasting.

	Body mass (g)	Temperature (°C)	Energy expenditure ($kJ.kg^{-1}day^{-1}$)
Freshwater species			
Goldfish	29	25	39
Common carp	80	28	41–50*
Sturgeon	230	18	30
Yellow pickerel	4–130	20	25
Sockeye salmon	10–20	15	23
Trout	100	8	29
	100	18	40–48*
Marine species			
Gilthead sea bream	10	20	79
	100	20	41
Red sea bream	35–70	20	64
Cod	30–80	7	30
	30–80	10	43
	30–80	15	52
Sea bass	10	20	81
	100	20	48
Plaice	5	20	163
	110	20	30

* Range of values found in the literature.

Table 5.5. Effect of water temperature on the utilisation of digestible energy and maintenance expenditure in trout. Diet contain 38% digestible protein and 9% digestible fat (from Cho & Slinger, 1980).

Temperature (°C)	7.5	10	15	20
Energy retained (% DE intake)	44	49	53	58
Maintenance ($kJ.kg^{-1\,0.824}/day$)	12	19	35	38

salmonids, energy deposited as lipid form becomes preponderant. The energy content of a fish's body can vary from 20–30 $kJ.g^{-1}$ DM, depending on the amount of lipids stored. This phenomenon probably increases the requirements for growth.

Thus, one type of food can result in different quantities and natures of energy deposition, in relation to the animal and the rearing conditions. Many experimental studies are still required to be able to accurately predict the fate of food energy, by integrating the different variation factors.

Energy retention usually represents 20–50% of GE intake, or 35–65% of DE intake. These values are higher than those observed in homeotherms. The difference can be attributed mainly to the low level of basal metabolism, or maintenance requirements, itself due to ectothermy and secondary factors such as ammonotely and buoyancy.

The role of temperature in the energy budget has not been studied in depth because nutritionists most often choose a temperature close to the optimum for the species. The energy budget is, however, strongly dependent on this factor (Table 5.5). Around and below the thermal optimum, intake and expenditure remain in a reasonably constant relationship. Retention is maximal around the optimal temperature for the species or a little above it. Above the optimum, intake level does not increase while maintenance expenditure continues to increase. At temperatures close to the lethal maximum, the budget deteriorates rapidly.

5.5 CONCLUSION

The various stages of energy production from food in fish are very similar to those known for mammals. Energy metabolism differs specifically from that of higher vertebrates from two points of view: firstly, at a cellular level, the nutrients providing energy are not used with the same priority as in omnivorous terrestrial vertebrates, and secondly, at the whole organism level, ectothermy results in a very marked influence of environmental temperature on metabolism. Nutrients used preferentially for energy production are the amino acids, the catabolism of which results in ammonia production. It is possible, however, in the salmonids at least, to substitute significant amounts of lipid for these nutrients in order to reduce protein catabolism. The consequences of ectothermy are often attenuated by thermoenzymes. Water temperature can have other repercussions on some components of energy metabolism, such as HiF or the proportions of protein or lipid retained. The low energy requirements resulting from ectothermy are also at the origin of the high DP/DE ratio which should be respected in fish diets, as well as the high efficiency of fish in retaining dietary energy as body energy. For several species, the recommendations relative to DPr/DE currently are under revision in view of protein economy and better energy efficiency.

5.6 BIBLIOGRAPHY

Blaxter, K. (1989) *Energy Metabolism in Animals and Man*. Cambridge University Press.

Brafield, A. E. (1985) Laboratory studies of energy budgets. In: Tytler, P. and Calow, P. (eds) *Fish Energetics. New Perspectives*. Croom Helm, London, 257–281.

Brett, J. R. (1972) The metabolic demand for oxygen in fish, particularly salmonids, and a comparison with other vertebrates. *Respir. Physiol.*, **14**, 151–170.

Brett, J. R. and Groves, T. D. D. (1979) Physiological energetics. In: Hoar, W. S. and Randall, D. J. (eds) *Fish Physiology*. Vol. III. Academic Press, London, 279–343.

Cho, C. Y. and Kaushik, S. J. (1990) Nutritional energetics in fish: energy and protein utilization in rainbow trout. *World Review of Nutrition and Dietetics*, **61**, 132–172.

Cho, C. Y. and Slinger, S. (1980) Effect of water temperature on energy utilization in rainbow trout (*Salmo gairdneri*). In: *Proceedings of 8th Symposium on Energy Metabolism*. Butterworth, London, 287–291.

Cowey, C. B. and Walton, M. J. (1989) Intermediary metabolism. In: Halver, J. E. (ed) *Fish Nutrition*, 2nd edition. Academic Press, San Diego, 259–329.

Guillaume, J. (1990) La nutrition des salmonidés. I – généralités, besoins qualitatifs. *La Pisciculture Française*, **100**, 47–59.

Guillaume, J. (1991) La nutrition des salmonidés. II – besoins quantitatifs. *La Pisciculture Française*, **103**, 4–20.

Kaushik, S. J. and Médale, F. (1994) Energy requirements, utilization and dietary supply to salmonids. *Aquaculture*, **124**, 81–97.

Legrow, S. M., Beamish, F. W. H. (1986) Influence of dietary protein and lipid on apparent heat increment of feeding of rainbow trout, *Salmo gairdneri*. *Can. J. Fish. Aquat. Sci.*, **43**, 19–25.

National Research Council (NRC) (1993) *Nutrient Requirements of Fish*. National Academy Press, Washington D.C.

Tytler, P. and Calow, P. (1985) *Fish Energetics. New Perspectives*. Croom Helm, London.

6

Protein nutrition

Protein nutrition is without doubt the most studied area within the whole of fish nutrition. Since the first attempts to use manufactured foods in fish culture, great efforts have been made to define the optimal protein content of diets. It very rapidly became evident that this level is much higher than for mammals or birds. Several complementary routes have been explored in order to explain the particular requirements of fish in this area. Whether the spectacular difference between fish and higher vertebrates results from particularly high requirements, either for maintenance or for growth, has been studied in depth. Great effort has been put into defining requirements in terms of essential amino acids (EAA), at least for some species. Explanations have also been sought in the fundamental mechanisms and, although most research has been carried out on protein metabolism, the role of proteins as well as sources of non-protein nitrogen compounds has also been explored.

In this chapter, the main characteristics of protein metabolism in fish will firstly be described, then protein and AA requirements, as well as the interrelations between proteins and energy.

6.1 CHARACTERISTICS OF PROTEIN METABOLISM IN FISHES

6.1.1 Amino acid metabolism

In animals, AAs are either in the free form or linked to proteins (i.e. incorporated within peptide chains). Free AAs have three possible origins: intestinal absorption of the products of hydrolysed food proteins, *de novo* synthesis and interconversions and lastly hydrolysis of body proteins. They can be used for synthesis of body proteins or other nitrogenous compounds (nucleic acids, amines, peptides, hormones etc.), provide a carbon source for intermediate metabolism or be oxidised to provide energy.

6.1.1.1 Synthesis

Fish, like higher vertebrates and many invertebrates, are incapable of synthesising certain AAs which must be provided by the diet. These AAs are termed essential (Chapter 2). To determine the identity of the essential amino acids, two types of method can be used.

The oldest and most classical methods consist of removing the compound assumed to be essential from the diet and monitoring the animals over a period of several weeks or months, looking for signs of deficiency. For amino acids, these experiments hardly ever provide evidence of specific deficiencies, except for methionine, lysine and tryptophan, deficiencies of which cause particular pathological symptoms in fish. In contrast, a deficiency in an essential AA leads immediately to a cessation in growth followed by a decrease in body mass. The reversibility of these symptoms after input of the missing substance is generally considered as sufficient proof. Thus, in 1974 Nose reviewed 18 AAs, substituting them one at a time in a mixture of AAs forming the dietary protein for carp. In ten cases, at the end of four weeks, a loss in body mass had been noted, while the addition of the removed AA re-established growth. Although they do not require any delicate biochemical measurements, these methods are long and relatively cumbersome.

More elegant methods, which are much more rapid and rely on biochemical techniques, in particular the use of radioactive labels, are preferred today. There are two variations on the method used: the fish receive an injection either of acetate or glucose, with labelled carbon ^{14}C and, after a few hours, they are sacrificed. A fraction of body proteins is taken and hydrolysed. The AAs obtained are then separated by chromatography and the radioactivity level assessed; this can only be found in non-essential amino acids (NEAA). This technique is very rapid and allows the simultaneous study of the essentiality of all the AAs. It does have its critics, however: after analysis, the acid hydrolysis of a given protein gives the same product for glutamine as for glutamic acid and the same for asparagine as for aspartic acid. In addition, this method does not allow distinction to be made between molecules which cannot be synthesised, and those which can be synthesised by the organism from a substance provided by the food, but which itself cannot be synthesised by the organism.

The two types of experiment lead to the same conclusions: the same ten AAs (Table 6.1) are essential for fish and higher, non-uricotelic, vertebrates. Arginine is one of the EAAs in fish, while in uricotelic animals it can at least partly be provided by the urea cycle. Two semi-essential AAs can also be identified in fish that can only be synthesised from EAAs. These are cysteine and tyrosine, which are derived from the 'coupling' of serine-methionine and phenylalanine respectively. Where AA synthesis is slow they are sometimes termed 'semi-indispensable': in fish proline and glutamine could fall into this category.

Under certain critical conditions, the rate of synthesis of NEAAs can be too slow and growth can be slowed down. Without being indispensable, they thus acquire an essential character. It is for this reason that the term 'non-essential' has been chosen. The NEAA can be synthesised from the corresponding α-ketoacids by

Table 6.1. Essential, semi-essential and non-essential nature of amino acids.

A. Classic distinction for fish.

Essential	Non-essential and semi-essential
Arginine	alanine
Histidine	asparagine
Isoleucine	aspartic acid
Leucine	glutamic acid
Lysine	glutamine
Threonine	glycine
Tryptophan	proline
Valine	serine ↓
Methionine →	cysteine*
Phenylalanine →	tyrosine*

* Semi-essential amino acids

B. Classification of amino acids taking into account the essential features of the carbon framework and main aminated group (after Jackson, 1983).

	Carbon framework	
Aminated group	Essential	Non-essential
Essential	lysine threonine	serine glycine cysteine
Non-essential	branched AAs* tryptophan phenylalanine methionine	glutamate, glutamine** alanine aspartate, asparagine**

* Isoleucine, leucine, valine
** These amino acids may be essential in fish.

transamination (reversible transfer of an amino group from an amino acid to an α-ketoacid, without release of ammonia). The classical concept assumes that the indispensable character is conferred by the carbon chain, which the animal no longer has the capacity to synthesise. It has now been shown, in the higher vertebrates, using ^{15}N, that there are several distinct cases. The EAAs can be divided into three categories (Table 6.1b, and Box 6.1):

- those which are synthesised within the framework of intermediate metabolism, without this synthesis leading to a positive energy balance, as a result of the low activity of the enzyme systems involved: arginine and histidine;
- those which can be synthesised by transamination, but at an insufficient rate, due to a lack of substrate: leucine, isoleucine, valine, tryptophan, phenylalanine, methionine;

Box 6.1 Formulae of the twenty protein amino acids and importance of their radical or amine group

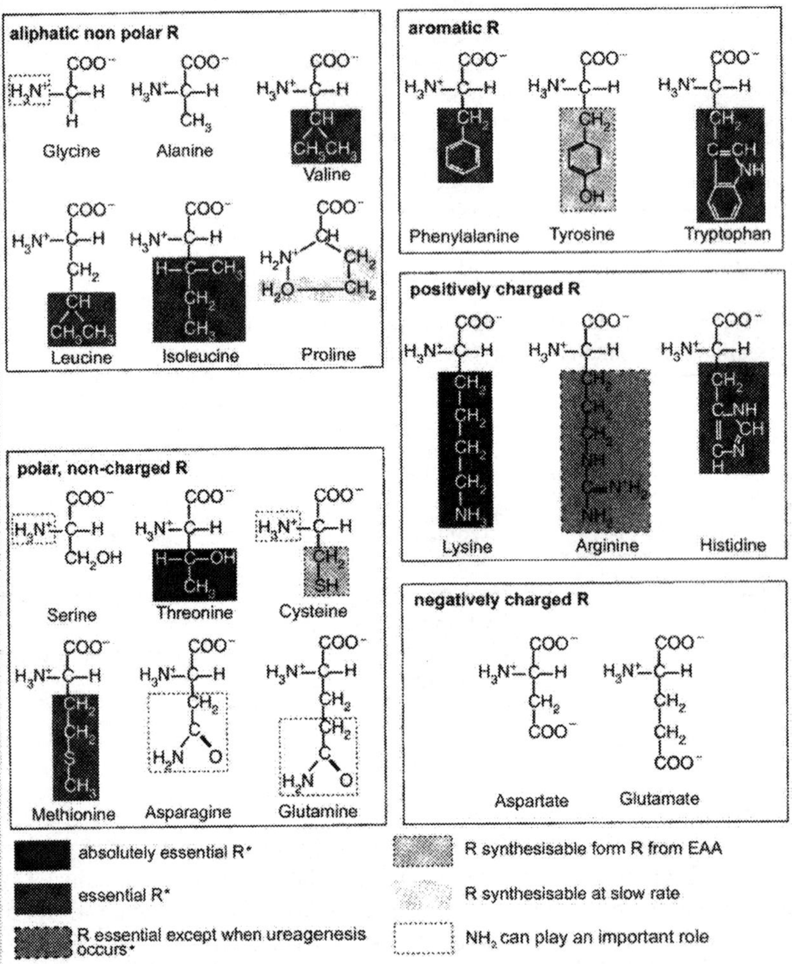

* In practice, and especially for food formulation, it is vital to consider as essential, the 10 amino acids listed, as those in Table 6.1A.

- those which cannot be synthesised from any intermediate metabolite, because of the absence of the required transaminases: lysine and threonine.

Consequently, only the last two AAs are absolutely essential (see Box 6.1). The eight other EAAs are not absolutely essential, in theory and in higher vertebrates at least. It could be assumed that the situation is the same in fish, in which the routes of synthesis appear to be similar. This, however, remains to demonstrated and for practical applications the essential nature of the ten AAs in Table 6.1a does not need to be questioned.

6.1.1.2 Utilisation

In fish, as in terrestrial vertebrates, body proteins are not static in character: protein synthesis and degradation occur all the time, drawing on the pool of free AAs and feeding it at the same time from the flow of food (Figure 6.1). The study of AA metabolism and, more particularly, the quantification of requirements, must take into account this dynamic aspect and especially the turnover rate of AAs in each compartment. In fish, studies have until now concerned two freshwater species, trout and carp, amphihaline species such as the Atlantic salmon or marine species such as the sea bass receiving only little attention.

Techniques for studying metabolic flux of amino acids These techniques utilise stable isotopes (^{13}C, ^{2}H, ^{15}N), which have still been little used in fish, or radioactive isotopes (^{14}C or ^{3}H). Fluxes are usually quantified after administration of labelled AA and measurement of the specific radioactivity (radioactivity/mole of relevant compound) of the AAs, free and linked to tissue proteins, as well as in excretory products. Analysis of the kinetic process follows the administration of label and is classically achieved using a two-part mathematical model, one representing the free AA pool, the other, the body proteins (Figure 6.1b). The AA can be administered using one of three methods: simple injection, perfusion, or simple injection associated with flooding using unlabelled AAs (see Box 6.2).

The simple injection technique requires laborious monitoring of the kinetics of appearance and disappearance of the labelled products. In addition, the reutilisation of the labelled AA as a result of protein turnover biases the results. In terrestrial vertebrates, perfusion is now the preferred method. In the aquatic environment, this method is difficult to apply without the risk of seriously stressing the animals, except with the use of subcutaneous micro-pumps which, however, do not guarantee ideal perfusion of products. These methodological difficulties especially limit measurements of oxidation rates of AAs (ko), while protein synthesis rate (ks) can be measured by the large dose, or flooding, method. The latter consists of flooding the free AA pool to reach equilibrium between specific activity of free AAs in the plasma (close to specific activity of the injected solution) and that of free AAs in the tissues.

None of these methods allows direct measurement of proteolysis or protein degradation (kd). Since protein deposition (or accretion) is the result of synthetic and proteolytic processes, the latter could be estimated from the difference between

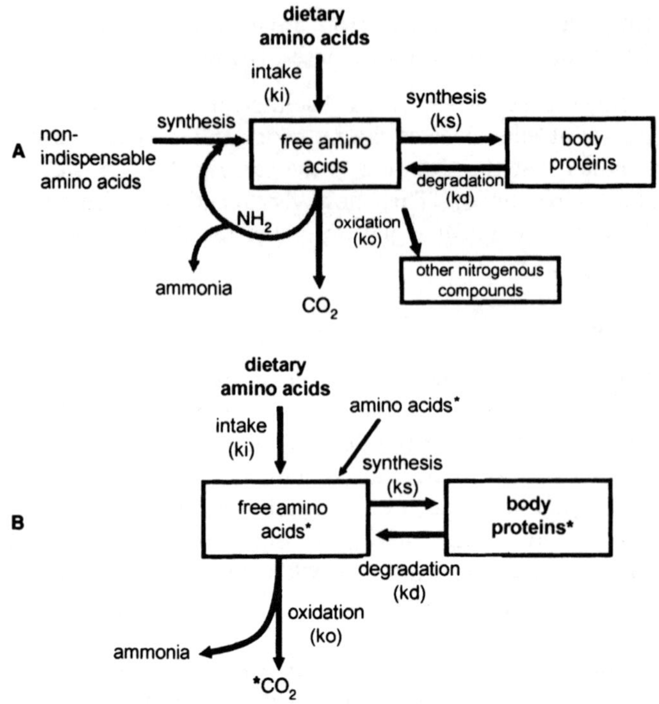

Figure 6.1. Diagram of total protein metabolism in fish. A: general model. B: two compartment model used after injection of labelled amino acid (indicated by an asterisk).

accretion and synthesis. However, accretion is measured from growth experiments over several weeks, while synthesis is usually estimated over a period of 40 minutes. Many criticisms have therefore been made and it would be desirable in the future to develop a specific technique to allow the rational study of amino acid fluxes.

Synthesis and protein accretion As in higher vertebrates, the rate of synthesis (ks) of proteins varies according to the nature of the tissue. In fish, the hierarchy is as follows: liver > gills > digestive tract > red muscle > white muscle (Table 6.2). The ks of the liver and gills is comparable to that of the liver and kidney in the rat, if the lower internal temperature of the ectotherms is taken into account. Levels in white muscle are very low in fish: 4–5 $g.kg^{-0.75}day^{-1}$, in contrast to 16 $g.kg^{-0.75}day^{-1}$ in mammals.

Most of the protein synthesised is retained in the white muscle of fish (50–70%, in contrast to 25–40% in mammals). These differences can be explained by the peculiarities of muscle growth in fish, which occurs by hyperplasia (continuous recruitment of muscle fibres) and not by simple hypertrophy, as in the higher vertebrates. Taking into account their relative proportions in the whole body, protein synthesis in the liver contributes to less than 2% of protein accretion in the fish, in contrast to 60% for white muscle and 65% for total muscle.

Box 6.2 Methods for measuring the dynamics of protein metabolism in fish; advantages and limitations.

Method	Measurement principle		Possible measurements	Advantages	Limitations
Simple injection		Series of samples taken over the period of change (6h), adjusted to estimate flux in amino acid utilisation	ks ko	simple minimal stress	uses many groups of fish reutilisation of label no state of equilibrium: significant variations in Sa
Perfusion	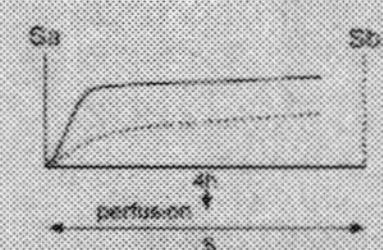	After checking equilibrium, one sample (4h) is sufficient; flux is measured using the Sa/Sb ratio	ks ko	few samples state of equilibrium individual measurements	long waiting time for equilibrium difficult to perfuse fish (stress) or use micro-pumps (reutilisation of label)
Large dose	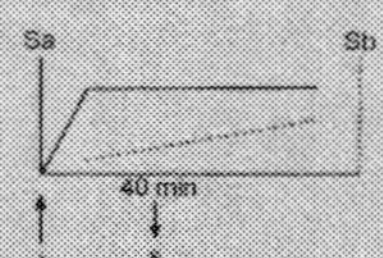	After checking linearity, one sample (40 min) is sufficient; synthesis rate is calculated from the Sa/Sb ratio	ks	minimal stress few samples rapid measurements little reutilisation of label individual measurements state of equilibrium	effect of large dose on protein metabolism(?) only measures synthesis rates

s = sampling
Sa = specific radioactivity in the free labelled AA
Sb = specific radioactivity in protein-bound labelled AA

Table 6.2. Synthesis rate (ks) of proteins (%/day) in various organs of fed fish, mininum and maximum found in the literature and accretion efficiency [(ks−kd)/ks × 100].

Organ	Synthesis rate	Accretion efficiency
Liver	5.3–20.0	5
Gills	2.4–23.0	–
Digestive tract	1.3–21.2	8
Stomach	7.8–18.3	–
Red muscle	0.3–7.7	–
White muscle	0.1–1.3	70

Table 6.3. Non-protein nitrogen compounds formed from amino acids.

Amino acid	Product
Aspartic acid	pyrimidines
Alanine	dopamine
Arginine	creatine, nitrous oxide, spermine
Glutamine	purines, pyrimidines
Glycine	creatine, purines, porphyrins
Histidine	histamine
Leucine	polyamines, spermine
Lysine	carnitine
Methionine/cysteine	taurine, choline, creatine, polyamines, glutathion
Phenylalanine/tyrosine	catecholamines, thyroid hormones
Serine	choline, ethanolamine, trimethylamine
Tryptophan	serotonin, niacin

Measurements of total AA flux show that, for protein synthesis, food constitutes a better source of AAs than the degradation of body proteins. Consequently, fish are much more dependent on dietary AA than higher vertebrates supply.

Synthesis of other nitrogenous compounds AAs can act as precursors for a large variety of other nitrogenous compounds (Table 6.3). In fish, these syntheses are difficult to quantify. Some metabolic pathways are minor (e.g. hormone synthesis), while others such as the synthesis of purine bases, creatine, or even trimethylamine can require significant amounts of AA precursors. Even qualitative information on this subject is still scarce, although many peptides are metabolic or physiological regulators and other molecules (nitrous oxide) play an important role in health. All these phenomena have still been little studied in mammals and are practically unknown in fish. However, in the latter, certain non-protein amino acids, for example taurine as well as trimethylamine, are involved in an additional area, namely osmoregulation.

Oxidation of amino acids Fish can be differentiated from terrestrial vertebrates by the fact that they oxidise a large amount of AAs to provide energy and that the final product of nitrogenous metabolism is ammonia. The different routes of AA catabolism leading to the provision of energy are described in Chapter 5. Only the general routes for AA catabolism are mentioned here.

During catabolism, the carbon chain of the AA can be used for the synthesis of other nitrogenous compounds (NEAA, nucleotides, etc.), but it usually enters the Krebs cycle to be oxidised and provide energy (Chapter 5). The total oxidation of AAs can be estimated by injecting a ^{14}C labelled AA and measuring the radioactivity of the expired CO_2. Injection of an EAA leads to a lower percentage of expired $^{14}CO_2$ than that observed when an NEAA is used. However, even with an NEAA, it remains high in comparison to that obtained for the rat. Measurement of ^{14}C labelled leucine flux indicates that more than 35–49% of the free leucine pool participates in oxidative metabolism, while in mammals this value does not exceed 20%. This phenomenon results from the propensity of fish for using AAs to provide energy. The provision of AAs by proteolysis is low (less than 45% for leucine) and so it is mainly those that are provided directly by the food which are oxidised.

The NH_2 groups of catabolised AAs can be used in transamination reactions, but the majority are released by deamination reactions before being eliminated. Reactions responsible for the release of ammonia are either direct deamination or the transfer of an amino group to a common receptor which will in turn be deaminated; this is called transdeamination. The second route appears to dominate since direct deaminations require very specific enzymes, concentrations of which are too low in fish tissues to form a significant amount of ammonia. Transaminases intervene, transforming AAs into α-ketoacids and glutamate, the latter being transformed into α-ketoglutarate and NH_3 through the action of glutamate dehydrogenase (Figure 6.2). This enzyme is found in higher concentrations in fish liver than in that of higher vertebrates and plays a major role in AA oxidation.

Excretion The great majority of fish excrete 80% of their nitrogen from nitrogenous catabolism in the form of ammonia; they are ammonotelic. Ammonia, NH_3, is extremely toxic but, at physiological pH, the majority of this compound is transformed into the ionic form of ammonia, NH_4^+. This form is less toxic and fish can easily eliminate it into the external environment either via the gills or in the urine (75 and 25% respectively). There is a direct relationship between nitrogen intake and ammonia excreted. Ammonotely differentiates fish from terrestrial vertebrates which have to incorporate circulating ammonia into non-toxic, but more complex, compounds, for example uric acid or urea. The whole range of urea cycle enzymes do exist in teleosts, but are only functional in a few exceptional examples. This is true for the elasmobranchs, in which ammonia accumulating in the egg would become toxic to the embryo, or in some African fish which live in a very alkaline environment where this compound would be very difficult to excrete. There is most often secretion of a certain amount of urea (about 5–20% of excreted nitrogen), but this comes from the catabolism of purines and arginine, and not from proteins. In contrast to ammonia excretion, that of urea

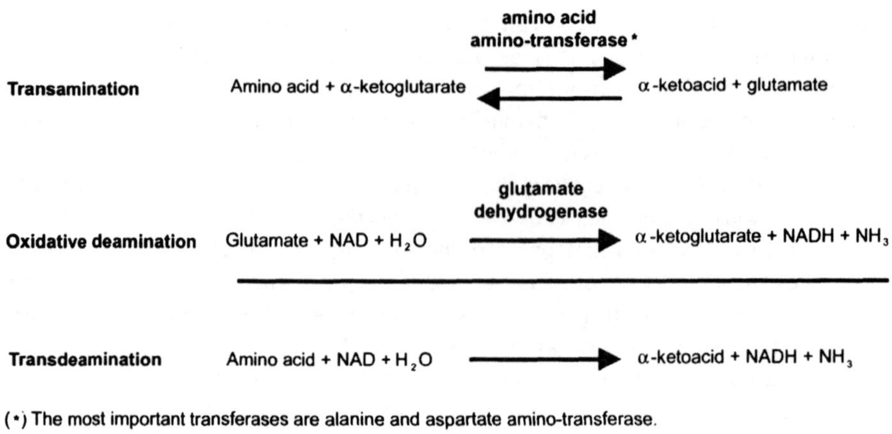

Figure 6.2. General reactions of transamination and deamination of amino acids.

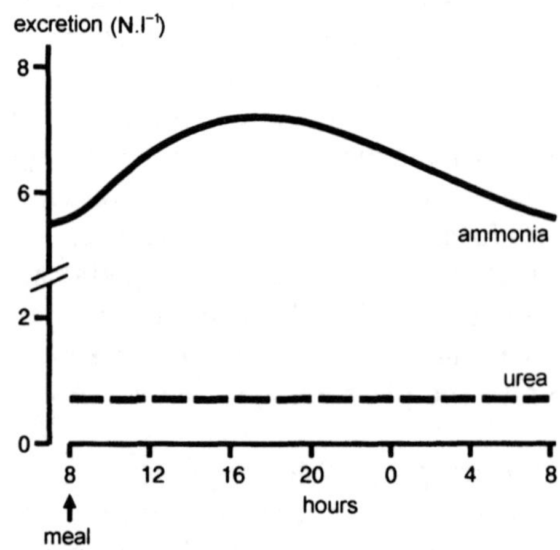

Figure 6.3. Kinetics of ammonia and urea excretion in trout (after Kaushik, 1980).

remains low and independent of meals (Figure 6.3), at least in general terms. It has recently been discovered that there are peaks of urea excretion in the turbot, but this phenomenon remains to be clarified.

6.1.1.3 Summary of features of metabolism

The main features of protein metabolism in fish are thus: low proteosynthetic activity in white muscle, low contribution of all body proteins to the AA precursor pool, significance of direct catabolism by the oxidative route of absorbed AAs and, lastly, excretion of nitrogenous waste in the form of ammonia.

Contrary to certain hypotheses, the significance of protein requirements of fish is not linked to hypertrophy of the white muscle mass. The low turnover of proteins in this tissue allows an apparent synthesis, the cost of which, for the whole organism, is of the same order of magnitude as in a higher vertebrate. However, the second and third characteristics described above explain why the protein metabolism of fish appears to be more dependent on dietary protein supply than that of the higher vertebrate. They also allow an understanding of why protein metabolism contributes more widely to covering the energetic requirements, except in particular nutritional contexts as described in Chapter 5.

6.1.2 Variations in metabolism

6.1.2.1 Age of the fish

The effect of biotic factors on nitrogen metabolism has been the subject of only a very few direct studies and, in addition, it is only possible to make very approximate comparisons between the different experiments since the methods used are far too variable. Only the effect of age has been determined: growth slows down with age but does not stop in fish; a simultaneous diminution in synthesis and degradation of body proteins is observed during ageing, accretion also progressively slowing down.

6.1.2.2 Abiotic non-nutritional factors: temperature

At first glance, the various researchers who have studied the effect of temperature on protein synthesis seem far from having reached the same conclusions, the Q_{10}s measured varying from 2.1 to 7. However, this involves an apparent rather than a real contradiction. In practice, a temperature difference does not have the same significance, depending on what is being compared:

- A: distinct fish species living in different climates to which they have become adapted during evolution and
- B: fish of one, relatively eurythermic, given species, acclimatised to certain temperatures or
- C: fish of the same species found at the same temperatures as in case B, but without having had time to acclimatise.

At the time of a variation in temperature with acclimation in a eurythermic species, Q_{10} is about 2.5. Species living in very cold waters have a relatively high synthesis rate in the muscle compared to that predicted from extrapolation from the curve showing temperature effects on temperate eurythermic species, and the inverse is observed for tropical species. Thus, fish from the Antarctic Ocean can have muscular protein synthesis rates three times higher than fish in tropical waters, at the same temperature. This difference leads to the thought that a direct compensation effect for temperature has appeared during evolution. Sudden changes in temperature without acclimation cause stress, which causes very marked effects on protein turnover, via cortisol, but does not correspond to a physiological effect.

6.1.2.3 Nutritional factors

The effect of feeding is exerted via several factors, particularly food intake, meal frequency and food quality. It has been incompletely studied and only the role of some factors will be mentioned here.

Fasting only slightly affects protein synthesis, but it clearly increases their degradation and raises aminoacidemia. Even after a long period of food deprivation, AA oxidation enzymes' activity remains high. These nutrients remain an important source of energy. The production of glucose from AAs, measured *in vitro*, increases during fasting.

When the animal is fed, synthesis and degradation increase with the amounts ingested, the former taking over from the latter once maintenance requirements have been met. As a result, accretion efficiency (ks – kd)ks increases from 0% for the maintenance ration to a value close to 50–60% for the maximal ration. It appears that the variation in this efficiency results mainly from the protein ration, although the true influence of food energy has still not been studied. As in higher vertebrates, synthesis appears to be a direct function of substrate availability, but it is certainly limited by the potential of the animal which depends on genetic and physiological factors.

Increasing meal frequency generally results in an increase in protein turnover in the muscle and liver.

The type of food modulates the protein metabolism in a complex manner. Only the role of protein supply has been studied. Increase in protein content, for a given level of feeding, leads to a simultaneous increase in protein synthesis and amino acid oxidation. The quality of food proteins can also influence turnover of body proteins, which is lower when the biological value is high.

6.2 REQUIREMENTS FOR PROTEINS AND ESSENTIAL AMINO ACIDS

6.2.1 Some definitions

The various definitions of requirements (Chapter 2) have neither the same significance nor the same usage. For example, during development, absolute requirements increase with animal size while relative requirements tend to decrease (Figure 6.4). Absolute requirements are the only ones to have any real physiological significance: the animal feeds on quantities, not percentages or ratios, and these quantities are a function of both external factors and factors linked to the animal itself. However, absolute requirements defined in this way vary rapidly throughout development, changing from a few milligrams for a small alevin to several tens of grams for a fish of several kilograms. Relative requirements are more stable, which is why they are used. Intermediate modes of expression have been proposed; thus, the quantity of proteins required per kg fish per day is of interest not only for scientific approaches but also for calculations of ration size. The replacement of body mass

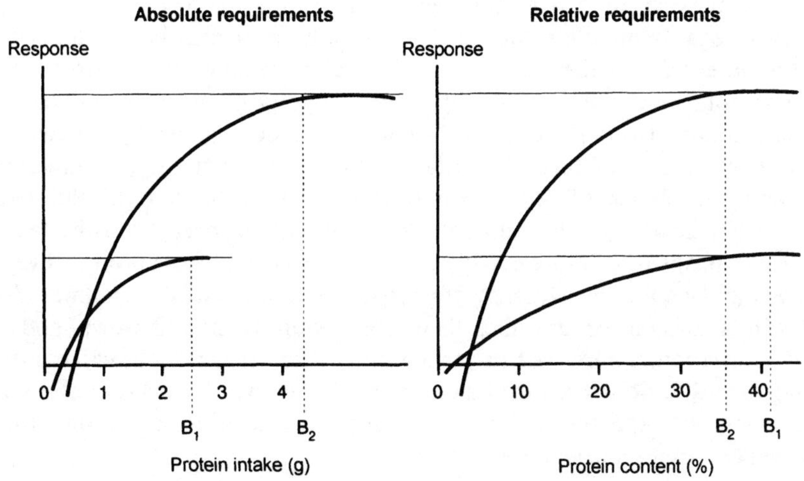

Figure 6.4. Absolute and relative protein requirements. Diagramatic representation for small (1) and large (2) fish of the same species.

M by 'metabolic size', $M^{0.75}$, which is *a priori* more valuable scientifically, is not easy to use. These different modes of expression can also be used for EAA requirements.

6.2.2 Study methods

The classic method is that of dose-response which results in the determination of either the minimal dose necessary or the optimal percentage of protein or of an EAA. The principle consists of comparing the performances or other characteristics of several groups of fish fed with feed containing increasing amounts of proteins or EAAs.

6.2.2.1 Experimental diets

The composition of experimental diets has to be rigorously controlled. To achieve a concentration gradient of dietary proteins, these are usually substituted for carbohydrates so that the energy level is hardly modified. However, the true energy value of the protein and carbohydrate source must be verified as far as possible, and the lipid or bulk (cellulose) content possibly adjusted. The possible input of vitamins, minerals, carbohydrates and fatty acids from the protein source must also be taken into account. In addition to the dietary characteristics described above, the equilibrium of EAAs plays a major role, to which we shall return. Ration size is also a factor that can markedly modify the response to a protein gradient.

When determining EAA requirements, it must firstly be ensured that the composition of the dietary protein covers the requirements for all EAAs except the one being studied. If, as is usually the case, these requirements are still unknown, a reference protein must be chosen. The EAA profile of a carcass of the species concerned, expressed in g.16 g nitrogen^{-1}, appears to provide the best guide (compared to egg or muscle proteins). EAA supply must also be maintained at a constant level, except for the one being studied, and the amount of NEAAs should be taken in total and not individually. These conditions are achieved either by formulating a diet from naturally deficient raw materials, or by replacing the actual protein with a mixture of purified AAs. In the first case, formulation is difficult and the choice of AA concentrations is very limited. In the second, growth is often slow. This slowness can result, at least in part, from the fact that purified AAs are absorbed more quickly than those produced from protein digestion, which tends to decrease the efficiency of protein synthesis. When crystalline AAs are coated with agar-agar or gelatine, their absorption is slowed down and their metabolic utilisation is improved to some extent. Two other reasons can contribute to limiting the utilisation of semi-synthetic diets: the acidity of the AA mixture which decreases the fish's appetite and the increased osmotic pressure of the stomach contents which upsets nutrient absorption.

6.2.2.2 Model

In theory, this method is simple if the response curve for the experimental diets can be described by two straight segments, with an inflection at the requirement level (Figure 6.5). In practice, the break point is not always well defined, especially in heterogeneous populations. In addition the chosen criterion does not inevitably result in a clear levelling off above the requirement level.

Other models have been proposed where the curve describing the relationship between dose and response is represented by various mathematical curves showing an asymptote. Requirement level is thus arbitrarily chosen as the value giving 95 or 98% of the maximum response. The equations for the curves giving the best

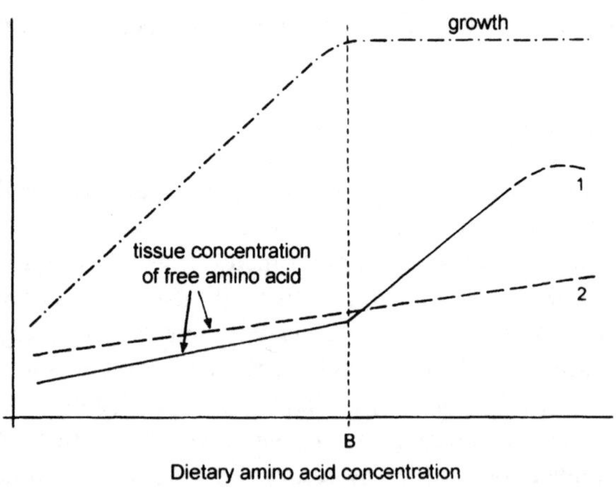

Figure 6.5. Influence of concentration of an essential amino acid on growth and on its concentration, in the free form, in the tissues. 1: amino acids for which a break point in the slope is observed (the abscissa B thus corresponds to requirements). 2: amino acids for which there is no break point.

description of the response generally provide higher estimates for requirements than the classical method of the break in the slope. These models can be criticised as being complex and not representative of reality, except in a range which can be difficult to define.

6.2.2.3 *Response*

The criteria chosen to show the development of body mass could be gain in body mass, specific growth rate, daily growth index, etc. (Chapter 2). Criteria relative to increase in length, which is less dependent on fatness, are hardly ever used, doubtless because they are less sensitive, changing approximately as a function of the cube root of the mass. Criteria related to protein accretion should, in theory, be more specific. They require supplementary measures (analysis of body proteins) and only slightly improve the significance of the results because of the constancy of the content of body proteins in fish under normal conditions.

Utilisation parameters or indices, either of food itself or its protein fraction, have sometimes been used for measuring requirements. The consumption index and its inverse, feeding efficiency, show optima for a protein input value very close to that which ensures optimal growth, but which is not necessarily the requirement (Figure 6.6a). In addition, this type of criterion is subject to errors in supplementary measurements. Utilisation criteria for protein itself, such as the protein efficiency ratio (gain in biomass/protein intake) or biological value, show an optimum for a food protein content clearly less than requirements (Figure 6.6b).

For AAs, the different dose-response methods for growth can be used, but growth can be relatively slow and the response differences take a long time to be demonstrated; the diets are also sometimes expensive. Thus 'metabolic' parameters have been sought that would be able to show instantaneously states of deficiency or satisfaction of requirements. The first measurement of this type was that of the change in free AAs in the blood. The curve obtained (Figure 6.5) shows, in general, a slow accumulation of EAAs up to the required level followed by rapid accumulation beyond it. For certain EAAs (leucine, isoleucine, valine, methionine and threonine), the break point is clear and corresponds reasonably well to the required value determined using the methods described above. In contrast, for other EAAs, no accumulation is observed beyond the required level, perhaps as a result of an increase in catabolism or interactions between AAs; the technique is therefore inoperable. Another method has been proposed which is based on the measurement of oxidation either of the AA being studied or an AA indicator. Having fed groups of fish with food containing increasing amounts of the AA being studied, the animals are injected with a solution containing the same AA or an indicator AA labelled with ^{14}C. The amount of $^{14}CO_2$ rejected, resulting from catabolism, is monitored during the following hours. Catabolism is increased above requirements for the AA being tested and below requirements for an AA indicator. The results obtained correspond well with those determined by the response method, but are less precise and require sophisticated equipment. Apart from certain cases, the application of this method thus appears to be very limited.

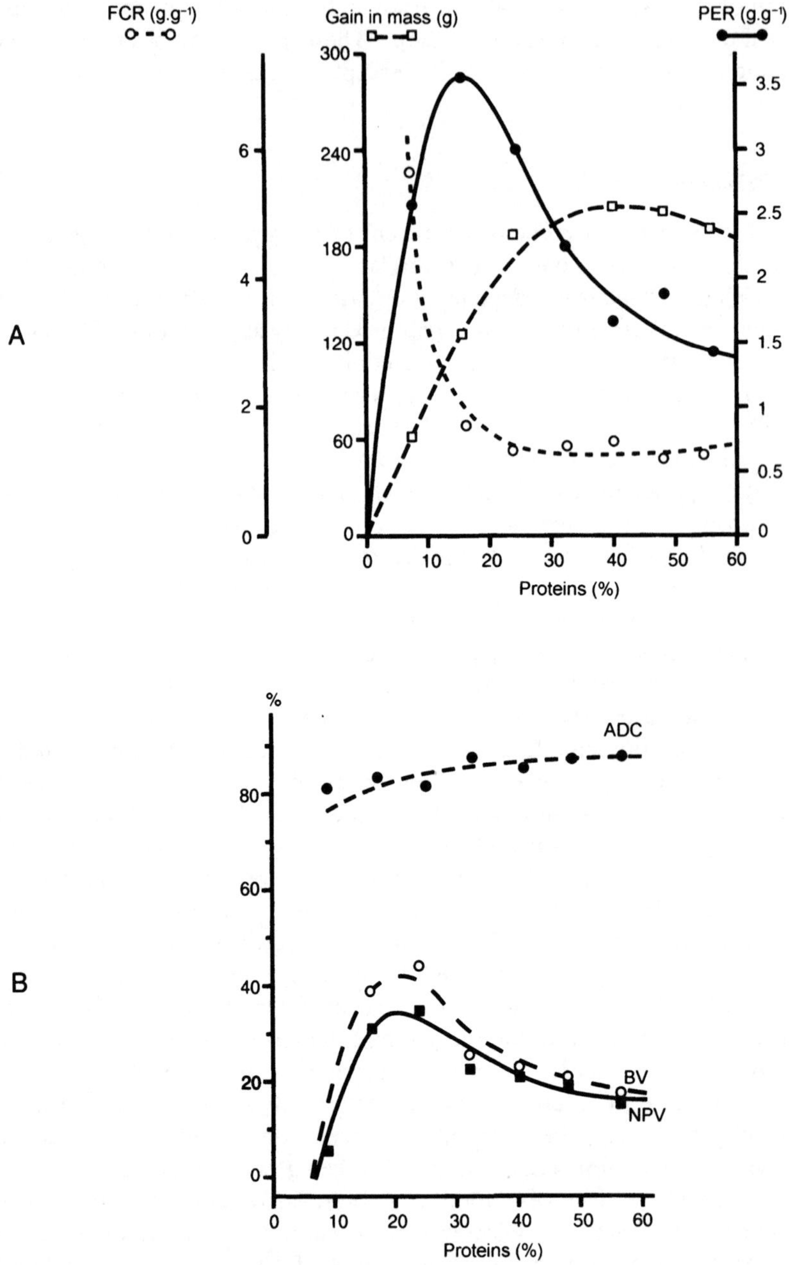

Figure 6.6. Influence of protein content of food on biological performances and various utilisation criteria of dietary proteins (after Shiau & Huang, 1989). A: influence on gain in body mass, feed conversion ration (FCR) and protein efficiency ratio (PER). B: influence on the apparent digestibility coefficient (ADC), biological value (BV) and net protein value (NPV).

In conclusion, the different methods for measurement or even the parameters chosen for a given method, most often provide coherent results, but they can be at the origin of significant variations in the quantification of requirements.

6.2.3 Nutritional quality of proteins

Quality has at least two aspects: digestibility and biological value.

6.2.3.1 Digestibility

The digestible coefficient (Chapter 4) is the first criterion to be taken into consideration. If requirements are expressed in gross proteins, they are inversely proportional to digestibility, while if they are expressed in digestible proteins, they are independent of it.

6.2.3.2 Biological value

Even when expressed in terms of digestible proteins, the requirements of a given animal remain, all other things being equal, a function of the protein source being utilised (Figure 6.7). There thus exists another characteristic of food protein, called by Thomas the biological value (BV) of the protein. This criterion corresponds to the ratio:

$$\mathrm{BV} = \frac{\text{corrected protein accretion}}{\text{proteins absorbed}} \times 100$$

where the two terms in the ratio are corrected to take endogenous and metabolic losses into account. It is measured using a balance not for proteins but for nitrogen.

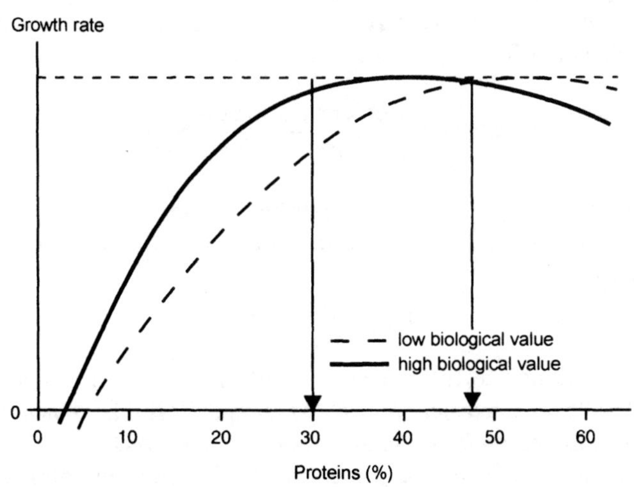

Figure 6.7. Determination of quantitative requirements of an animal using two proteins with distinct biological values.

Two groups are required for its determination, one fed with a diet containing the protein to be tested, the other with a protein-free diet (0% protein). The first allows the determination of faecal losses (Nf) and urinary and gill (branchial) excretion (Nub). The second allows the determination of so-called metabolic faecal losses (Nf_o) and so-called endogenous urinary and gill losses (Nub_o). Knowing the quantity of nitrogen intake (Ni), BV can be calculated from the following formula:

$$BV = \frac{Ni - [(Nf - Nf_o) + (Nub - Nub_o)]}{Ni - (Nf - Nf_o)} \times 100$$

BV is not constant for a given protein source. In fact, it changes very markedly in relation to the protein content of the diet (Figure 6.6b). However, it remains very characteristic of a given source of food proteins. It allows a hierarchy to be established between different sources of food 'proteins' (i.e. between sources or mixes of proteins) as long as they are measured under the same conditions and, in particular, at the same level of incorporation in the food. This hierarchy is practically the same throughout the vertebrate groups, including the fishes.

In contrast to the ADC which, by definition, only measures digestive utilisation, BV gives the metabolic utilisation of proteins. The curves in Figures 6.6a and 6.6b show that this utilisation increases at first and then decreases in relation to protein intake above and below the requirements.

6.2.3.3 *Net protein value (NPV) or net protein utilisation (NPU)*

Another criterion of food protein quality, the net protein value (NPV), has been defined. It is measured by the simultaneous determination of nitrogen accretion in a group receiving a diet containing the source of proteins being tested (1) and the nitrogen loss from another group receiving a similar diet, but lacking in proteins (2). NPV is calculated from:

$$NPV = \frac{\text{accretion}_1 + \text{losses}_2}{\text{proteins absorbed}_1} \times 100$$

where the indices relate to the groups described above.

It can be shown that NPV is linked to BV and to nitrogenous digestibility using a simple formula:

$$NPV = BV \times ADC.N$$

NPV changes in relation to the protein content of the food in a similar way to BV (Figure 6.6b), allowing verification of the simple formula which links the two criteria, taking into account the relative constancy of ADC.N (Chapter 4). NPV shows both digestive utilisation and metabolic utilisation, in contrast to BV in its true sense. It should be noted that NPV is sometimes confused with BV, but the latter is often used in a broad sense (not as according to Thomas).

BV and NPV are hardly ever measured nowadays, even for research into protein metabolism in fish. The Terroine practical utilisation coefficient of nitrogen (PUC) is generally preferred, which is simply an approximation of NPV:

$$\text{PUC} = \frac{\text{protein accretion}}{\text{protein intake}} \times 100$$

The concept of BV, to which frequent reference is made, has however led to significant progress in knowledge. This criterion possesses quite a rare characteristic in being non-additive: the BV of a mixture is not always equal to the weighted mean of those of its constituents; in certain cases, it is possible to obtain a good BV by mixing protein sources with low BVs. This characteristic, which is very interesting in itself, has been put to good use for a long time, in an empirical manner, in the feeding of both terrestrial animals and man. It has, however, rendered the use of BV difficult, if not impossible, in feed formulation. Due to complete analysis of proteins, this property has become clearly explicable by the deficiencies in EAAs of various sources, and their possible complementarity.

6.2.3.4 Chemical indices

Simple quantitative indices for deficiency in one or more EAAs of a tested protein in comparison to a reference protein, can be linked to BV. The two main ones are Mitchell and Block's Chemical Score and Oser's Essential Amino Acid Index (EAAI).

Like BV, these two chemical indices have the disadvantage of not being additive. They thus cannot be used for linear formulation. It is for this reason that EAA requirements are substituted preferentially for protein requirements for a given BV, at least in species whose nutritional requirements are well known.

6.2.3.5 Availability of amino acids

The concepts of protein digestibility and metabolic utilisation described above have their equivalents for AAs. Unfortunately, they are often still designated by the same term, availability. The digestive utilisation of each AA in a given raw material is measured as for proteins, except for the fact that it requires the measurement of the amount of each EAA. Criticisms of this technique are connected to deaminations, interconversions and other reorganisations resulting from the actions of the intestinal flora. Deaminations in particular lead to systematic overestimation of availability. In poultry and pigs, where determinations are carried out on a large scale, flora activity is limited by using fistulas that short-circuit the distal parts of the intestine where the flora is most active. In fish, where digestibility is starting to be measured, this precaution is *a priori* rarely necessary owing to the low activity of the bacterial population.

As for protein digestibility, *in vitro* methods have been proposed. These are based on hydrolysis using enzymes such as pepsin and measurement of the hydrolysed residue. Criticisms can be made: pepsin action is much less complex than that of the proteolytic enzymes of the animal and absorption is not taken into account. Although they are rapid, these methods are rarely employed in fish.

Some *in vivo* methods concern the availability of a particular AA. This is the case for Carpenter's method for available lysine which can be determined for any protein source, independently of its destination.

Overall *in vivo* methods that can be used for measuring EAA availability can only be used for one EAA at a time and this must be limiting in the diet. The response measured is either the level of free AAs in the tissues, blood or, most often, muscle, or a criterion which shows animal growth. In both cases, several dietary regimes must be used, one of which is supplemented with increasing amounts of the EAA being studied. The latter is input either in its pure form (it is assumed to be 100% available), or via a tested protein source. If the responses obtained are P and T with the pure source and the tested source respectively, the availability corresponds to the ratio T/P. Neither of these methods is currently employed for fish.

6.2.4 Total protein requirements

6.2.4.1 General requirements of fishes

Fish's requirements for proteins appear to be very high (Table C1, p. 379), compared to those of higher vertebrates. The optimal protein content of the diet is about two times higher than that for mammals or birds during the first few weeks of life, i.e. at their most demanding stage. However, these differences concern the total relative requirement. Maintenance requirements are much lower in fish than in mammals: 320 in contrast to 625 mg protein.$kg^{-0.75}day^{-1}$. In contrast, growth requirements or the quantity of food proteins taken in above maintenance requirements and necessary for the synthesis of one gram of body protein, is of the same order of magnitude in fish and higher vertebrates. The high requirement of fish for proteins is thus relative. It derives from, in addition to the points mentioned above, the fact that energy requirements are low, resulting in low 'requirements' for non-nitrogenous nutrients (energy sources), and thus in a high optimal dietary protein concentration.

6.2.4.2 Requirements of spawners

These requirements are very poorly understood, even in qualitative terms. It appears, however, that the protein inputs that ensure good growth also allow good synthesis of reproductive products in the female and, *a fortiori*, in the male. Protein deficiency in future spawners can delay sexual maturity in the female but, in Nile tilapia at least, this phenomenon results only from the slowing down of growth. Elsewhere, protein input does not influence the protein or amino acid content of the oocytes.

6.2.4.3 Effect of biotic factors

The total protein requirement of fish presents quite large variations from one species to another. The true variability is certainly less than that which appears from differences arising from measurement methodology. However, it is clear that there are large differences between, for example, mullet (low requirements), salmon and especially turbot, a species in which requirements appear to be particularly high. Such interspecific variability, which is found in other vertebrate classes, remains quite difficult to explain. It is usually attributed to the metabolic specificities as well as other physiological characteristics (activity, energetic metabolism, etc.). It is tempting to link these differences to the fish's natural feeding regime, with carnivorous species having higher protein requirements than 'herbivorous' species. This hypothesis has been put forward many times but has never been verified for fish.

Among the other factors likely to influence requirements within a species, genetic inheritance should be mentioned. In domestic animals where growth rate has been the subject of powerful genetic improvement, absolute requirements have changed greatly: high production animals consume much more than unimproved animals. In contrast, the repercussions of genetic improvement on relative requirements are moderate, or even negligible. In fish, it is almost certainly the same but, in any case, genetic improvement has so far only caused (except in a few cases) modest accelerations in growth.

In addition to genetic inheritance, age influences growth rate markedly. Specific growth rate decreases exponentially with time. The ratio between growth requirements and maintenance requirements decreases simultaneously and a decrease in relative protein requirements should follow. Some authors have indicated that the optimal protein level of diets can be lowered by about 25% in fish such as trout, carp or catfish between the alevin stage and one year old, i.e. during the period in which growth rate falls most rapidly. However, in older animals, for example between 100 g and 1 kg, it is generally difficult to demonstrate any relationship between animal size and optimal protein content. This quasi-independence contrasts with the characteristics of terrestrial mammals and birds; it could be explained by the absence of the adult stage in fish and the persistence of growth throughout ageing.

6.2.4.4 Effect of abiotic factors

The most significant abiotic factors for fish physiology are temperature and salinity. It is uncertain from some studies carried out on euryhaline species such as trout or tilapia that salinity has any effect on protein requirements, other than the weak and indirect effect derived from the stimulation or slowing down of growth or appetite, or even a slight influence on digestive utilisation. Ambient temperature has a very marked effect on absolute requirements, but not on relative requirements, which do not appear to depend on this factor. In other words, optimal protein concentrations are practically independent of rearing temperature and there are no fundamental differences between freshwater and saltwater fish. However, for individual EAAs, more care must be taken since, in the salmonids, arginine requirements are influenced by salinity.

6.2.4.5 Effect of dietary factors

Ration is one of the nutritional factors that can influence protein requirements. In theory, absolute requirements being a quantity, relative requirements should be an inverse function of the ration distributed. This reasoning assumes that input of other nutrients remains unchanged; as this is not generally the case, the true relationship between relative requirements and ration is quite complex. However, it is possible for both rainbow and brown trout to economise on food by reducing the ration if the diet is rich in proteins.

Many characteristics other than dietary quality can modify protein requirements. Energy input is very important. When feed intake is voluntary, it is mainly energy intake that determines food consumption, and thus the protein for a given protein/energy ratio. In other terms, the relative protein requirement should be expressed in relation to energy, using the ratio DP/DE, for example, rather than in relation to the mass of food (Chapter 5). This practice is currently used in terrestrial animal nutrition, but has only recently been applied to fish.

6.2.5 Amino acid requirements and metabolism

6.2.5.1 General

The requirements for EAAs have only been determined (usually by the dose-growth-response method) for seven species of teleost; two species of carp (Indian, major and common), the Channel catfish, two species of salmon (chinook and chum), the Japanese eel and the Nile tilapia. More limited data are available for six freshwater, brackish water or amphihaline species: the trout, two species of salmon (coho and sockeye), bandeng or milkfish, Mozambique tilapia, an American bass hybrid, as well as for three marine species: the red drum, royal sea bream and sea bass. For these different species only one single determination is often available and it should be added that, without harmonisation of the methodologies, the measurements are often spoiled by marked imprecision.

Requirements for EAAs, expressed in g.16 g nitrogen^{-1}, are quite similar from one species to another. In addition, they are comparable to those of non-ureotelic higher vertebrates. For ureotelic vertebrates the only difference relates to arginine which can be produced by the urea cycle. Expressed in this way, these requirements are strongly correlated with the AA profile of the proteins from the whole carcass (Table 6.4). The provision of NEAAs should not, however, be forgotten: for a given intake of EAAs, protein accretion is optimal if the proportion of NEAAs is close to 50%, which is generally the case when the diet is rich in proteins of animal origin.

For certain EAA precursors of biologically-active molecules, specific physiological conditions can render the actual requirements significantly different from the requirements estimated for the profile of body AAs. We shall therefore review in this chapter not only requirement values but also some metabolic particularities of EAAs that can explain such divergences as well as interspecific differences.

Table 6.4. Essential amino acid composition (g.16 g nitrogen^{-1}) of proteins in egg, muscle and whole carcass of different species of fish [mean and coefficient of variation (%)] compared to relative requirements also expressed in g.16 g nitrogen^{-1} (minimum and maximum).

	Eggs	Muscle	Carcass	Requirements
Arginine	6.1 (19.4)	6.3 (9.2)	6.0 (16.5)	2.2–6.0
Histidine	2.6 (18.4)	2.7 (24.0)	2.4 (26.8)	0.9–2.7
Isoleucine	5.7 (22.8)	6.3 (16.3)	4.3 (22.6)	1.2–4.4
Leucine	8.9 (13.4)	8.6 (8.3)	7.3 (9.9)	2.7–8.4
Lysine	7.6 (14.6)	9.9 (12.6)	7.3 (12.5)	3.8–6.6
Methionine	2.5 (23.1)	3.3 (9.0)	2.7 (16.4)	1.3–3.6
Phenylalanine	4.8 (14.4)	4.4 (13.0)	4.1 (11.7)	2.0–6.5
Threonine	5.4 (14.8)	5.2 (9.9)	4.3 (12.9)	0.6–5.0
Tryptophan	1.1 (26.6)	1.1 (14.4)	1.0 (25.2)	0.2–1.1
Valine	6.4 (17.6)	6.7 (16.7)	4.7 (11.8)	0.6–5.5

6.2.5.2 Quantitative requirements of growing fish

In this section the requirements (see Table C2, p. 380) will be expressed as a percentage of proteins, i.e. in grams per 16 g nitrogen.

Arginine As mentioned above, this AA is essential for fish, with the exception of some species in which the urea cycle is truly active. Requirements are between 2.0 and 6.0 g/16 g nitrogen inclusive. Variations can in part be explained by the variable capacity of each species for urea production. Thus, the salinity of the water as well as an increase in ambient ammonia are likely to increase urea synthesis (which certainly acts as an osmotic regulator). In the chicken or the rat, there is a lysine-arginine antagonism, with an excess of lysine increasing arginase activity and thus arginine requirements. However, such an effect has not been demonstrated in either trout or catfish; only a negative effect of arginine on intestinal absorption of lysine has been observed in trout.

The catabolic route favoured by arginine is the transformation into urea by arginase. The concentration of urea in the plasma is correlated, more or less directly, depending on the species, with the arginine input from the feed. The amount of urea in the plasma has been used successfully to determine the arginine requirements of trout and sea bass. Arginine can also be used for the synthesis of creatine which is an energy-rich compound that regenerates muscle ATP. This molecule also plays a greater role in crustaceans than in vertebrates. Arginine triggers the secretion of insulin and can also increase that of growth hormone. It participates in the synthesis of polyamines, molecules which intervene in the regulation of growth. It is also a precursor for the production of nitrous oxide, a central mediator in cell functions and their communications, a phenomenon recently discovered in higher vertebrates. The presence of such a molecule and *a fortiori* its multiple effects found in mammals have not yet, to our knowledge, been demonstrated in fish.

Lysine The requirements for this AA are between 3.8 and 6.6 g.16 g nitrogen^{-1}. In higher vertebrates, taking into account the possible inhibitory effect of arginine on lysine absorption, they must be increased where there is an excess of dietary arginine. Such information is not available for fish. Lysine acts as a precursor for carnitine (Chapter 9), but conversion occurs from peptides derived from protein catabolism and not from free lysine. A lysine deficiency can cause erosion of the caudal fin in freshwater fish (trout and carp), which has not been found in sea bass or milkfish.

Histidine Histidine requirements are between 0.9 and 2.7 g.16 g nitrogen^{-1}. The concentration of free histidine is always high in the muscle, where the imidazole nucleus acts as a buffer. This AA also plays a role in the tertiary structure of proteins. It is capable of attaching to metallic ions, for which reason it is usually found at active sites of enzymes which use metals as coenzymes.

Threonine The requirement for threonine is between 0.6 and 5.0 g.16 g nitrogen^{-1}. Glycine formation from threonine, by action of threonine aldolase, has not been detected in the trout, while it represents 80% of threonine oxidation in the pig. Other metabolic pathways are, however, identical.

Tryptophan Tryptophan requirements are between 0.2 and 1.1 g.16 g nitrogen^{-1}. This AA is the precursor of a neurohormone, serotonin. In contrast to higher vertebrates, fish appear to be incapable of converting tryptophan into niacin. Tryptophan deficiency leads to scoliosis, lordosis and abnormal deposition of calcium in trout and sockeye salmon kidneys. These effects have not been observed in catfish or carp.

Branched amino acids: isoleucine, leucine and valine Requirements are 1.2–4.4, 2.7–8.4 and 0.6–5.5 g.16 g nitrogen^{-1} for isoleucine, leucine and valine respectively. These three AAs have very similar formulae and are antagonists to transport mechanisms: a high concentration of leucine can inhibit isoleucine absorption by the intestine or even facilitate its catabolism. Food lacking in leucine causes an increase in isoleucine and valine concentrations in the plasma in salmonids and catfish, as found in higher vertebrates. However, these phenomena have been demonstrated using purified AAs while with normal diets a significant amount of AA is undoubtedly absorbed in the form of peptides; under these conditions, competition phenomena are limited and it is probable that isoleucine and valine requirements would only increase slightly when there is excessive input of leucine from the food.

The branched AAs are largely metabolised in the muscle, as in the higher vertebrates. Leucine oxidation occurs mainly in this tissue, especially during swimming or after ambient temperature changes. This AA is supposed to play a role in the control of protein synthesis in ruminants and monogastric animals, but such an effect has not been demonstrated in fish.

Sulphated amino acids: methionine and cysteine Requirement values for methionine given in the literature are between 1.3 and 3.6 g.16 g nitrogen^{-1}. However, the input from food of cysteine or cystine (in the hydrolysates, only cystine is measured; it is a dimer of cysteine and has a strong disulphide bond, hence only cystine is mentioned in the tables), is not always given by authors. However, methionine can lead to the synthesis of cysteine, but not the reverse, the total reaction being almost irreversible. In other words, the methionine requirement can only be met in practical terms by methionine input from the food, while cysteine requirements can be met by either of the sulphur AAs (SAA). In practical terms, both methionine requirements alone and SAA requirements, which some authors call 'methionine requirements in the absence of cystine' must be considered. The possibility of converting methionine into cysteine has been evaluated at an average of 50% (based on mass) in trout and, in the catfish, this value is comparable to that observed in higher vertebrates.

Methionine is a source of methyl groups. It plays an important role in the initial stage of protein synthesis. For this same reason, it also constitutes a precursor of choline (Chapter 9). It is equally necessary for the synthesis of creatine and polyamines. Cysteine is a constituent of glutathione, a peptide that protects the organism against free radicals, in particular those produced from lipid peroxidation (Chapter 7). Methionine deficiency causes cataracts as a result of the limited synthesis of glutathione. Through decarboxylation, cysteine can be transformed into taurine which, conjugated with bile acids, participates in lipid digestion and absorption. In aquatic animals, this non-protein AA acts as an osmoregulatory agent, its concentration in the plasma increasing with salinity.

D-methionine does not occur naturally but is easily isomerised by animals, unlike other EAAs, except phenylalanine, leucine, valine and tryptophan, in certain higher vertebrates at least. As a result, the synthetically manufactured DL form is almost as efficient as the L isomer which is not produced industrially for animal feeds. However, it is not possible to use the D form alone, without observing falls in performance. In contrast, the methionine hydroxy-analogue (MHA), a non-nitrogenous synthetic compound used as a methionine precursor, appears to be significantly less efficient in fish than in higher vertebrates. In trout, growth is slower than with L- or DL-methionine. Inorganic sulphur cannot be used, any more than taurine, to spare SAAs in the feed.

Aromatic amino acids: phenylalanine and tyrosine Phenylalanine requirements are between 2.0 and 6.5 g.16 g nitrogen^{-1}. The variations can be partly explained by the variable concentrations in the feed of tyrosine, a semi-essential AA, of which phenylalanine is the precursor. In the presence of tyrosine, phenylalanine input can be reduced to strict requirement level, hydroxylation of phenylalanine becoming negligible. This 'phenylalanine sparing' by tyrosine appears to be between 40% and 60%. The aromatic AAs give rise to hormones (thyroid hormones and dopamine) and pigments (melanin).

6.3 SATISFYING REQUIREMENTS

6.3.1 Total proteins and essential amino acids as percentage of food

6.3.1.1 Total protein requirements and the equilibrium of essential amino acids

In this old method the formulae for feed were based mainly on total proteins. Secondly, the EAAs must be equilibrated by imitating the profile of a reference protein and proceeding either by supplementing limiting EAAs with input in a pure form, or by complementation (Figure 6.8). There are several disadvantages to this process:

- total protein requirement is generally derived from experiments in which the protein source has a different BV from that used for formulation. This can only be an approximation of requirements as they would appear with the proteins available;
- the mode of expression of EAA requirements (in g.16 g nitrogen^{-1}) does not lend itself directly to linear programming, in which only variables expressed in quantities per kg of food are taken into consideration.

6.3.1.2 Requirements for essential and non-essential amino acids

For some species for which requirements for each EAA are known, even if only in summary, it is possible to express the EAA requirements directly as a percentage of

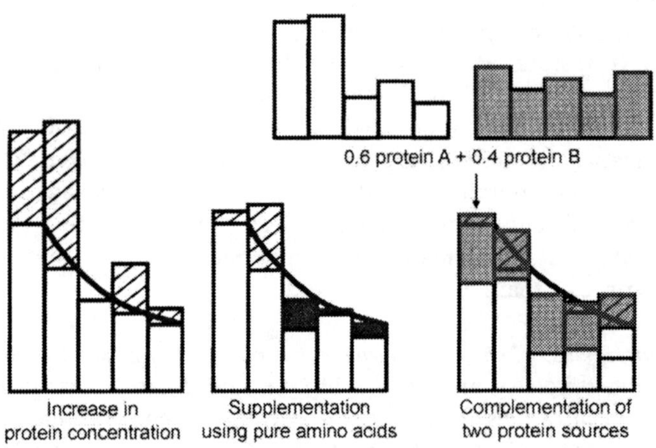

Figure 6.8. The three methods of meeting an animal's requirements using a protein with low biological value: increasing protein concentration, supplementation using pure AAs or complementation of two proteins. The hyperbola-shaped curve corresponds to the requirements for 5 EAAs (the only ones included in this theoretical example). The first protein source (in white) is deficient in EAAs nos. 3 and 5. EAAs in excess are hatched. EAAs input in their pure form are indicated in black and those of a second protein with a very different profile, in grey.

the diet (see Table C2, p. 380). In theory, it should also be possible to refer to an optimal input (requirement in its broad sense) of NEAAs taken globally and to formulate foods taking into account all these requirements. In practice, the sum of the NEAAs is never taken directly into consideration. It is, however, dependent on this fact since, if one considers the requirements for total proteins on the one hand and the requirements for each EAA on the other, the difference between total proteins and the sum of EAAs corresponds to the NEAAs.

Working in this way for a species whose EAA requirements have been determined, theoretically it should be possible to obtain the same precision in meeting requirements as is currently found in pig and poultry nutrition. This is still not the case. In practice, meeting requirements is still approximate, which requires significant safety margins. There are several reasons for this: few species have been studied in comparison to the large number of aquaculture species, and there is imprecision of requirement estimations and lack of information about certain dietary characteristics, in particular about their energy content.

6.3.2 Total proteins, essential amino acids and energy level

As indicated in Chapter 5 and above, consideration of the energetic level, i.e. calculation of the ratios DP/DE and EAA/DE, constitutes definite progress. While the EAA/DE ratio is still not used in fish nutrition, it has been for a long time in avian and porcine nutrition, where otherwise an energy level (DE/kg) is chosen. The latter standard does not only arise from nutritional considerations. In terrestrial species, in the days before the preoccupation with environmental protection, it was based on energy ('calorie') cost. This cost had a minimum for a given energetic level, which is why it was chosen. In aquaculture (especially for salmonids) attempts are made to maximise the sparing effect of lipids with regard to proteins: the DP/DE ration is decreased by inputting more lipids. The search for foods that give both excellent FCRs and little pollution pushes the choice towards higher and higher energetic contents and thus greater and greater lipid concentrations (Chapters 4 and 7). In these, the EAA/DE ratios are lowered as a result of the optimisation of EAA equilibrium and the use of highly digestible protein sources. The ratio of available EAAs/DE should remain about the same for 'classic' and 'low pollution' foods, in contrast to the NEAA/DE ratio which can be lower for the latter since the contribution of the NEAAs to energy provision or carbon decreases with high lipid inputs. However, the EAA/NEAA ratio cannot drift far from its ideal value without causing a decrease in growth. This point requires complementary research for species in which it appears more difficult to make use of the sparing effect of lipids on proteins (brown trout, turbot).

Currently, it is still difficult to utilise EAA/DE ratios, or even the DPr/DE ratio, to formulate feeds for all fish. Despite this fact, it is in this direction that many food manufacturers appear to be heading.

6.3.3 Some other approaches for the future

The considerations discussed above concern relative requirements, while absolute requirements are more significant in scientific terms. As a result, it could be said that the best modes of expressing protein requirements are amount of protein intake/ unit body mass and EAA intake/ unit mass. Tacon and Cowey (1985) had the idea of expressing the former ratio in relation to specific growth rate for a series of data from experiments carried out on different species. A strongly positive correlation thus appeared between requirements defined in this way and growth rate. Interspecific variability is very low, with absolute requirements being, for practical purposes, proportional to growth. It would be convenient to establish similar graphs or tables allowing absolute requirements for each EAA to be related to fish growth rate, and to gain in biomass (Figure 6.9).

The distance of the requirements of a given species away from the regression line would result from specific characteristics. While differences in requirements have been clearly indicated for several EAAs, it cannot be confirmed that these do not result simply from measurement imprecisions or artefacts. It is still possible that, in a first step, one single interspecific relationship between absolute requirements and growth rate will have to be used. In practice, specific requirements are only necessary for EAAs that have been found to be limiting and costly. This has the advantage of allowing the fish farmer to ascertain if a fish of a certain species, being

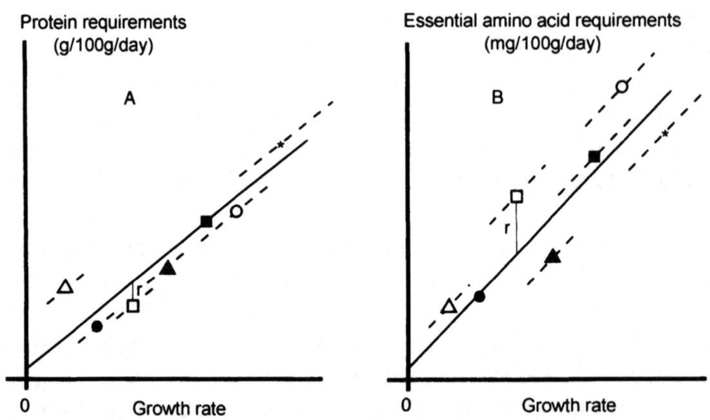

Figure 6.9. Relationship between requirements (expressed in daily amount per 100 g of biomass) and growth rate for whole proteins (A) and a given EAA (B). The curve in A is from Tacon and Cowey (1985). It shows a high correlation between the two variables, each symbol corresponding to a distinct species. It provides a first approximation of requirements (protein ration) for a fish whose growth rate is being monitored. The curve (B) shows that a species may have a particular requirement for a certain EAA. The residual r shows that, compared to the mean, the species indicated by a white square has less requirement for proteins, but more for the EAAs being considered, taking into account its growth rate.

fed a certain ration and the growth of which is being monitored, is really taking in the EAAs which it requires.

Wherever possible, it is better to replace feed composition tables and standards for crude AAs by corresponding values for *available* AAs. This mode of expression can, in theory, only bring advantages, in particular the better assessment of AA sources of multiple quality and origin. The work involved, however, takes a long time: first of all the methodology must be tested in depth, made reliable and rapid, verified on distinct species under different abiotic conditions, before going on to do repeated tests on primary material of economic interest.

6.4 CONCLUSION

Although the optimal protein level is particularly high in fish, the myth that fish require more protein than higher animals must be abandoned. It disappears when absolute requirements are examined. Fish have quite a number of metabolic characteristics (especially those concerning protein turnover, the destiny of AAs raising, either from catabolism or from food or even the mode of excretion of nitrogenous waste), but their repercussions on EAA requirements do not appear to be spectacular.

These repercussions are even more difficult to assess when the requirements are only imprecisely known. In the short term, it is convenient to choose an expression that takes into account the food energy and intake of each EAA (or that of each of the most important EAAs) per unit of body mass. It would also be desirable, in the future, to take EAA availability into account.

6.5 BIBLIOGRAPHY

Cowey, C. B. and Walton, M. J. (1989) Intermediary metabolism. In: Halver. J. E. (ed) *Fish Nutrition*, 2nd edition. Academic Press, San Diego, 259–329.

Fauconneau, B. (1983) La biosynthèse des protéines chez les poissons. *Ichtyophysiologica Acta*, **7**, 34–75.

Houlihan, D. F. (1991) Protein turnover in ectotherms and its relationships to energetics. In: Gilles, R. (ed) *Advances in Comparative and Environmental Physiology*, **7**. Springer-Verlag, Berlin, 1–43.

Jackson, A. A. (1983) Amino acids: essential or not essential? *The Lancet*, **I** 8332, 1034–1036.

Kaushik, S. J. (1980) Influence of nutritional status on the daily patterns of nitrogen excretion in the carp (*Cyprinus carpio* L.) and the rainbow trout (*Salmo gairdneri* R.). *Repr. Nutr. Develop.*, **20**, 1751–1765.

Shiau, S. Y. and Huang, S. L. (1989) Optimal dietary protein level for hybrid pilapia (*Oreochromis niloticus* × *O. aureus*) reared in sea water. *Aquaculture*, **81**, 119–127.

Tacon, A. G. J. and Cowey, C. B. (1985) Protein and amino acid requirements. In: Tytler, P, and Calow, P. (eds) *Fish Energetics. New Perspectives*. Croom Helm, London, 155–183.

Wilson, R. (1989) Amino acids and protein. In: Halver, J. E. (ed) *Fish Nutrition*, 2nd edition. Academic Press, San Diego, 111–151.

7

Lipid nutrition

The most important classes of lipids in animal nutrition are the triacylglycerols (TAG) or triglycerides and phospholipids (PL). TAGs are triesters of glycerol and three fatty acids (FA) and PLs are esters of two fatty acids and one phosphoric acid, which is itself linked by a diester bond to another alcohol (usually ethanolamine, choline, serine or inositol). In the marine environment, another class of lipids, the waxes, can be found in large quantities. These are esters of an FA and a fatty alcohol. Free or esterified sterols are also present and play a particular role in crustacean nutrition.

Lipid supply in the feeding of fish, as in that of mammals, is vital firstly to satisfy essential fatty acid (EFA) requirements, FAs which are not synthesised by the organism are necessary for cellular metabolism (for the synthesis of prostaglandins and similar compounds) as well as for the maintenance of membrane structure integrity (via their fluidity). Lipids also serve as a vector during intestinal absorption of liposoluble vitamins and carotenoid pigments (Chapters 9 and 11). Lastly, lipids also play a major role in energy supply, a role which is all the more important in fish, the majority of which poorly digest complex carbohydrates (Chapter 8).

The main stages of lipid metabolism in fish are now known and show major similarities with those of mammals. However, lipid nutrition in fish differs from that of higher vertebrates in several ways. The aquatic environment is characterised by a great richness in polyunsaturated fatty acids (PUFA) and in particular, long chain PUFAs (20 carbon atoms or more) or HUFA. HUFAs of the n-3 series (n-3 PUFA) are those for which fish have the highest requirements, in contrast to higher terrestrial vertebrates. Many studies have been carried out on the quantitative requirements of lipids in fish food, in order to improve production yields as well as protect the environment. Product quality, which is also influenced markedly by dietary lipids, was taken into consideration much later.

After a description of the principal stages of lipid metabolism in fish, this chapter is devoted to EFA requirements and the role of lipids in energy supply followed by the effect of dietary lipids on body composition and product quality.

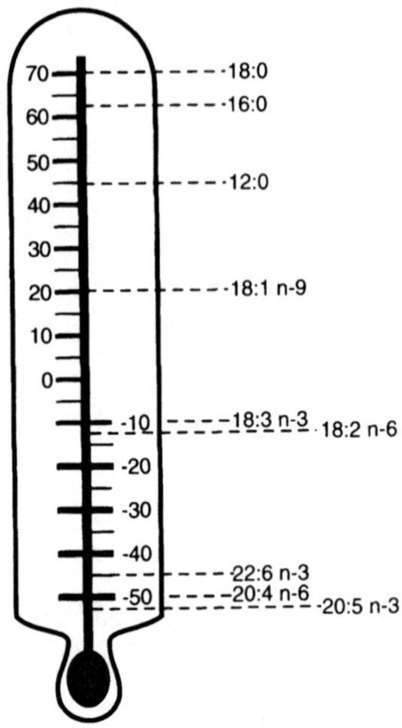

Figure 7.1. Comparison of melting points of some fatty acids.

7.1 DIGESTIVE UTILISATION AND LIPID METABOLISM IN FISH

7.1.1 Lipid digestibility

Lipids are usually either well or very well digested. The exception is when, as a result of their high melting point, they are solid at the temperature at which the fish lives (Chapter 4). Thus the hydrogenation of oils, while improving their resistance to oxidation, renders them less digestible. The melting point of triacylglyceride depends mainly on that of the constituent FAs, as shown in Figure 7.1. PUFAs are particularly well digested. In the Atlantic salmon, their ADC is 90–98% when they are ingested as free TAGs or FAs. For saturated FAs, digestive utilisation is lower and decreases with the length of the carbon chain, changing from 70% for myristate (14:0) to 50% for stearate (18:0) (see Box 7.1). The same type of response has been observed in other species such as trout and carp.

Interspecific differences in aptitude for digesting large amounts of lipids have been little studied, although they do exist. Thus, in turbot, a decrease in ADC and a slowing down in growth rate are observed when the food contains more than 15% lipid. In contrast, diets containing more than 30% fat give excellent results for trout or Atlantic salmon, implying a good digestive utilisation.

7.1.2 Transport

Absorbed FAs are re-esterified in the enterocyte into triacylglycerides and phospholipids. In the enterocyte there are large lipid droplets formerly called 'spread-out

Box 7.1 Main fatty acids present in fish lipids

The denomination 'n-a', or historically 'ωa', indicates the number of double bonds as well as the position of the one closest to the terminal methyl group.

Thus, for nutritionists, the abbreviated denomination of linoleic acid C18:2 (n-6) or 18:2 n-6 (historically, 18:2 ω6) indicates that this acid possesses 18 carbon atoms and 2 double bonds, the first positioned between the 6th and 7th atoms of carbon away from the methyl group.

Note that only biochemists describe the positions of the double bonds (or Δ) in relation to the carbonyl group and state the position of each double bond. Thus, biochemical notation would be abbreviated as 18:2 Δ9,12.

Series	Formula	Denomination	
		abbreviation	common or scientific
Saturated	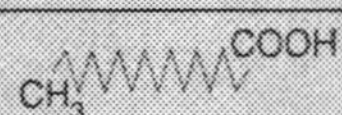	16:0	palmitic acid
		18:0	stearic acid
n-9		18:1(n-9)	oleic acid
n-6	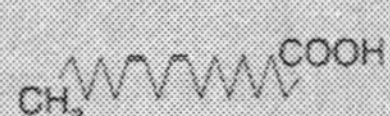	18:2(n-6)	linoleic acid
		20:4(n-6)	arachidonic acid
n-3		18:3(n-3)	linolenic acid
		20:5(n-3)	eicosapentanoic acid (EPA)
	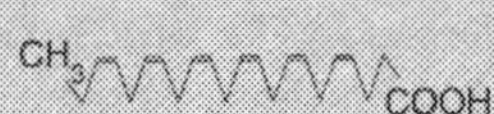	22:6(n-3)	docosahexanoic acid (DHA)

lipids', a form of temporary storage specific to fish. However, resynthesised lipids are mainly incorporated into lipoprotein particles analogous to the chylomicrons of mammals and to very low density lipoprotein (VLDL). The size of these particles is related to the nature of the dietary lipids. The presence of chylomicrons, containing mainly TAG (80%), has been clearly demonstrated in trout, 2–4 hours after a meal. These lipoproteins are transported to the liver mainly by the lymphatic system, but also in the hepatic portal system in certain species (carp and trout). Part of the FA absorbed can also be transported via the circulatory system, bound to albumin. This route only seems to be significant in certain situations: refeeding after fasting and when glycerol concentration is limiting.

Lipids reaching the liver in this way, as well as those that are synthesised *de novo* by this organ are complexed with particular proteins, apoproteins and free cholesterol, and esterified to form new lipoproteins. As in mammals, lipid transport from the liver to other tissues is carried out by three categories of plasma lipoproteins: VLDL, LDL (low density lipoprotein) and HDL (high density lipoprotein). The structure and nature of apoproteins as well as the lipid composition of these different classes of lipoproteins are quite close to those of mammals. The presence of apoproteins analagous to A and B apoproteins in mammals has been demonstrated in HDLs and VLDLs respectively. However, fish lipoproteins contain much higher percentages of n-3 HUFAs. In many species (carp, turbot, etc.), the main lipoproteins are HDLs. The lipoprotein profile varies, however, in relation to the habit and lifestyle of the species, age, nutritional condition and stage of the reproductive cycle. Thus, in immature juvenile trout, VLDL and especially LDL predominate, while in a sexually maturing trout, the HDL are most abundant (up to 80% of lipoproteins, while LDL only represents 10% and VLDL even less). A small part of the lipids (5–10%) is transported in the form of free FAs, or bound simply to albumen. In most fish, lipid concentration of the plasma is very high (1.85 to 2.40 g/100 ml in salmonids).

In the circulation, these lipoproteins are subjected to attack from various enzymes: lipoprotein lipase whose presence has been demonstrated in the plasma and in various tissues in trout (liver, muscle, heart, ovaries, adipose tissue); a 'salt resistant' lipase, analagous to hepatic lipase in mammals, but which is also found in extra-hepatic tissues; lecithin-cholesterol-acyl-transferase, which has a very high activity in the plasma. The action of these enzymes leads to the intra-vascular transformation of lipoproteins. Chylomicron VLDL and triacylglycer are hydrolysed and the liberated FAs are either stored in the tissues or used to provide energy. Triacylglycer hydrolysis, accompanied by modifications in apoprotein composition results progressively in the formation of LDL. Exchange of surface material is also observed (phospholipid, cholesterol, apoproteins) between LDL and HDL, as well as an enrichment of the latter in cholesterol esters. The processes of conversion and catabolism of lipoproteins are thus very close to those of mammals.

Another class of lipoproteins, vitellogenin, exists in female fish during and at maturation. This is a very high density lipoprotein which is synthesised by the liver and transported to the ovaries during the first stages of oogenesis. It is made up of about 80% proteins and 20% lipids, mainly PLs. It is not directly involved in the transport of dietary lipids.

7.1.3 Storage

The phospholipids, mainly localised in cell membranes and cellular organelles, represent a relatively constant fraction of the tissues and consequently the biomass. Stored lipids are generally triacylglycerides and sometimes waxes. They are resynthesised in the tissues from free circulating FAs liberated by lipase action. In mammals, lipid storage is carried out mainly in quite specific tissue – adipose tissue. In fish it can occur in several tissues: the liver, perivisceral adipose

Table 7.1. Lipid content of muscle and liver of different fish species (% wet weight). From Cowey & Sargent (1972), Henderson & Tocher (1987).

Species	Muscle	Liver
Haddock	0.3	50–75
Cod	0.4	50–75
Tuna	4	4–28
Halibut	5	4–28
Ayu	1–5.4	3–9
Atlantic salmon	4–10	10
Coho salmon	2.5–4.6	4–6
Trout	2.5–5.7	3.5–6
Carp	1.5–12.5	4.8–8.8
Mackerel	13	8
Herring	11	2
Eel	22	–

tissue and muscle, or even subcutaneous tissue. In white muscle, lipids are stored in small formations of adipose tissue outside the myomeres, while in red muscle they are also stored within the fibres themselves. The localisation of lipid deposits is very different depending on species and serves as a criterion for differentiating between several categories of fish. Thus 'oily' fish such as herring and mackerel which have muscle lipid contents of more than 10%, are distinguished from a category of whitefish such as the cod, whose flesh contains less than 2% lipid, this being stored in the liver (Table 7.1). Between these two categories are 'intermediate' fish that store lipids in the muscle (levels of 2.5 to 6%) and in other sites such as the perivisceral adipose tissue in salmonids. The lipid content of fish varies in relation to age and physiological stage. Generally, it increases with age and size of the animal, while protein content varies little. In wild fish in cold or temperate climates, it fluctuates widely in relation to seasonal cycles. Numerous other factors (genetic, nutritional, endocrine and environmental) are likely to influence the magnitude of lipid deposits and, of these, diet plays a preponderant role. Overall, reared fish are clearly fatter than their wild congeners, although this difference becomes less obvious if feeding is controlled both qualitatively and, especially, quantitatively.

In addition to the deposition sites and size of reserves, lipid composition differs considerably between fish and higher terrestrial vertebrates, both for transport or membrane phospholipids and for reserves of neutral lipids (triacylglycerols and waxes). While mammalian and avian lipids consist mainly of saturated and mono-unsaturated FA, those of fish always contain a high percentage of HUFA. Among the latter, a preponderance of n-3 series FAs is observed, especially in cold water species, a phenomenon linked, as mentioned in the introduction, to the maintenance of membrane fluidity at low temperatures.

7.1.4 Mobilisation

In mammals, lipid mobilisation is dependent on a hormone-dependent TAG lipase that hydrolyses TAG. Free FAs in the circulatory system are then oxidised, mainly in the mitochondria, but also in the peroxysomes, leading to the provision of energy via β-oxidation (Chapter 5). In fish, data concerning lipid mobilisation are fragmentary and still the subject of debate.

However, the existence of a TAG lipase has been clearly demonstrated in the liver and adipose tissue of the trout. This enzyme, localised in the cytosol, has an optimal pH of between 6.5 and 7.5, the same as for mammalian lipase. Recent studies have shown that it is activated by a phosphorylation/dephosphorylation process that is dependent on cyclic AMP. Data related to hormonal regulation of this enzyme are few and far between and they differ depending on which tissue is being studied. Thus, TAG hepatic lipase activity seems to be stimulated by glucagon and catecholamines, whereas for TAG lipase in adipose tissue (when it exists) there appears to be no clear hormonal influence, with the exception of thyroid hormones action. Elsewhere an acid lipase (optimal pH between 4 and 4.5) has been found in red muscle lysosomes in trout, but its role has not yet been clarified. It could act in the post mortem degradation of intra-cellular lipids.

During the biological cycle in fish, periods can be observed during which reserve lipids are actively mobilised, i.e. periods of fasting during the winter months and the gonad development phase. In anadromous species, gonadal maturation occurs during the non-feeding period corresponding to migration. In all cases it is the TAG at the deposition sites (firstly from the viscera and then from the muscle) that is mobilised to provide energy (in the case of starvation and migration) and gonad formation. Various studies have shown the importance of lipid reserves for reproduction. Thus in the trout, an enforced period of starvation preceding gonad development leads to a decrease in fecundity as a result of the insufficiency of reserves for gonad development (egg production).

7.2 ESSENTIAL FATTY ACID REQUIREMENTS

7.2.1 Endogenous synthesis and bioconversion of fatty acids

In fish, FA synthesis occurs mainly in the liver, by means of the FA synthetase complex. This enzymatic complex shows many similarities (molecular mass, kinetic properties) to that of mammals. The main newly-synthesised fatty acids are palmitate (16:0), stearate (18:0) and myristate (14:0), in different proportions depending on species. Thus, in the trout, the 16:0 is the dominant FA (57%) followed by the 14:0 (29%), while in the carp these two FAs are synthesised in similar proportions (41 and 37% respectively). In contrast, in several marine fish species such as plaice and flounder the 16:0 and the 18:0 are in the majority. Like other vertebrates, fish cannot synthesise linoleic acid (n-6, 18:2) or linolenic acid (n-3,

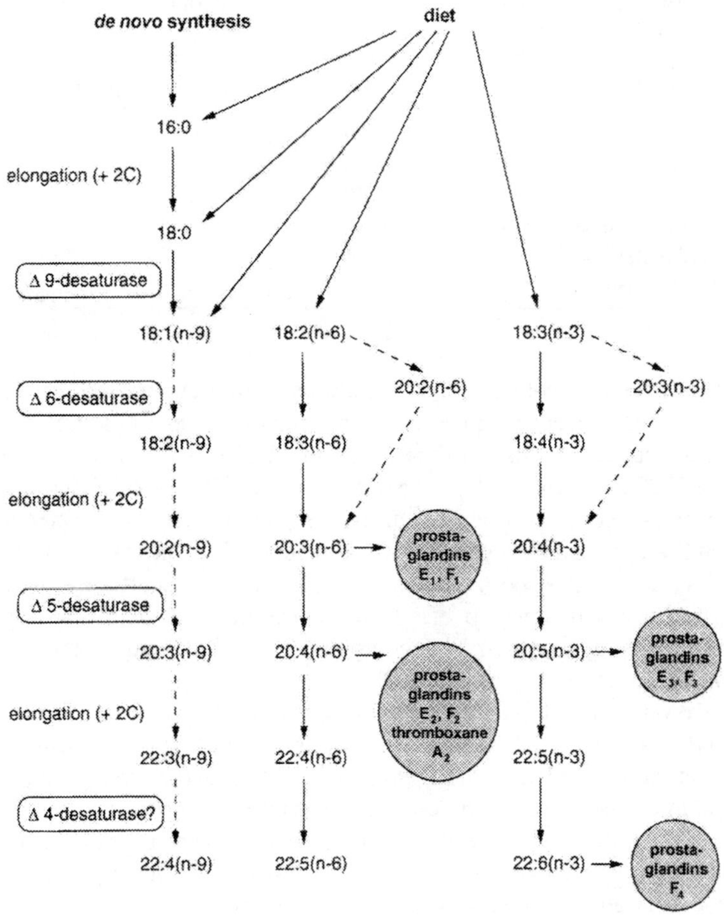

Figure 7.2. Diagram of main routes for bioconversion of fatty acids and prostaglandins (after Bell *et al.*, 1986). ——; main routes; - - - -: minor routes. n-9 series; transformations occuring in freshwater fish, where EFAs are deficient. By kind permission of Elsevier Science.

C18:3) *de novo*. These two PUFAs, or their long chain derivatives (HUFA), must be supplied in the food; they thus are essential.

The neosynthesised FAs or those supplied by the food can be transformed (bioconverted) into FAs with longer or more unsaturated chains, at least to a certain extent. The mono-unsaturated (palmitoleic acid, 16:1, n-7 and oleic acid, 18:1, n-9) are synthesised from the corresponding saturated FAs by the action of a Δ9 desaturase. Bioconversion of PUFAs includes elongations (addition of two carbons) and desaturations (addition of a double bond), ensured successively by Δ6, Δ5 and perhaps Δ4 desaturases (Figure 7.2). Desaturation and elongation systems in fish have not been completely characterised. However, the desaturases,

Table 7.2. Bioconversion capacity of 18:3 n-3 (from Kanazawa *et al.*, 1979).

Species	Relative bioconversion capacity
Trout	100
Ayu	36
Eel	20
Gilthead seabream	15
Japanese puffer	13
Red fish	7

like the other enzymes involved in lipid metabolism, are not very specific. They have affinities for the FA series which decrease from the n-3 series to the n-9 series. Thus the preferential substrates for Δ6 desaturase are: 18:3 n-3 > 18:1 n-6 > 18:1 n-9. In addition, the HUFAs such as 22:6 n-3 exert a retroinhibition on Δ6 desaturase. The appearance of n-9 HUFA, which is hardly ever observed in freshwater fish, only occurs when there is a deficiency in 18:2 n-6 and 18:3 n-3; this constitutes a deficiency criterion. Lastly, as has been shown in the rat, Δ4 desaturase is probably not a true enzyme but a group of enzymes including Δ6 desaturase and chain elongation–shortening enzymes. This complex allows the change from 22:5 n-3 to 22:6 n-3 via the intermediary of 24:6 n-3.

There are very great differences between species in their capacity to convert C18 FAs into HUFAs, which is high in freshwater fish but low in marine fish. The bioconversion capacity of n-3 18:3 is given in Table 7.2, the trout being used as a reference point (value of 100). The quasi-incapability of marine fish to convert C18 FAs is due to the absence of or very low activity levels of certain enzymes involved in these reactions, in particular Δ5 and Δ4 desaturases. These interspecific differences lead to large differences in essential fatty acid requirements.

7.2.2 Essential fatty acids

7.2.2.1 Role of essential fatty acids

EFAs are required by the organism for three reasons.

- Their main role is as a constituent, as they are important components of phospholipids which in turn are major constituents of cell membranes and transport lipoproteins. Phospholipid composition varies little, although it is not as constant as the amino acid (AA) composition of proteins. It is for this reason that a total deficiency in EFA, in contrast to total lack of EAAs, does not immediately block tissue synthesis; instead it simply slows down while causing a reorganisation and disturbance of triacylglycerides and then phospholipids.
- They serve as a substrate for the synthesis of a whole family of molecules which are hormonal in character (*sensu lato*): the prostaglandins (PG) and similar compounds, leukotrienes and thromboxanes. These compounds provide the

link between the hormones and the cellular sites where the reactions regulated by these hormones occur. They have multiple functions and act on the nervous system, circulatory system, digestive tract, reproductive organs, kidneys, etc. In fish, the best known functions of these compounds are involved with reproduction (induction of ovulation), renal and branchial excretion and osmoregulation. The main prostaglandins are PG1 and PG2 which are derived from n-6 FA, but there are also others which are derived from the n-3 series (Figure 7.2). The prostaglandins of these two series often have antagonistic effects and their respective synthesis is directly influenced by the equilibrium of the substrates. The amount of EFA metabolised by the organism for the synthesis of PG is low compared to that incorporated into phospholipids.

- The third role, only recently and incompletely discovered, is the role of second messenger played by at least one EFA: arachidonic acid (C20:4 n-6). In its free state, this FA acts as a mediator on protein kinases and the two enzymes regulating the synthesis and degradation of cyclic AMP. In this way, it contributes to the very complex regulation of cellular multiplication.

7.2.2.2 Essential fatty acids in fish

According to a simplification only recently formulated, mammalian EFAs belong to the n-6 series, those of fish to the n-3 series. Taking into account the multiple roles of EFAs described above, this assertion is not acceptable and it is certain that the two EFA series are essential to both fish and mammals. It is only the quantities required that differ radically between these two vertebrate classes. In mammals, the quantitative requirements for n-6 EFA prevail over requirements for n-3 while the converse is true in fish, at least for cold water fish which are well known for their n-3 HUFA-rich tissues. In warm water fish species, however, quantitative requirements for EFAs are similar for the two series, or putatively even larger for the n-6 series (Table C.3, p. 459).

In fish, the capacity for bioconversion of C18 FAs into HUFAs allows two categories to be distinguished: freshwater fish that have capacities similar to those of mammals and in which the only EFAs which are truly essential to the animal are the two C18 FAs (linoleic and linolenic) and marine fish. In the latter, bioconversion rates are low or very low, and the truly important EFAs are the HUFAs such as eicosapentanoic acid (n-6 20:5 or EPA), docosahexanoic (n-3 22:6 or DHA) and arachidonic (n-6 20:4). This classification is based on the study of only a few species; it has undoubtedly been wrongly simplified since a freshwater species, the pike, has recently been found to have bioconversion capacities close to those of marine fish. It should also be made clear that the existence of a HUFA requirement does not signify the disappearance of all roles of the C18 FAs in the same series. In one species at least – the turbot – linoleic acid retains an essential character that can be explained by a residual bioconversion capacity which allows the synthesis of 'intermediate' HUFAs such as the n-6 20:3 precursor of the prostaglandins (PG 1).

7.2.2.3 Essential fatty acid deficiencies

Diets that are EFA-deficient lead to a slowing down in growth and a decrease in feeding efficiency. These phenomena have been observed in several species of freshwater fish (trout, salmon, eel, common carp and tilapia) and marine fish (turbot, gilthead sea bream). At the end of a certain period of time (sometimes several months), pathological signs appear, such as hepatic degeneration with lipid accumulation, fin erosion, gill lesions or decrease in haemoglobin levels. In trout, a deficiency in EFAs over several months can lead to a 'shock syndrome' with loss of movement in response to stress. In spawning fish, if this deficiency occurs during the period preceding spawning, it causes a significant reduction in egg production, the percentage of eyed eggs and thus the hatching rate. In addition, the majority of larvae show various morphological deformities and have limited survival rates.

7.2.2.4 Requirements for essential fatty acids and meeting these requirements

Quantitative requirements for EFAs are very difficult to determine precisely. As mentioned above, the criteria of growth rate are less sensitive to EFA supplies than inputs of essential AAs (EAA). In contrast, tissue composition varies in relation to FA inputs, essential or not, while they do not vary depending on dietary AA. The composition of body lipids constitutes a criterion for requirements that is not easy to manipulate. The appearance of an n-9 HUFA may be a criterion for n-3 EFA deficiency in freshwater fish, but in dose-response studies, TAG composition of the tissues provides almost no useful information. Under the same conditions, PL composition is much more stable. However, the concentration of HUFAs increases with increasing food supply and even an excess of n-3 HUFA depresses growth. Requirement values determined by taking EFA content of the tissues into account thus correspond more to overestimation while those based on growth criteria correspond to underestimation. Some purely practical difficulties can be added to those of interpretation: for cost reasons, it is difficult to determine requirements with purified HUFAs and so mixtures of FAs, more precisely, TAGs or other esters, are generally used which complicates the interpretation of results.

As a result, requirements are only known very imprecisely. Requirements shown in Table C.3 vary from 0.5 to 1% of the diet. At first sight these values appear to be very high compared to the requirements of terrestrial animal (with the exception of a laying hen), but they should be considered as orders of magnitude rather than quantitative requirements in the strict sense. According to recent data, they have been widely overestimated. Elsewhere, some authors have proposed expressing EFA requirements as a percentage of dietary lipid rather than as a percentage of the food, since high concentrations of lipid would increase the EFA requirements. With regard to n-6 EFAs, quantitative requirements are still not known for most cold water species.

In order to formulate a diet that meets the EFA requirements of a given species of fish, it is necessary to know the dominant EFA series for the species (n-3 or n-6) and its capacity for bioconversion of C18 FA. The choice of dietary lipids thus depends on species characteristics. Fish oils are very rich in n-3 HUFAs: EPA and

DHA (about 10% and 12% respectively). Fish meals also contain lipids (about 10%) which are rich in n-3 series HUFAs. In addition, lipids of marine origin always contain a lower but not insignificant amount of n-6 HUFA. In contrast, vegetable oils are totally lacking in HUFAs and contain mainly the following FAs: C18:0, C18:1 n-9 and C18:2 n-6. However, certain oils, such as soya or rapeseeds, contain 8–10% of n-3 18:3.

In freshwater fish, oils lacking in n-3 PUFAs, such as maize oil, cannot contribute to meeting requirements for n-3 EFAs. However, this requirement can be met by the incorporation into the diet of 10% soya or rapeseed oil (i.e. an input of 1% of n-3 C18:3) or even better, 4% fish oil. These experiments have shown that HUFAs are twice as efficient as C18 PUFAs in meeting requirements, even in the salmonids. In practice, aquaculture feeds almost always contain fish meal and a minimum of HUFA is supplied by this protein source. In marine fish, only the incorporation of fish oil (5–8%) or another material of marine origin whose lipids have not been removed (fish or shrimp meal) can satisfy this requirement. Unicellular marine algae, the only plants that synthesise HUFAs, play this role in the feeding of live prey for marine fish larvae (Chapter 13 and 14). When the EFA requirements are met, the lipids serve as an energy source that can be supplied by other types of fat such as vegetable oils.

7.2.2.5 Problems of balance and excess

In theory, it is not sufficient to input quantities of n-3 and n-6 HUFAs which are at least equal to the requirements; a certain equilibrium must also be maintained between n-3 and n-6 EFA, and the relative digestive efficiencies of the different HUFAs, at least for the n-3 series, must be taken into account. The problem of equilibrium between EFAs is quite complex as has been shown by studies carried out with trilinoleine and trilinolenine in trout and coho salmon. When there is minor imbalance, everything occurs as if the two series are interchangeable, at least at the phospholipid level and up to certain limits. When the imbalance is too pronounced, a harmful effect is observed, the excess supply of one series having an effect on the utilisation of the other. This phenomenon is probably due to competition between desaturases on the one hand and, on the other, to the synthesis of prostaglandins. The problem of the relative efficiency of the two main n-3 HUFAs probably results from a difficulty in converting EPA into DHA. It does not appear to have any notable consequences in practical terms other than in the case of marine fish larvae.

An excess of n-3 EFA can also have harmful consequences for fish. In trout, diets containing 4% 18:3 or 2% 20:5 and 22:6 in the form of methyl esters, i.e. four times the required amount, slows growth and decreases feed efficiency. An overall increase in water content and a decrease in liver size have also been observed. These effects have sometimes been attributed to the form of supply of EFAs in the case of methyl esters instead of TAGs. In the turbot, a harmful effect of excesses of EPA and DHA has been observed even when these compounds were supplied as TAG.

7.3 IMPORTANCE OF LIPIDS IN FISH DIETS

7.3.1 General

As well as meeting EFA requirements, lipids provide an energy source. They are very rich in energy: 1 g of PL provides on average around 33 kJ (8 kcal) of gross energy and 1 g of TAG, 40 kJ (9.5 kcal). Assuming a digestibility of 95%, the energetic value (in terms of digestible energy) of 1 g of dietary lipids is of the order of 37 to 38 kJ, i.e. approximately 2.5 times that of carbohydrates and 1.8 times that of proteins.

7.3.2 Protein sparing effects and waste reduction

The current trend in feeding fish, and salmonids in particular, is to increase the lipid content of the feed. This is justified by numerous studies carried out on several species. Thus, in the trout, increasing the lipid content from 14% to 20% improves growth and feed efficiency even when protein content is decreased (Table 7.3). This is the sparing effect of proteins by lipids, already described in Chapters 5 and 6. In the salmonids, it is estimated that an increase in lipid level from 15% to 20% allows the protein content to be lowered from 48% to 35% without altering growth performances. In parallel, an increase in protein efficiency is found, as well as better utilisation of energy. It should be noted that the beneficial effect of lipids on feeding efficiency results firstly and mainly from their energetic density. With the sparing effect of proteins, the improvement in feeding efficiency contributes to the decrease in aquaculture pollution and participates in the maintenance of the quality of the aquatic environment.

Data concerning the optimal level of dietary lipid for several species currently exist (Table 7.4). These values could however, rapidly become obsolete; during the last decade, the lipid content in salmonid feeds has continued to rise, now often exceeding 30%. It should also be noted that the 'standards' cannot be used alone but together with the DP/DE ratio, the importance of which was emphasised in Chapters 5 and 6. Lastly, while the optimal value of this ratio has been determined for the trout, Atlantic salmon, sea bass and yellowtail, it seems premature to extend these

Table 7.3. Effect of dietary lipid content on nutrient utilisation efficiency in trout.

Diet	Diet A 'classic'	Diet B lipid enriched
Lipids (%)	14.0	20.0
Proteins (%)	44.0	37.0
Digestible energy ($kJ.g^{-1}$)	17.0	19.0
Growth ($\%.day^{-1}$)	1.91	2.21
Consumption index	1.31	1.10
CPE	1.74	2.43
Proteins (g) required per kg of production	575.0	412.0
N waste (g)/kg of production ($N.kg^{-1}$)	63.5	38.7

Table 7.4. Optimal dietary fat incorporation level for some species.

Species	Lipid content (%)
Trout	18–20
Carp	< 18
Tilapia	< 10
Gilthead sea bream	12–15
Japanese sea bream	10
Turbot	< 15
Sea bass	12–15
Yellowtail	11
Red drum	7–11
Tropical sea bass	13–18
Japanese puffer	< 6
Milk fish	7–10
Grouper	13–14
Sole	5

results to species for which no such research has been undertaken and in particular to species that appear to be difficult to feed using feeds containing high levels of lipids, such as the turbot or some tropical freshwater omnivorous species.

7.4 INFLUENCE OF DIETARY LIPIDS ON BODY COMPOSITION AND FISH QUALITY

7.4.1 Relationship between dietary lipids and tissue lipids

As indicated previously, the main storage sites for lipids in fish are the liver, perivisceral adipose tissue or the muscle, which is the only truly edible part. In many species (trout, carp, plaice, Channel catfish, turbot, gilthead sea bream, etc.), high-fat diets, especially when distributed *ad libitum*, lead to modifications in body composition: as a percentage of fresh weight, lipid deposits increase while water content decreases, with protein content generally remaining unchanged.

However, the relationship between the dietary lipid content and lipid content of the carcass is less simple than it appears. If the diets are compared on the basis of the same energy content, the influence is relatively reduced. However, if they are compared on the basis of the same food ration, the diets richer in lipids lead to higher energy and lipid consumptions and can lead to a marked increase in fat deposition. The different body parts do not all react in the same way. In the trout, the lipid content of the muscle can change from 5% to 15%. The perivisceral lipids increase even more dramatically (Figure 7.3A). It is these tissues that are responsible for the overall fattening of the animal. In marine fish such as the cod, and also the sea bass or sea bream, the liver acts as the storage site, leading to steatosis. However, this state (which is reversible and not pathological), can also result from simple

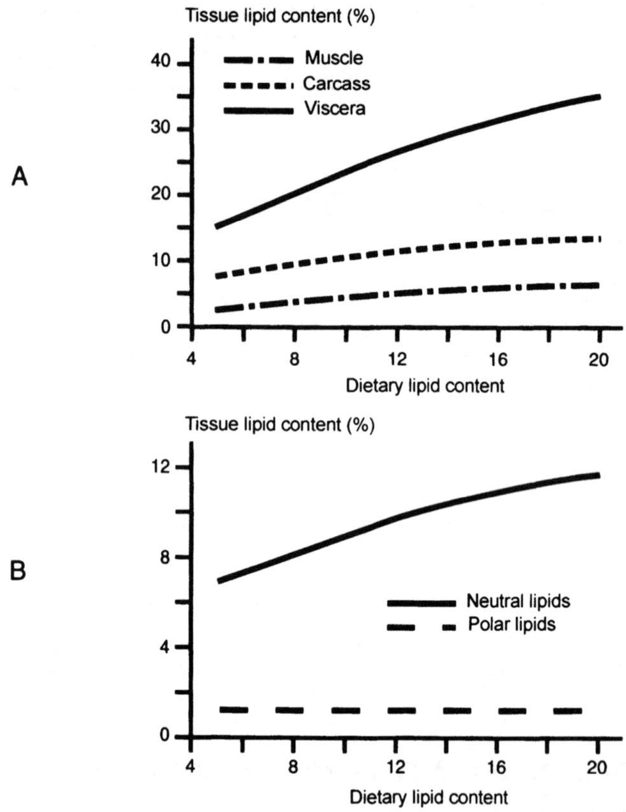

Figure 7.3. Influence of dietary lipid content on lipid content of tissues (above) and on the types of lipid in the muscle (below) in trout (after Takeuchi *et al.*, 1978).

overconsumption of energy and not an excess of dietary lipids. In other words, it is the preferred storage site which is subjected to the most marked effect. There are also differences with age: lipid contents of dorsal and ventral muscles are higher in large fish. Increases in tissue lipids are not reflected in the same way for the different classes of lipids (Figure 7.3B): phospholipid concentration remains almost constant while TAGs are responsible for almost all the observed increase.

The utilisation of diets that are rich in lipids, called 'high energy' or 'environment friendly' (Chapter 6), requires many precautions, in particular the management of feeding. This is based mainly on the quantity of digestible energy (not the amount of food) distributed; the aim is not solely to optimise biological performances but also to preserve the environment and to maintain the quality of the flesh (Chapter 22). It must therefore be ensured that the food provided does not cause excessive lipid deposition.

In contrast to the quantitative lipid dietary supply, their nature does not have any significant repercussions on overall fat deposition in the fish, or on the lipid

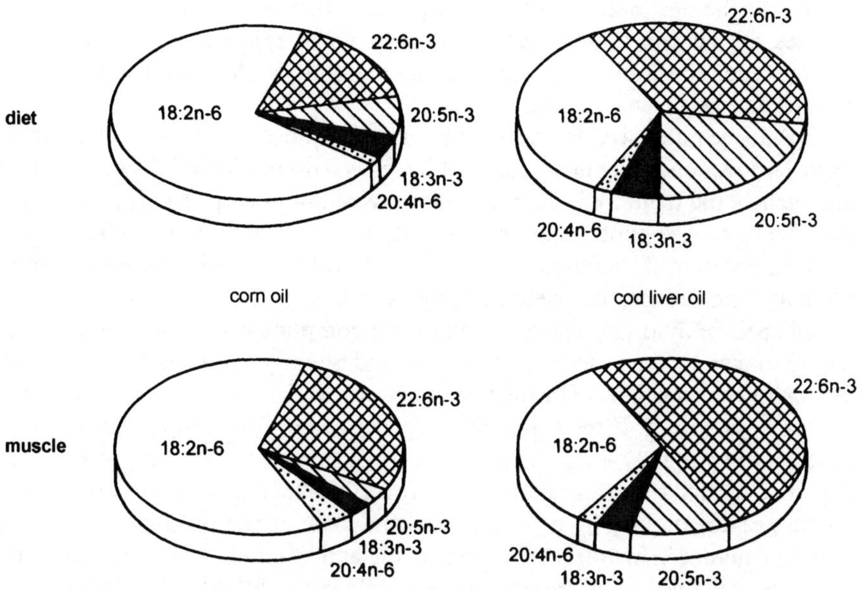

Figure 7.4. Influence of dietary lipid source on the composition of fatty acids in trout muscle.

content of the muscles. The nature and content of fat have a notable influence on the FA composition of body lipids. The higher the supply of dietary lipids, the more the *de novo* synthesis of FA is suppressed and is replaced by deposition of exogenous FAs. As in other species, the composition of body lipids thus reflects the composition of dietary lipids. In trout fed with diets containing a significant amount of corn oil (8%) (Figure 7.4), high concentrations of n-6 18:2 are found in the flesh, while n-3 HUFAs predominate when the feed is based mainly on fish oil. In more general terms, reared fish fed commercial diets containing vegetable oils or simply non-delipidated vegetable products have higher levels of n-6 FAs than their wild counterparts. However, variations in FA composition in the tissues are less pronounced than in feed. Due to elongation-desaturation reactions and selective retention of certain FAs, fish modulate the lipid content of their own tissues, in particular depositing HUFAs very efficiently. This regulation is all the more efficient when fattening is limited since, under these conditions, the fraction of PLs, which are relatively constant in composition, is larger.

7.4.2 Influence of dietary lipids on fish quality

Fish quality cannot be defined by any single criterion. It has many facets that differ in significance during processing, distribution and consumption. Among the main components are: yield at slaughter and processing, human health and food safety aspects, organoleptic quality (colour, texture, flavour, etc.), and quality. Organoleptic criteria are sometimes assessed very subjectively and their significance varies depending on the destination of the final product (fresh, frozen or smoked) or the preferences of the consumers. Many components of quality depend on multiple factors (genetic, environmental, nutritional) and this is the case for fattening, in which feeding plays a major role.

Fish shape is influenced by genetic factors and is affected as much by thinness as excessive fatness, especially in species that have been domesticated over a long period of time such as the trout and, even more so, the common carp. In practical terms, it is advisable to ensure good adjustment of rations in order to avoid deficiencies and excesses. In addition, fattening increases perivisceral or hepatic deposits more than muscle mass and reduces the yield at slaughter.

In all species, lipid content of the flesh is a component of organoleptic quality. Fattening makes the raw flesh more tender and slightly modifies the overall taste perception of the consumer. During taste tests of raw fish, neither very thin nor very fat fish are appreciated. If the fish are to be processed, too-high lipid levels also have a harmful influence: they are prejudicial to salting (loss of water and microbiological quality) and to smoking (loss of oils). A correct balance must therefore be found. The nature of the dietary lipids also affects the taste of the products, but in a less pronounced manner. In fillets of brown trout and salmon, tasters are capable of detecting the influence of oil (marine or vegetable in origin) added to the food; their preference is for the fish that have received marine oils. In contrast, no clear difference is perceived in the rainbow trout. No precise data are available at present for other species.

The nutritional quality of the flesh and especially its energy value ('calorific', according to the commonplace terminology still used in dietetics) is obviously influenced by fattening. However, two facts must be emphasised: firstly, in thin fish such as the turbot, even if the lipid content of the flesh could be doubled by excessive lipid or energy intake, it remains low compared to that of fatty fish, or even meat of terrestrial farm animals. In addition, in humans, consumption of oily fish as opposed to fatty meat does not aggravate the risks of cardiovascular diseases. In contrast, the richness of aquatic products in n-3 HUFAs is favourable to health, decreasing cholesterol levels and the subsequent occurrence of cardiovascular disease. It is thus of the utmost importance to ensure that feed lipids are high in n-3 HUFAs since the dietetic value of reared fish is to be preserved.

7.4.3 Significance of lipid peroxidation

In contact with oxygen, unsaturated FAs (and especially the more unsaturated of them), the n-6 HUFAs and still more the n-3 HUFAs, are easily altered by peroxidation. This phenomenon occurs readily *in vitro* (in food during storage or in the flesh after slaughter) as well as *in vivo*. In oils and fats, this is generally called rancidity (purely chemical reactions). *In vivo*, natural phenomena regulated by chemical or enzymatic mechanisms can be observed, as well as abnormal phenomena that rapidly assume a pathological character. The phenomena occurring *post mortem* are strongly influenced by the liberation of enzymes involved *in vivo*. The difficulty in preserving fish flesh, even frozen, also results largely from these phenomena.

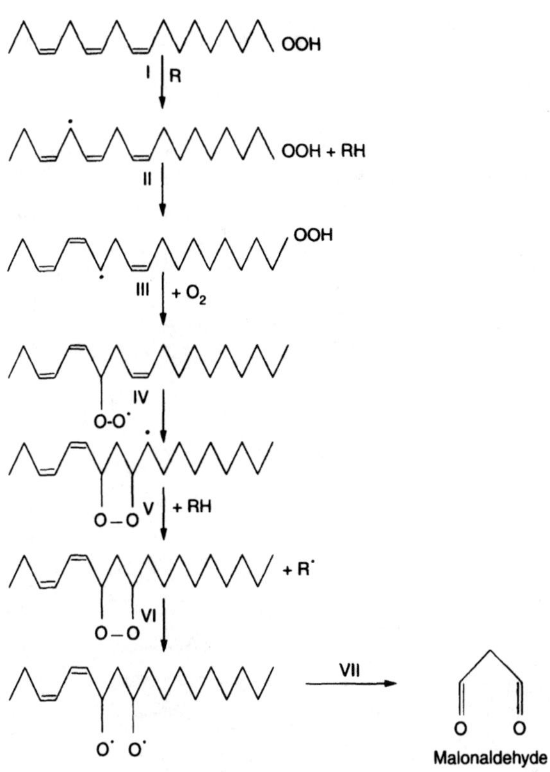

Figure 7.5. Diagram of the main stages involved in fatty acid peroxidation (after Dianzani *et al.*, 1978 and Slater, 1972) By kind permission of Academic Press.

The overall peroxidation process includes three main phases that are summarised in Figure 7.5: initiation (extraction of an electron from a double bond and formation of an unstable radical); propagation with the formation of a chain of a series of free radicals acting sometimes as reducers and sometimes as oxidisers; and, lastly, the cessation of the reaction with the formation of terminal compounds such as aldehydes (malonaldehyde in particular) and ketones. Free radicals are very harmful to cell organelles and even to macromolecules. The terminal compounds, although stable, are toxic: malonaldehyde is carcinogenic. The organisms possess two lines of defence against this phenomenon: the trapping of free radicals and an enzyme system allowing the destruction of certain intermediate compounds. Free radical traps are antioxidants such as vitamin A and especially the tocopherols (vitamin E). Ascorbic acid (vitamin C) is often attached to this group; in fact it is a hydrosoluble sequestrator of oxygen while tocopherol is a proton donor which is particularly active in a liposoluble environment (Figure 7.6). In the second defence mechanism, the first enzyme involved is superoxide dismutase (SOD) which transforms the superoxide anion into hydrogen peroxide. This compound is very toxic to the cell and must be reduced by catalase which transforms it into water and oxygen, or by glutathione peroxidase (Figure 7.6). These different defence mechanisms are

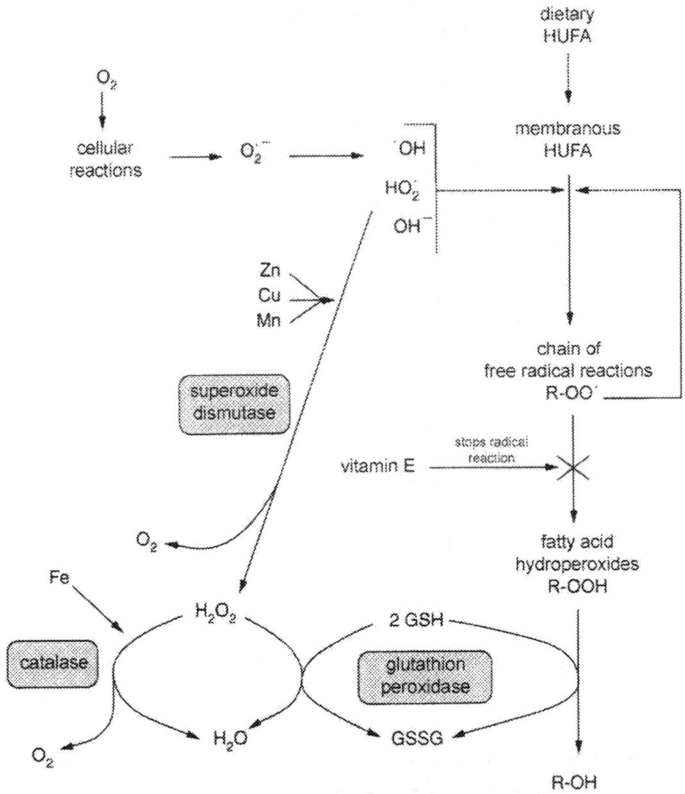

Figure 7.6. Possible interactions between the active metabolites of oxygen, cellular compounds and the anti-oxidant systems (after Sargent *et al.*, 1989). By kind permission of Academic Press.

not well understood in fish and can also be very active against exogenous peroxides and free radicals, i.e. those in the food. It is possible to prevent the peroxidation of FAs in foods and to economise on natural antioxidants by adding authorised synthetic antioxidants (Chapter 18). However, synthetic antioxidants are inactive *in vivo*.

Lipid peroxidation has many harmful effects. It leads to the destruction of labile elements (vitamins A, E, riboflavin, folic acid, etc.) in foods. In fish it can cause diarrhoea as a result of the decrease in the activity of enzymes such as amylase, lipase or trypsin. At a metabolic level, problems vary from hepatic steatosis to the inhibition of certain enzymes in the TAC cycle. However, the symptoms most frequently encountered correspond to vitamin E deficiency which has been totally utilised in free radical destruction. Thus, muscular dystrophy and lysis of red blood cells are observed (Chapter 9). The deposition of lipofuschins (altered lipoproteins which have become insoluble in solvents) is also characteristic of this state; it is easily demonstrated by histology and allows the pathological effects of peroxidation to be described. For the consumer, in addition to the muscular dystrophy which under extreme cases can render the texture of the flesh totally unacceptable, there is reduced coloration and an alteration of its flavour (smell and taste) may happen. Added to this is a decrease in dietary quality as a result of the partial destruction of the FAs which are most important to human health as well as vitamin E and other vitamins sensitive to oxidation. In practice it appears that it is the reduction in storing time or shelf life which should receive the most attention from researchers.

The simplest method of solving these problems consists of adding to the foods (which already contain antioxidants and ascorbic acid), doses of vitamin E which greatly exceed the vitamin requirement. Several studies have shown that supplementing diets with vitamin E leads to an increase in the concentration of this vitamin in the muscle and, consequently, a decrease in oxidation of muscle lipids after slaughter. In trout, Channel catfish or turbot, doses of the order of 300–350 $mg.kg^{-1}$ of feed have been shown to be effective. This is especially recommended when the fish is to be frozen or fast-frozen, as at these temperatures peroxidation is slowed down but does not stop. These considerations argue in favour of supplementation with synthetic antioxidants, as well as vitamins E and C, at higher levels for fish than for terrestrial animals, the flesh of which is lower in PUFAs.

7.5 CONCLUSION

Lipid nutrition is one of the most studied areas of aquaculture nutrition. It has revealed original features (significance of the requirement for n-3 HUFA) and especially a diversity in the area of HUFA requirements, and the aptitude for bioconversion of short chain EFAs. The quantitative requirements for these FAs appear to be high, but they may have been overestimated. With the proteins, the lipids constitute an energy source which is particularly well adapted to the physiology and metabolism of fish. They are increasingly employed in the manufacture of very efficient feeds. More than other nutrients, they exert a direct and an indirect influence on the quality of the flesh, a parameter whose importance will grow rapidly. However, the lipids, and especially those of fish rich in n-3 HUFAs, have the inconvenience of being very sensitive to peroxidation which renders necessary certain precautions both during feed manufacturing and during the storage or processing of the fish.

7.6 BIBLIOGRAPHY

Bell, M. V., Henderson, R. J. and Sargent, J. R. (1986) The role of polyunsaturated fatty acids in fish. *Comp. Biochem. Physiol.*, **83B**, 711–719.

Corraze, G. (1994) Nutrition lipidique des poissons: importance et conséquences. *La Pisciculture Française*, **117**, 25–36.

Cowey, C. B. and Sargent, J. R., (1972) Fish nutrition. *Adv. Mar. Biol.* **10**, 383–492.

Dianzani, M. U. and Ugazio, G. (1978) Lipid peroxidation. In: Slater, T.F. (Ed.) *Biochemical Mechanisms of Liver Injury*. Academic Press, London, 669–707.

Greene, D. H. S. and Selivonchick, D. P. (1987) Lipid metabolism in fish. *Prog. Lipid Res.*, **26**, 53–85.

Henderson, R. J. and Tocher, D. R. (1987) The lipid composition and biochemistry of freshwater fish. *Prog. Lipid. Res.*, **26**, 281–347.

Kanazawa, A., Teshima, S.-I. and Ono, S. (1979) Relationship between essential fatty acid requirements of aquatic animals and the capacity for bioconversion of linolenic acid to highly unsaturated fatty acids. *Comp. Biochem. Physiol.*, **63B**, 295–298.

Sargent, J., Henderson, R. J. and Tocher, D. R. (1989) The lipids. In: Halver, J.E. (Ed.) *Fish Nutrition*. Academic Press, San Diego, 153–218.

Sheridan, M. A. (1988) Lipid dynamics in fish: aspects of absorption, transportation, deposition and mobilisation. *Comp. Biochem. Physiol.*, **90B**, 679–690.

Slater, T. F. (1972) *Free Radical Mechanism in Tissue Injury*. Pion Ltd, London.

Takeuchi, T., Watanabe, T. and Ogino, C. (1979) Digestibility of hydrogenated fish oil in carp and rainbow trout. *Bull. Jpn. Soc. Sci. Fish*, **45**, 1521–1525.

Watanabe, T. (1982) Lipid nutrition in fish. *Comp. Biochem. Physiol.*, **73B**, 3–15.

8

Carbohydrate nutrition: importance and limits of carbohydrate supplies

Carbohydrates form the most widespread organic compounds in the biosphere. The term 'carbohydrate' covers a large number of molecules, mainly formed of carbon, oxygen and hydrogen atoms in proportions which almost always correspond to the crude formula $(CH_2O)_n$. These are polyalcohols with one or several aldehyde or ketone functions. Simple sugars (monosaccharides) include a single unit of hydroxyaldehyde or hydroxyketone. The oligosaccharides contain between two and ten monosaccharide molecules linked by glycoside bonds. Polysaccharides are polymers with either straight or branched chains. Although very many carbohydrates exist, only a few have any nutritive value in animal nutrition. These mainly involve hexoses (glucose), disaccharides (formed from two hexose molecules) and some homopolysaccharides, including starch. Other carbohydrates can play a secondary nutritional role, or even a negative one; this involves fermentable carbohydrates and especially fibre.

There are two types of starch, amylose and amylopectin. The former is formed from long, unbranched chains of glucose units linked by $\alpha 1 \rightarrow 4$ bonds while the latter is very branched with $\alpha 1 \rightarrow 6$ bonds. These molecules are hydrolysed into maltose by α amylase and the residues of hydrolysis (maltose and short chain dextrins) are hydrolysed by maltase and sucrase-isomaltase (Chapter 3). In animal cells, the main form of carbohydrate storage is glycogen, a similar compound to amylopectin but more branched. Structural polyosides such as peptidoglycan in bacteria, cellulose in the plant kingdom and chitin in the fungi and the animal kingdom are distinguished from the starches by their $\beta 1 \rightarrow 4$ bonds. In the laminarines from algae, the bonds are $\beta 1 \rightarrow 3$ type. The distinction between α and β allows the classification of glucose polymers into two main categories – available carbohydrates and fibres – but there are also digestible disaccharides with $\beta 1 \rightarrow 3$ links such as lactose, or the indigestible oligosaccharides such as the α-galactosides (Table 8.1 and Box 8.1).

Table 8.1. Classification of different types of carbohydrate. Those in bold have a nutritional role.

Monosaccharides	
trioses ($C_3H_6O_3$)	glyceraldehyde, dihydroxyacetone
tetroses ($C_4H_6O_4$)	erythrose
pentoses ($C_5H_{10}O_5$)	ribose, arabinose, xylose ...
hexoses ($C_6H_{12}O_6$)	**glucose, galactose, mannose, fructose** ...
Oligosaccharides	
disaccharides ($C_{12}H_{22}O_{11}$)	**maltose (2 glucose),**
	saccharose or sucrose (glucose + fructose)
	lactose (glucose + galactose)*
trisaccharides ($C_{18}H_{32}O_{16}$)	raffinose (galactose + glucose + fructose)*
tetrasaccharides ($C_{24}H_{42}O_{21}$)	stachyose (2 galactose + glucose + fructose)*
pentasaccharides ($C_{30}H_{52}O_{26}$)	verbascose (3 galactose + glucose + fructose)*
Polysaccharides	
homopolysaccharides	pentosanes*
	hexosanes: starch, glycogen, cellulose*, laminarine*
	fructosanes: inuline**
	galactanes
	mananes
heteropolysaccharides	pectin, hemicellulose*, mucopolysaccharides
Others	
chitin	(polymer of N-acetyl-glucosamine)

* Compound with at least one β type bond
** Polysaccharide also containing glucose units.

Although carbohydrates are not indispensable in fish feed, they constitute an inexpensive source of energy. In the absence of carbohydrates, there is increased utilisation of proteins and lipids as an energy source. In many species, a dietary carbohydrate supply appears to be necessary as it improves growth and especially protein utilisation. In the wild, fish food is low or sometimes even completely lacking in carbohydrates, with the exception of chitin which is not very digestible. The poor ability of fish to make use of dietary carbohydrates is a reflection of the scarcity of carbohydrates in the aquatic environment.

There are two types of problem in the carbohydrate nutrition of fish. Firstly, as a general rule, digestion of complex carbohydrates such as starch is low in aquatic animals. Secondly, these animals appear to be poorly adapted to the metabolic utilisation of simple sugars (mono- or disaccharides): intolerance phenomena seem to manifest themselves when a certain dietary incorporation level is exceeded.

8.1 DIGESTION – DIGESTIBILITY

Although amylase activity is always found in fish from early larval stages onwards (Chapter 12), there is great variability in amylase activity linked to feeding habits in

Box 8.1 Structure of some carbohydrates which are significant in fish nutrition

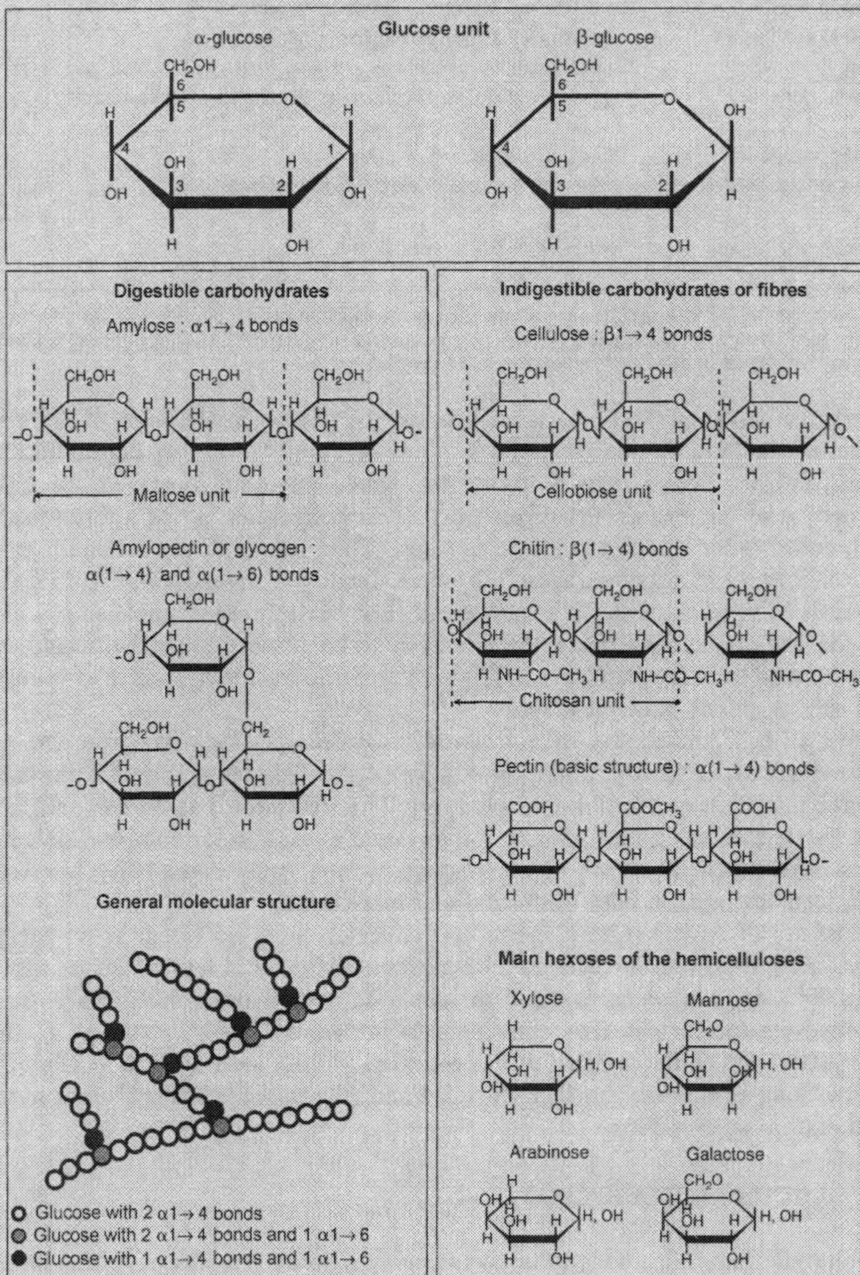

Table 8.2. Comparative efficiency of different carbohydrate sources in some teleosts. Note that highest efficiency does not always correspond to highest digestibility.

Species	Utilisation efficiency of carbohydrate sources
Common carp	gelatinised starch* > dextrin* > glucose
Japanese sea bream	gelatinised starch > dextrin > glucose
Catfish	corn starch > dextrin > glucose, maltose, fructose, saccharose
Chinock salmon	glucose, maltose, saccharose, dextrose, cooked starch > fructose, galactose, lactose
Siberian sturgeon	cooked starch > raw starch
American sturgeon	glucose, maltose > dextrin, raw starch
Trout Sea bass, sea bream	cooked starch > raw starch

* The term 'gelatinised' indicates that the product has been subjected to heat treatment of variable intensity. The term 'dextrin' (in a commercial sense) corresponds to a product which has undergone dry thermal treatment; the degree of depolymerisation is thus very variable.

juvenile fish (Chapter 3). For a given species, the utilisation efficiency of different sources of starch is not the same either. Thus, in 'carnivorous' fish, digestibility of carbohydrates is lower than in those fish which are omnivorous in the wild. Irrespective of the species, the digestibility of carbohydrates is, for a given osidic bond, linked to the complexity of the molecule. Thus, simple sugars such as glucose or saccharose have a higher digestibility than dextrin or raw starch (Figure 4.11). However, in practical terms, it is not worthwhile incorporating simple sugars into diets and the only carbohydrates that are likely to be included in the formulation of manufactured feeds are plant carbohydrates with complex structures such as starches from cereals, pulses, roots or tubers.

For all fish, preliminary hydro-thermal treatment (cooking-extrusion, flaking, toasting) improves the digestibility of complex carbohydrates (Tables 8.2 and 8.3), increasing the dietary digestible energy supply. This treatment is all the more efficient when the species has a low capacity for digestive utilisation for complex carbohydrates. The beneficial effects of cooking on carbohydrate digestibility has been clearly demonstrated in both freshwater and marine fish.

The use of cereals (wheat, maize and triticale) or pulses (peas) as an energy source in fish feeds thus involves technological pretreatment. A level of gelatinisation above 70% appears to be required to maximise carbohydrate digestibility. Such carbohydrate sources can thus contribute to protein sparing (Chapter 6). In this way, pulses such as the pea can be incorporated up to a level of 25% in salmonid feeds, as long as they are dehulled, extruded and micronised, and the level of starch gelatinisation reaches 80%.

8.2 GLUCOSE METABOLISM

The total efficiency of carbohydrates as an energy source is not always directly linked to its digestibility. Even when fish digest carbohydrates, the metabolic utilisation of

Table 8.3. Digestibility data (apparent digestibility coefficients, ADC) for carbohydrates in trout (after Bergot, 1979; Kaushik & Médale, 1994).

	ADC of starch (%)
Effect of nature of ingredient	
Raw starches	
rice	39
wheat	54
potato	<5
cassava	<15
maize (28% amylose)	33
'waxy' maize (99% amylopectine)	54
amylo-maize (65% amylose)	<20
Treated starches	
extruded wheat starch	96
gelatinised wheat starch	96
Effects of incorporation level of raw maize starch	
(level of incorporation/feeding rate/daily intake)	
10%/*ad lib.*/0.24 g	38
25%/*ad lib.*/0.75	26
40%/*ad lib.*/1.6	22
40%/rationed/0.28	36
40%/rationed/0.78	25
Effect of water temperature	
Raw starch (>30%)	
8°C	31
18°C	41
Gelatinised starch (>30%)	
8°C	64
18°C	75

the absorbed glucose can be low. Intolerance phenomena are reported in the literature, when more than 20% simple sugars (glucose, saccharose) are incorporated in the diet. In this particular case, this intolerance to high dietary carbohydrates is not explained by poor digestibility; instead, it is generally clearer when the simple sugars produced from digestion are rapidly absorbed.

8.2.1 Hyperglycaemia and control by insulin

In metabolic terms, monitoring post-prandial glycaemia (after a test meal) allows the estimation of poor tolerance to carbohydrates. In contrast to monogastric terrestrial animals, the blood glucose level remains high after a carbohydrate-rich meal

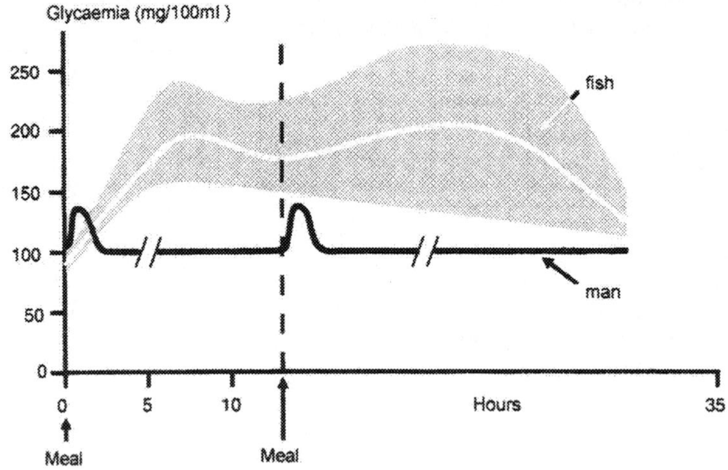

Figure 8.1. Comparison of glycaemia in fish and man. The shaded area corresponds to fluctuation in glycaemia levels due to variation in digestible carbohydrate input.

in all teleosts. This very marked (reaching levels of 17 mmol.l^{-1} or 300 mg.100 ml^{-1}) and prolonged (it can exceed 24 h) hyperglycaemia after the administration of glucose is observed in most teleosts (Figure 8.1), with slight differences between species. In comparison, the post-prandial peak in man occurs at less than 7 mmol.l^{-1} and only lasts for a few hours. This prolonged hyperglycaemia after glucose administration or after a carbohydrate-rich meal results from a lack of control of glycaemia. To start with, this phenomenon was attributed to an absence of insulin secretion and led some authors to consider fish (trout, Channel catfish, yellowtail, etc.) as 'naturally diabetic' animals.

The development of specific assays (RIA) for fish insulin has demonstrated that they secrete insulin. Values vary from 1–3 ng.ml^{-1} in starved fish. In fed fish, secretion levels are much higher, plasma values varying from 5–48 ng.ml^{-1}. Thus, it now appears clear that in most fish insulin secretion is not a limiting factor. In addition, insulin administration lowers glycaemia in trout just as effectively as in other species; the respective roles of insulin (hypoglycaemiant) and of glucagon (hyperglycaemiant) seem to be the same in fish as they are in terrestrial animals or alloxane, although data on glucagon are limited. In addition, insulin receptors are present in several fish tissues. There are certainly notable interspecific differences, in particular in tissue receptivity and in tyrosine kinase activity, which are higher in certain species such as the common carp than in the trout. However, in general terms, the intake of large quantities of digestible carbohydrates results in hepatic malfunction associated with hepatomegaly and other effects on lipid or carbohydrate metabolism.

8.2.2 Glucose phosphorylation

The main stage of glucose phosphorylation (conversion of glucose into glucose-6-phosphate) is carried out by the hexokinases. It was initially suggested by some authors that the activity of these enzymes was low and that the capacity for glucose phosphorylation was limited. However, recent works have clearly shown that glucokinase is very much present and, both at the biochemical and molecular levels, their activities and expression reflect dietary carbohydrate levels (Panserat *et al.* 2000). The existence of a mammalian-type glucose transporter (Glut) in fish is still a matter of debate; studies using mammalian anti-Glut antibodies suggest a very limited distribution of Glut-1 and a total absence of Glut-4 in tilapia. On the other hand, studies with rainbow trout and brown trout have demonstrated the existence of both Glut-1 and Glut-4.

8.2.3 Carbohydrate catabolism

8.2.3.1 Glycolysis

Glycolysis is the main catabolic pathway for glucose in fish tissues. Glycolysis transforms glucose into pyruvate with ATP formation and produces oxidisable substrates for the tricarboxylic acid cycle under aerobic conditions. Under anaerobic conditions, which prevail in fish white muscle, it leads to the formation of lactate. All enzymes involved in glycolysis (Chapter 5) have been found in the tissues of various species of fish. The highest glycolytic activities are found in cardiac and skeletal muscles, while they are low in fish kidney and liver. In addition, in the liver, glycolysis acts more to provide precursors for biosynthesis than in producing pyruvate for oxidation.

As in other animals, all glycolytic reactions are reversible and can thus function in the direction of gluconeogenesis, with the exception of the three stages catalysed by hexokinase (formation of glucose-6-P), phosphofructokinase (formation of fructose 1,6 diphosphate) and pyruvate kinase (transformation of phosphoenolpyruvate into pyruvate) respectively. These are all limiting stages in glycolysis, especially in the salmonids.

8.2.3.2 Fate of pyruvate resulting from glycolysis

Lactate formation Under anaerobic conditions, such as during intense muscular exercise, pyruvate is converted into lactate via a lactate dehydrogenase. This reversible reaction occurs in the cytosol; it generates NAD^+ which allows the maintenance of the cell's redox potential and stimulates the formation of oxaloacetate. This pathway for pyruvate utilisation is preponderant in comparison to the oxidative pathway in the white muscle of fish. As in mammals, lactate dehydrogenase is present in the form of several isozymes which differ between tissues. However, fish isozymes have particular characteristics. The temperature required to obtain maximal activity is much lower than that required by the mammalian enzyme owing to a different enzyme-substrate link. In carp, muscle lactate dehydrogenase is only slightly sensitive to pH and maintains a high activity level even at low pH, which can be considered as an adaptation to severe hypoxic conditions. The activity of this enzyme in white muscle is very variable from one species to another, being higher in fast-swimming fish.

Formation of acetyl-CoA: Pyruvate is converted by decarboxylation to acetyl-CoA, which can then be oxidised in the TCA cycle. The decarboxylation stage is carried out in the mitochondrial matrix by the pyruvate dehydrogenase complex, to thiamine pyrophosphate which coexists in the inactive phosphorylated form and the active dephosphorylated form. The activity of the enzyme complex is regulated by the concentration of mitochondrial ATP in the oxidative tissues. In white muscle, citrate inhibits the appearance of the active form.

8.2.3.3 Other catabolic routes for glucose

The pentose phosphate pathway, also called the phosphogluconate pathway or shunt, leads to the synthesis of ribose 5 phosphate from glucose 6 phosphate, with NADPH generation. This metabolite is an energy transporter with powerful reducing potential and is required for syntheses, in particular of fatty acids and steroids. The pentose pathway also provides the ribose necessary for the synthesis of nucleic acids. In fish, the enzymes for this oxidation pathway are mainly found in the liver and their synthesis is temperature sensitive.

Alcoholic fermentation is generally considered only to occur in microorganisms under anaerobic conditions (yeasts), but has been described in the goldfish under prolonged anoxic conditions. This evidence is, however, only anecdotal.

The degradation pathway for glucose into glucuronic acid and ascorbic acid is incomplete in fish. The absence of gulonolactone oxidase prevents ascorbic acid (vitamin C) synthesis (Chapter 9). However, the first reactions leading to the synthesis of glucuronate are important, since the glucuronyl residues are used for the synthesis of acid polysaccharides during detoxification reactions.

8.3 SYNTHESIS AND MOBILISATION OF GLYCOGEN

Few data are available relating to the glucose 'requirement' of the whole animal. Current estimations show that it is low (about 500 $mg.kg^{-1}$ body $mass.day^{-1}$). The turnover rate of circulating glucose is slow (more than 5–6 hours) compared to that of mammals (less than 30 min). Glucose turnover varies from 2–55 $mmole.kg^{-1}$ body $mass.min^{-1}$ depending on species, environmental conditions and nutritional state.

8.3.1 Gluconeogenesis

The metabolic pathways for gluconeogenesis (formation of glucose from non-carbohydrate substrates) are quite similar to those found in other animal species. The liver and kidney are the main sites concerned. Amino acids (AAs) play a large part in gluconeogenesis in most teleosts (carp, trout, catfish, bass), glucogenic AAs such as alanine, serine or glycine being preferred as substrates to glutamic or aspartic acid. Gluconeogenesis from AAs from the degradation of body proteins is high in fish. The activities of enzymes involved in these syntheses increase and those linked to glycolysis decrease when protein content of the diet increases.

8.3.2 Glycogen reserves

In the same way as higher animals, fish have carbohydrate stores in the form of glycogen. In the fish, hepatic glycogen consists of 10,000 to 14,000 units of glucose, depending on species. There is significant variation between tissues: the brain, liver and red muscle contain more than white muscle. In the brain, concentrations can reach $350\,mg.100\,g^{-1}$. In white muscle, these vary between 40 and $200\,mg.100\,g^{-1}$ and in red muscle, between 300 and $900\,mg.100\,g^{-1}$. Values in the liver vary from 1–12% of fresh tissue, in direct proportion to dietary carbohydrate supply: there is a linear relationship between carbohydrate intake and hepatic glycogen content.

8.3.3 Mobilisation of reserves (glycogenolysis)

Breakdown of glycogen is carried out by the action of glycogen phosphorylase and has been observed in many species. In trout, this enzyme is localised not only in white muscle but also in the liver. In contrast, there does not appear to be any relationship between glycogen catabolism and phosphorylase activity, the function of which in fish is the subject of controversy. For glycogen, as for the fatty acids, there is also a lysosomal degradation pathway which appears to be preponderant in the muscle.

In general, glycogen is an energy substrate which is used very little in fish tissues. White muscle is the exception: mobilisation of this compound, observed notably under conditions of stress (hypoxia, exercise), leads to an increase in lactate concentrations. Mobilisation of reserves is very rapid in white muscle compared to red muscle or to the liver. In the latter, glycogen content and its mobilisation depend on the previous nutritional state. In fish fed with diets rich in digestible carbohydrates, hepatic glycogen reserves are high, but are rapidly depleted during starvation. Conversely, in fish fed a diet less rich in carbohydrates, mobilisation during starvation is lower. In the wild, even after long periods of starvation, as observed in salmon that starve in synchrony with migration, or in carp during the winter months, glycogen reserves are only slightly reduced. This is very different from the situation encountered in higher vertebrates in which mobilisation of hepatic glycogen is very rapid and that of muscle glycogen is slow. Despite this fact, hypoglycaemia is rarely observed in starved fish, probably because there is a decrease in metabolic activity in general (and consequently in glucose utilisation) and also there is activation of gluconeogenesis.

Glycolysis and glycogenolysis, the main catabolic pathways for carbohydrates, appear to function at low levels in fish. For example, the injection of glucagon has a hyperglycaemic effect in fish, just as in other animals. However, this hyperglycaemic effect is more as a result of an increase in gluconeogenesis than of mobilisation of hepatic glycogen. The injection of a gluconeogenesis inhibitor effectively cancels the hyperglycaemic action of glucagon. Increasing carbohydrate supply results in an increase in enzyme activity involved both in glycolysis and the pentose pathway (notably glucose 6 phosphate dehydrogenase) and hence lipogenesis. An inverse relationship is often observed with increasing lipid supply (Chapter 7).

8.4 IMPORTANCE AND LIMITS OF THE ENERGETIC ROLE OF CARBOHYDRATES

The mean value for crude energy content of carbohydrates is close to $16.7\,kJ.g^{-1}$ (15.9 for glucose and 17.6 for starch). When feeding fish, it is important to consider the values for digestible energy of carbohydrates as these values can vary considerably (see above). In addition, in terms of net energy the energy value can also differ depending on species. The incorporation of simple sugars into foods for aquaculture is not economic and the only carbohydrates likely to be retained in manufacture of compound feed are the complex carbohydrates. Various studies have shown that, at an equivalent protein concentration, the growth performances and feed conversion ratio of trout can be improved by including polysaccharides such as pretreated starch in the diet.

Starches differ depending on their botanical origins. In common carp and trout, raw wheat starch is better utilised than that of maize, triticale, or potato. For the last two, heat treatment appears to have a marked effect. The presence of amylase inhibitors in wheat does not appear to have any significantly harmful effect, at least in species such as the common carp. It has also been demonstrated that increasing meal frequency can sometimes improve carbohydrate utilisation in some species. This phenomenon can be explained in part by the possible effects of unavailable carbohydrates (fibre) on the rate of digestive transit. Some indications of optimal limits of incorporation for different species are given in Table 8.4.

The protein sparing effect of digestible carbohydrates is also subject to much variability. However, the supply of digestible energy by inessential carbohydrates appears to have an overall beneficial effect in terms of improving growth and protein utilisation in most species.

8.5 FIBRE AND FILLING AGENTS

8.5.1 Definition

Sometimes termed as cellulose compounds, wall compounds (of plant cells) or even plantix (substance X, always coming from plants), fibres have never been satisfactorily defined. They mainly include carbohydrates of varying degrees of complexity (Box 8.1), which cannot be hydrolysed by higher vertebrate enzymes. However,

Table 8.4. Suggested maximal incorporation levels (% diet) of digestible carbohydrates in different species (ranges).

Salmonids	25–30
Cyprinids	40–45
Siluriforms (catfish)	30–35
Acipenserids (sturgeons)	25–30
Eels	25–30
Cichlids (tilapias)	35–40
Turbot, plaice	15–20
Sea bass	25–30
Sea bream	25–30
Marine tropical fish	
Milk fish	35–40
Tropical sea bass	20–25
Red drum	20–25
Yellowtail	10–15

lignin, which is classed among the fibres, is a polyphenol and not a polysaccharide. These compounds are generally fibrous in structure, which is not the case for the pectins.

In monogastric animals, these compounds mainly play a role as filler or bulk, facilitating digestive transit, but they can be partially degraded which means that they play a small role in nutrition.

8.5.2 Assays

The usual measuring methods do not always give constant results: the very classical and ancient Weende method allows the evaluation of the sum of cellulose + lignin (crude fibre), but with significant underestimation. The Van Soest method allows the measurement of three main fractions (including lignin), obtaining satisfactory estimations for cellulose and the hemicelluloses (Table 8.5). Pectins are not, however, measured using this method which also results in large errors when it is applied to compounds rich in starch. It is convenient in this case to utilise the modified method of Giger *et al.* (1987). If the sum of lignin, cellulose, hemicellulose and pectins can be measured, then it is possible to measure the total cell wall compounds (Carré and Brillouet, 1989). For cereals alone, the measurement according to Carré and Brillouet can be replaced by that of NDF (neutral detergent fibre) of Van Soest, since monocotyledons, in contrast to dicotyledons, contain almost no pectin.

Table 8.5. Current methods for the measurement of the main components of dietary fibres.

PCW* (plant cell walls)	pectins + hemicelluloses + cellulose + lignin
NDF** (neutral detergent fibre)	hemicelluloses + cellulose + lignin
ADF** (acid detergent fibre)	cellulose + lignin
ADL** (acid detergent lignin)	lignin
Thus, the difference:	
PCW − NDF	pectins
NDF − ADF	hemicelluloses
ADF − ADL	cellulose

* Method of Carré & Brillouet, 1989
** Modified method of Van Soest: Giger *et al.*, 1987.

8.5.3 Nutritive roles

Fibrous compounds (except lignin) are subject to partial digestion by hydrochloric acid in the stomach and especially by enzymes in the bacterial flora and to even more digestion in species with herbivorous tendencies. In increasing order of digestibility, the classification is:

$$\text{lignin} < \text{cellulose} < \text{hemicelluloses} < \text{pectins}$$

The digestibility of polyholosides of aquatic origin (alginates and carrageenans) has been little studied in fish. It is certainly nil or very limited, since these binding substances are sometimes used to stabilise not only food particles but also the faeces. Some qualitative studies mention high polysaccharide digestibility as being extremely rare, only found in some true herbivorous species. In omnivorous fish (carp, tilapia, etc.), there are few data, but all lead to the conclusion that fibre digestibility is negligible. Even in herbivorous carp (e.g. grass carp), it is practically nil for cellulose. In carnivorous fish, it can be assumed that all cell wall compounds, including pectin, are indigestible.

8.5.4 Other roles

In terrestrial vertebrates there is a negative correlation between fibre content and energy digestibility. The total effect results mainly from a dilution effect of the digestible components, sometimes with an added negative effect of the cell wall components (mainly lignin and cellulose) on the digestive processes. The overall effect appears to be similar in fish.

Fibre has a stimulant effect on digestive transit and allows compensation for digestible energy input by an increase in intake in higher animals. In contrast, 'crude fibre' components acting as gels (gums and pectin) tend to slow down stomach emptying. Many other effects of fibre, both beneficial and detrimental, have been described or are still the subject of studies with higher vertebrates (desquamation of intestinal cells, increased elimination of cholesterol and minerals). Most of these are still poorly understood or unknown in fish. Even the incorporation limits in carnivorous or omnivorous fish are impossible to determine precisely; the value of 10% which is sometimes used appears to be rather arbitrary.

In experimental diets, cellulose can be used conveniently to dilute diets without altering nutrient utilisation. Some binding agents (alginate, agar, guar gum and carrageenans) or even bran improve faeces consistency and make measurements of digestibility more reliable without interfering with digestive utilisation, at least up to a certain level of incorporation which can be fixed at 4 or 5%. Lastly, carboxymethylcellulose, chemically transformed cellulose, is a binding agent with gelling properties which appears to limit post-prandial hyperglycaemia in fish.

8.6 CONCLUSION

In general, fish do not have as great a capacity for utilising carbohydrates as the terrestrial higher vertebrates. This difference is true both for available carbohydrates and the carbohydrates that fall into the category of fibres. The amounts of starch currently incorporated into fish foods far exceed the limits imposed two decades ago. For certain species, it is possible to supply an even larger amount of carbohydrate energy when the economic context is favourable. Before doing this, however, it is advisable to target certain species and study their physiological characteristics which are still far from being known in detail.

8.7 BIBLIOGRAPHY

Bergot, F. (1979) Effects of dietary carbohydrates and of their mode of distribution on glycaemia in rainbow trout (*Salmo gairdneri* Richardson). *Comp. Biochem. Physiol.*, **64A**, 543–547.

Carré, B. and Brillouet, J. M. (1989) Determination of water unsoluble cell walls in feeds: inter-laboratory study. *J. Assoc. Off. Annal. Chem.*, **72**, 463–467.

Giger, S. (1985) Revue sur les méthodes de dosage de la lignine utilisées en alimentation animale. *Ann. Zootech.*, **34**, 85–122.

Giger, S., Thivend, P., Sauvant, D., Dorleans, M. and Journaix, P. (1987) Etude de l'influence préalable de différentes enzymes amylolytiques sur la teneur en résidu NDF d'aliments du bétail. *Ann. Zootech.*, **36**, 39–48.

Kaushik, S. J. and Médale, F. (1994) Energy requirements, utilization and dietary supply to salmonids. *Aquaculture*, **124**, 81–97.

Mommsen, T. P. and Plisetskaya E. M. (1991) Insulin in fishes and agnathans: history, structure and metabolic regulation. *Rev. Aquatic Sci.*, **4**, 225–259.

Palmer, T. N. and Ryman, B. E. (1972) Studies on glucose intolerance in fish. *J. Fish Biol.*, **4**, 311–319.

Panserat, S., Médale, F., Blin, C., Brèqu, J., Vachot, C., Plagne-Juan, E., Gomes, E., Krishnamoorthy, R., Kaushik, S. J. (2000) Hepatic glucokinase is induced by dietary carbohydrates in rainbow trout, gilthead seabream and common carp. *Am. J. Physiol.*, **278**: R1164–R1170.

Shiau, S. Y. (1989) Role of fibre in fish feed. In: Shiau, S.-Y. (ed.) *Progress in Fish Nutrition. Proceedings of the Fish Nutrition Symposium*, Keelung, Taiwan, ROC, 6–7 September 1989.

Walton, M. J. and Cowey, C. B. (1982) Aspects of intermediary metabolism in fish. *Comp. Biochem. Physiol.*, **73B**, 59–79.

Wilson, R. P. (1994). Utilization of dietary carbohydrate by fish. *Aquaculture*, **124**, 67–80.

9

Vitamin nutrition

The concept of vitamins appeared at the end of the nineteenth century, lending some weight to the 'accessory factors' to which nutritionists previously referred when certain nutritional effects were not explained by the nutrients known at that time. In 1948, vitamin B_{12} was isolated in the form of cyanocobalamin and its formula was determined in 1955. The list of vitamins has been closed, perhaps in a rather arbitrary manner. Studies on vitamin nutrition in fish have mainly taken place after this date. Numerous questions have been posed: are the vitamins required by fish the same as found in higher vertebrates? What are the symptoms of deficiency? What are the quantitative requirements? While currently there are no satisfactory answers to all these questions, many aspects have been added to by fundamental and applied research that has benefited from knowledge of higher vertebrates. This chapter is centred on the role of each vitamin and effects of deficiencies on fish; quantitative requirements are given in Appendix C.

9.1 DEFINITIONS

The term 'vitamin' was coined in 1910 by Funk, to describe 'the amine necessary for life', which we now call thiamin or vitamin B_1. However, it very rapidly became essential to lengthen the list of compounds 'necessary for life' in very small quantities, which were also called vitamins, even though many of them contained no nitrogen. All these studies followed the observation of pathological states, described nowadays as states of deficiency, in animals fed with purified diets (vitamin deficiencies seem to be unknown in wild animals). Input of the vitamin involved relieves deficiency symptoms; the naming of anti-pellagra, anti-beriberi and anti-rickets vitamins followed on although it is more logical to define vitamins by their positive role in normal situations than by their aptitude for repairing damage caused by a deficiency. Vitamins can be defined as organic compounds which are necessary for an organism, which cannot be synthesised and are different from essential amino acids or essential fatty acids. To this definition is often added the fact that vitamins – which are catalytic, non-plastic compounds, unlike amino acids or fatty acids – are required in minuscule amounts. Within the group, however, this definition is not very easily applied to two classes of compound among the vitamins in the broad sense (quasi-vitamins according to some authors): choline and inositol.

It should also be noted that incapacity for synthesis is not an absolute character of the compounds figuring in the 'official' list: niacin can in fact be synthesised from an amino acid, tryptophan (admittedly at very low levels in fish); cholecalciferol can also be made from cholesterol if ultraviolet radiation is sufficient (however, it penetrates little through water), and retinol from certain precursors (mainly β-carotene), supplied by the diet. The concept of inability to synthesise is thus relative. In contrast, there are compounds such as carnitine, the synthesis of which is also possible but at an insufficient rate, which have not been included in the list of vitamins. These 'growth factors' will be mentioned only briefly.

Vitamin nomenclature still causes confusion. It was customary to use letters, sometimes followed by a suffix. Following current conventions, only the name of the molecule should be used, but this is often cumbersome and obscure for non-specialists. However, care should be taken in that in certain cases the letter and suffix terminology is very ambiguous: pantothenic acid, for example, is called vitamin B_3 in Germany, B_5 in Britain and B_3 or B_5 in France; the designation B_7, recently given to choline, corresponds historically to adenine, a purine base which often acts as a growth factor, also called B_5! It is thus important to reserve the use of nomenclature using letters and suffixes to cases where there is no ambiguity.

9.2 GENERAL

9.2.1 Vitamin characteristics and species

Ascorbic acid is the best example of a molecule whose vitamin character varies between species: it is perfectly synthesised from glucose by most vertebrates, but not by primates (including man), nor by some species such as the guinea-pig. Fish form the largest group of vertebrates for which this compound is a vitamin; vitamin C.

The presence of an active intestinal flora sometimes causes confusion, most bacteria effectively synthesising large quantities of B group and K group vitamins, which the host organism can recover to a certain extent. An animal with an active flora can appear indifferent to the dietary supply of most vitamins while in fact its body requires them. In fish, the role of the intestinal flora is very limited, except in very few cases, and dietary intake constitutes the quasi-exclusive source of vitamins.

9.2.2 Classification

The vitamins form a group which is neither chemically nor functionally homogeneous (Table 9.1). Their classification is therefore not simple. Two groups are usually

Table 9.1. Classification of vitamins.

	Vitamers		
Vitamin family	Chemical names	Biologically active forms	Metabolic functions
A	retinol retinal retinoic acid retinyl palmitate* or acetate β-carotene (pro-vitamin)	retinol retinal retinoic acid	co-enzymatic pro-hormonal
D	ergocalciferol (D_2) cholecalciferol (D_3)	dihydroxyergocalciferol [$1.25(OH)_2D_2$] dihydroxycholecalciferol [$1.25(OH)_2D_3$]	pro-hormonal
E	tocopherols α-acetate-tocopherol*	α-tocopherol	H^+/e^- transfer (membrane protection)
K	phylloquinone (K_1) menaquinones (K_2) menadione (K_3)*	phylloquinone menaquinone	co-enzymatic, H^+/e^- transfer
B_1	thiamine thiamine chlorohydrate*	thiamine pyrophosphate (TPPH) thiamine triphosphate (TTPH)	co-enzymatic
B_2	riboflavin	flavine mononucleotide (FMN) flavine adenine dinucleotide (FAD)	co-enzymatic, H^+/e^- transfer
PP	niacin nicotinic acid nicotinamide	nicotinamide adenine dinucleotide (NAD) nicotinamide adenine dinucleotide phosphate (NADP)	co-enzymatic, H^+/e^- transfer
[B_5]	pantothenic acid calcium pantothenate*	coenzyme A(CoA) acyl protein transportase (ATP–ACP)	co-enzymatic, H^+e^- transfer
B_6	pyridoxine pyridoxal pyridoxamine pyridoxine chlorohydrate*	pyridoxal phosphate	co-enzymatic
[B_8]	biotin	biotinyl-AMP	co-enzymatic
[B_9]	folic acid polyglutamates	tetrahydrofolates (THF)	co-enzymatic
B_{12}	cobalamines	methylcobalamine adenosylocobalamine	co-enzymatic
C	ascorbic acid dehydroascorbic acid ascorbyl phosphates*	ascorbic acid dehydroascorbic acid	H^+/e^- transfer (co-enzymatic?)

[] ambiguous terms, to be avoided
* vitamers

distinguished according to their physical properties: the four liposoluble vitamins (vitamins A, D, E and K) or A group vitamins which are soluble in oils and their solvents and the 11 hydrosoluble vitamins sometimes assigned to the B group – in fact, there are only eight, in the strict sense, but vitamin C as well as choline and inositol are added since they are also soluble in water.

9.2.3 Stability and availability

The fragility of certain vitamin molecules, along with the simplification of feeding formulae, is one of the main causes of deficiencies observed in fish farming. Some vitamins are not fragile at all and some are astonishingly stable; this is true for pantothenic acid and also for niacin which not only resists boiling in an alkaline environment but often becomes even more available after this treatment. Vitamins are, however, generally fragile, which makes them similar to EFAs. In practical terms, certain precautions must be taken with this family of nutrients. The problems are complex because of the distinct actions of the different aggressors (heat, humidity, light, pH) on each vitamin. Thus, retinol (vitamin A) is sensitive to oxidation, riboflavin (vitamin B_2) to light, ascorbic acid to alkaline pH, tocopherol (vitamin E) to heat and ultraviolet rays, and so on. In practical terms, vitamin stability under defined conditions will not be the same for natural compounds made from raw materials as for synthesised molecules. This is especially true for commercial products which are usually salts or other forms chosen for their stability: the palmitate or acetate of retinol, the acetate of tocopherol, and the pantothenate of calcium are much more stable than retinol, tocopherol and pantothenic acid respectively. Vitamin C constitutes an extreme case, ascorbic acid being the most labile form, while the sulphate forms and especially the phosphate forms are very stable. Commercial products are not always active in the form in which they are incorporated into diets. Thus α-tocopherol acetate or ascorbyl phosphate must firstly be hydrolysed in order that the actual vitamins can be absorbed. Ascorbyl sulphate, despite its good stability, is less easily hydrolysed and thus less available.

9.2.4 Vitamers and antivitamins

For a given vitamin activity, several vitamers can exist. These may be a whole family of compounds present in nature, compounds slightly modified during their metabolism (biological forms which remain active) or purely synthetic molecules, sold commercially for reason of cost of synthesis or stability (see above). Provitamins, a term applied mainly to the carotenoids, correspond to precursors which are vitamers in the broad sense of the term.

The different vitamers, compared on a molecular basis, do not have the same activity level and nutritionists have introduced the concept of bioavailability. This is applied in principle to the absorption or bioconversion of natural substances and in particular to cases where the molecules are present in the form of complexes.

A whole series of compounds have been defined as antivitamins. These can be enzymes which hydrolyse vitamins, substances blocking their intestinal absorption or natural or synthetic antagonists. An example from the first group is thiaminase in raw fish. The second group includes antibiotics and other medicines. The last, even larger group, includes compounds used experimentally and also a series of medicines.

9.2.5 Absorption

Certain vitamers, and in particular those that correspond to active forms of group B vitamins in the tissues, must be subjected to hydrolysis in the digestive tract before being absorbed. For these compounds there thus exists a true digestive phase, but this usually corresponds either to the action of proteases or that of non-specific enzymes such as esterases or phosphatases.

Vitamin absorption is only well understood in the higher vertebrates in which it results from several distinct mechanisms, some of which are quite specific. A-group vitamins are firstly incorporated into micelles, as are free FAs and mono- and diacylyglycerols, before being absorbed with these nutrients. B-group vitamins generally have active transporters, whose action is sometimes completed by facilitated diffusion phenomena, or even passive diffusion. Thiamine, for example, is absorbed both by passive diffusion and active transport. In the case of active transport, there is regulation of transporter activity by nutritional state: this activity is higher when deficiency or sub-deficiency occurs, while it is reduced when input is excessive. This very important phenomenon contributes to ensuring a certain 'vitamin homeostasis' (Figure 9.1).

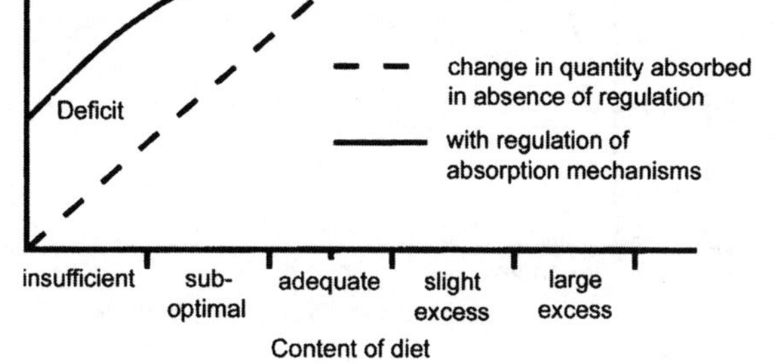

Figure 9.1. Diagrammatic regulation of tissue concentrations.

9.2.6 Metabolic role

It is impossible to attribute a specific role to the two groups, A and B. It is true that group B corresponds to molecules which are fundamental to intermediate metabolism (co-enzymes in general) in both animals and plants, while the group A vitamins have a more varied role, sometimes present only in animals. Combs (1992) proposed a new classification based on four functions:

- coenzyme function (11 vitamins);
- transfer of protons or electrons (6 vitamins);
- pro-hormonal function (2 vitamins);
- membrane protection (1 vitamin).

Of these vitamins, five fulfil several functions and some mechanisms remain only partially understood, even in mammals. The classification used is based entirely on Combs', while taking into account the known multiple roles and the remaining gaps in knowledge.

9.2.7 Requirements and feeding standards; sparing effect and excess; deficiency symptoms

In theory, determination of a vitamin requirement is carried out in the same way as for any nutrient, using a dose-response curve. It differs, however, in the multitude of criteria that can be used to quantify the response: growth rate, state of reserves (vitamin status), biochemical functions and even survival rate or suppression of deficiency symptoms (Figure 9.2); it is also complicated by a series of other considerations: availability or stability of vitamers, multiple interactions between nutrients and even analytical difficulties. Among the interactions to be taken into

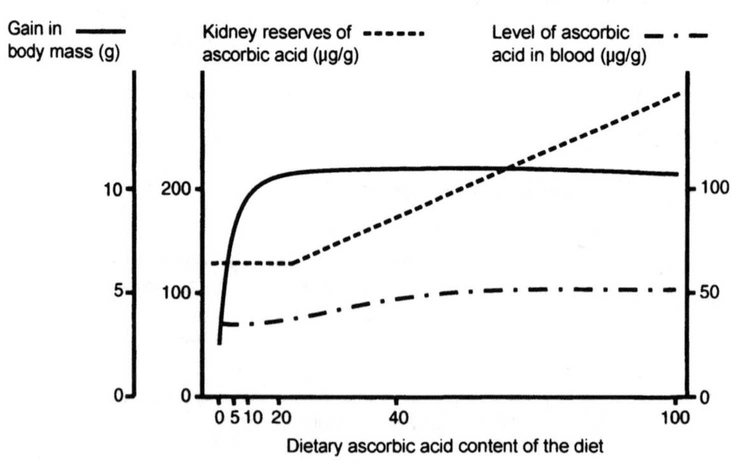

Figure 9.2. Dose-response curves obtained for ascorbic acid in trout, indicating the differences between chosen criteria (after Halver, 1972).

account, those that occur between the vitamins themselves – called sparing effects – are particularly important. They can lead to underestimation of requirements for a given vitamin if others are input in excess; this has been clearly demonstrated in fish. For these various reasons, vitamin requirements are far less precisely known than those for AAs, for example. In addition, although reserves can meet requirements for a period of several weeks, or even months, security margins of the order of 100% are very common for amounts in the food. Excessive amounts of vitamins can be harmful, except for group B vitamins which are easily excreted in the urine. In fish, the harmful effect of excess has been demonstrated mainly for menadione, an industrial vitamer of vitamin K, and retinol. Certain vitamins possess a pharmacodynamic role at levels greatly exceeding nutritional standards. This is the case for vitamin C which is sometimes used in 'megadoses' in aquaculture.

Deficiency symptoms, which are more or less characteristic of each vitamin, are summarised in Table 9.2.

9.3 SIMPLE COENZYME VITAMINS

The formulae for these vitamins are given in Box 9.1 (p. 154).

9.3.1 Thiamine (vitamin B_1)

This vitamin plays a major role in carbohydrate metabolism at several levels of intermediary metabolism. Thiamine can be phosphorylated, producing thiamine mono-, pyro- and triphosphate respectively. Only thiamine pyrophosphate (TPP) and thiamine triphosphate (TTP) are significant in metabolic terms. PPT is a coenzyme that acts as a transporter for aldehyde groups in oxidative decarboxylation reactions of α-ketonic acids such as pyruvic acid and α-ketoglutaric acid, on the one hand, or other ketoacids arising from branched AAs on the other. This sequence of reactions is catalysed by a group of enzymes that bring into play other vitamins in the form of cofactors, including FAD (riboflavin) and NAD (niacin). Similarly, TPP participates in the transfer of glycoaldehyde radicals during the two transketolation reactions that occur in the pentose phosphate pathway. TTP plays an essential role in neurotransmission.

Thiamine deficiency has been observed in the eel, turbot, catfish, carp and especially the trout. Nervous problems, similar to those found in mammals, constitute the most typical signs. Taking into account response criteria such as gain in body mass or transketolasic activity, requirements have been evaluated at 1 $mg.kg^{-1}$ of feed for young trout, from purified diets; it is of the same order of magnitude for other species such as catfish, carp and turbot. Requirements does not seem to be correlated with potential for carbohydrate utilisation in the different species.

9.3.2 Pyridoxine (vitamin B_6)

This is the key enzyme of AA metabolism. Under the name vitamin B_6 there are three different vitamer derivatives of pyridine which differ in function, i.e. alcohol

Table 9.2. Main signs of deficiency and response criteria in different species.

Vitimins	Main signs of deficiency	Response criteria
Thiamine	Nervous problems (Sa, CF, E, C, T), melanism (CF, Y), skin discoloration (C), subcutaneous haemorrhages (E, C, SB), congestion of fins (E, Y, SB)	Gain in body mass and tissue reserves (liver, erythrocytes), activity of transketolases
Pyridoxine	Nervous problems: erratic swimming, convulsions, hyper-irritability (Sa, CF, E, C, Y), blue-green coloration (CF)	Gain in body mass, hepatic reserves, activity of transaminases
Biotin	Degeneration of gill lamellae, muscular atrophy, degeneration of pancreatic acini (Sa), skin discoloration (CF), melanism (E), lethargy (C, Y)	Gain in body mass, hepatic reserves, acetyl-CoA carboxylase and pyruvate carboxylase activity, resistance to swimming
Folic acid	Lethargy and anaemia (Sa, CF), melanism (Sa, E), increased bacterial infections (CF), fin congestion (Y)	Gain in body mass, hepatic reserves, haematocrit
Cobalamines	Microcytic and hypochromic anaemias with erythrocyte fragmentation (Sa), low haematocrit (CF), fin congestion and anaemia (Y), anorexia (E)	Hepatic reserves
Retinol	Ocular lesions with degeneration of the retina (Sa), skin discoloration and exopthalmia (Sa, C, CF), oedema (Sa, CF), haemorrhages (CF, C, Y), melanism (Y)	Gain in body mass and hepatic reserves
Riboflavin	Lethargy (Sa, E), melanism (Sa, Y), photophobia (Sa, C), fin haemorrhages (Sa, E, C), opacity or vascularisation of the cornea (Y, Sa), nanism (CF), dermatitis (E), renal necrosis (C)	Gain in body mass, hepatic reserves, glutathione reductase and DAAO activity
Niacin	Lesions and haemorrhages in skin and fins (Sa, CF, C, Y), anaemia (Sa, CF), photosensitivity (SA), exopthalmia (CF), abnormal swimming (E), fin deformities (CF), mortality (CF, Y)	Gain in body mass and hepatic reserves

Pantothenic	Atrophy of pancreatic acini (Sa), club-shaped gills (Sa, CF, Y), skin haemorrhages (E, C), exopthalmia (C), dermatoses (E), mortality (Sa, CF, Y, SB)	Gain in body mass and hepatic reserves
Menadione	Delay in blood clotting with anaemia, haemorrhages in gills and eyes (Sa), skin haemorrhages (CF)	Haematocrit
Ascorbic acid	External and internal haemorrhages (Sa, CF, E, Y), lordosis, scoliosis (Sa, CF, Y), distortion of gill filaments (Sa), ascites (Sa, T), melanism (Y, T), lower jaw erosion (E), renal granulomatosis (T)	Gain in body mass, hepatic and renal reserves, vertebral collagen content, tyrosinaemia
Tocopherols	Muscular dystrophy (Sa, CF, C), skin discoloration (Sa, CF), melanism (Y), erythrocyte fragility (Sa), exudative diathesis and ceroid deposition (CF), exopthalmia and lordosis (C)	Gain in body mass and hepatic reserves
Calciferol	Disturbance of calcium homeostasis and tetanus (Sa), decalcification of bones (CF)	Gain in body mass and feeding efficiency
Choline	Fatty liver (Sa, C), haemorrhages of kidney and intestine (Sa, CF), vacuolisation of hepatic cells (C), melanism (Y), exopthalmia (Sa), intestine greyish in colour (E)	Gain in body mass, hepatic reserves, lipid content of liver
Inositol	Melanism (Sa, Y), anaemia with fin erosion (Sa), intestine greyish in colour (E), loss of cutaneous mucus (C)	Gain in body mass, hepatic reserves, choline esterase and transaminase activity

Sa: salmonids, CF: catfish, E: eel, Y: yellowtail, C: carp, T: turbot, SB: sea bream.

(pyridoxol or pyridoxine), aldehyde (pyridoxal) or amine (pyridoxamine). All three are transformed in liver cells into an active coenzyme: pyridoxal phosphate which acts in many catalytic systems such as transaminases, transulphurases and transferases. Pyridoxal phosphate also allows, by AA decarboxylation, the formation of biogenic amines that act as neuromediators (histamine, serotonin or 5-hydroxytryptophan, tyramine, dopamine, etc.) and also acts as a constituent of dehydratases in the deamination of certain AAs. Vitamin B_6 is also involved in many other

Box 9.1 Formulae of coenzyme vitamins

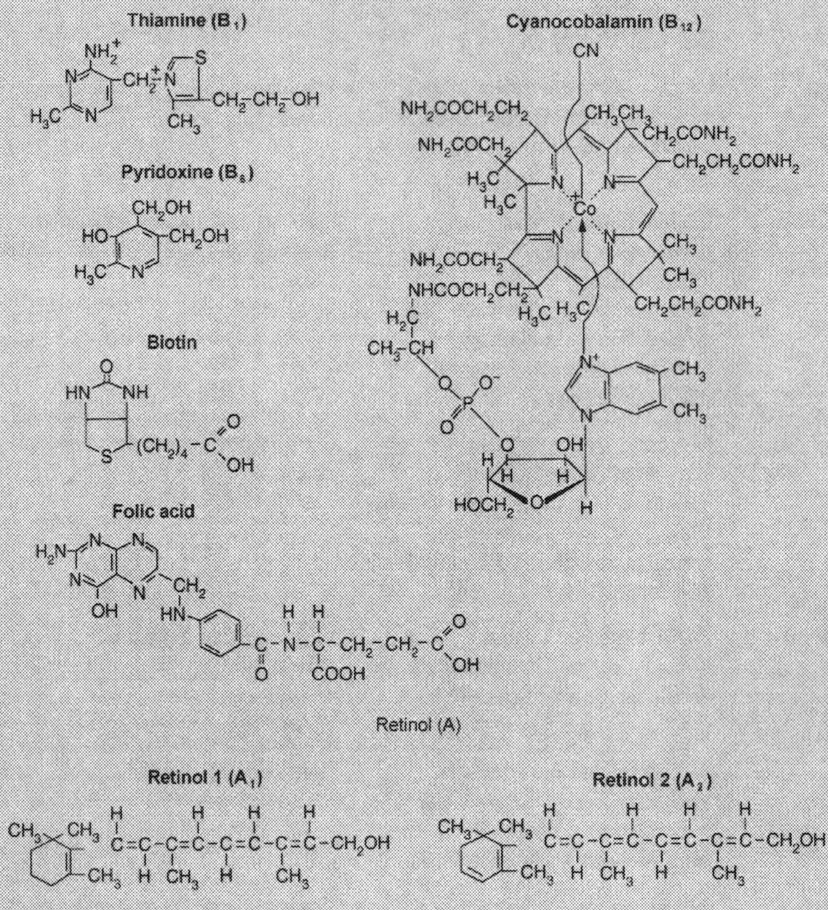

functions which vary from the synthesis of pancreatic enzymes to glycogen metabolism and even immune functions.

Fish have a marked carnivorous tendency, especially at the start of their lives, and rapidly manifest signs of deficiency in pyridoxine, particularly in nervous system problems (convulsions, erratic swimming and hyper-irritability) due to a defect in mediator synthesis. Growth of young trout is improved by input of protein in the diet at a level of $2\,mg.kg^{-1}$ of food, i.e. $0.5\,mg.100\,g^{-1}$ of protein. In other species, such as turbot, sea bream or Channel catfish, requirements are of the same order of magnitude, between 1 and $3.2\,mg.kg^{-1}$ of feed. The activities of the two commonly-measured transaminases, glutamate-oxaloacetate transaminase (GOT) and glutamate pyru-vate transaminase (GPT) in various tissues (liver, muscles and erythrocytes) have been used as response criteria in the determination of requirements for trout, turbot and sea bream. In the trout, requirements are three times higher if results are based on obtaining maximal GOT activity in white muscle rather than on growth. On the basis of an interspecific comparison, it does not appear that pyridoxine requirements are correlated with dietary proteins, in contrast to what might be predicted from the role of this vitamin in protein metabolism.

9.3.3 Biotin

Biotin acts as a coenzyme in carboxylation and decarboxylation (transfer of CO_2) reactions, after being activated into biotinyl-AMP. The main enzymes involved are: pyruvate carboxylase which is involved in glucose metabolism (glycolysis and neoglucogenesis); acetyl-CoA carboxylase, key enzyme in the synthesis of FAs or lipogenesis; propionyl-CoA carboxylase which catalyses the transformation of propionyl-CoA, produced by the degradation of FAs and certain AAs into methylmalonyl-CoA; β-methylcrotonyl-CoA carboxylase which acts during leucine breakdown.

Various signs of deficiency have been described in several species, but few seem to be specific and, moreover, they have never been observed in fish fed on a diet based on natural ingredients. This could be due to the low level of requirement. This has been estimated from purified diets at 0.08 $mg.kg^{-1}$ of feed for trout alevins, using gain in body mass as the criterion, and at 0.14 $mg.kg^{-1}$ of feed for maximal pyruvate carboxylase activity in the white muscle.

9.3.4 Folic acid and polyglutamates

Folic acid acts as a coenzyme in monocarbon transfer reactions; its molecule is composed of a pteridine nucleus linked to para-amino-benzoic acid (sometimes considered as a vitamin or a vitamer of folic acid), thus forming pteroic acid. This acid is linked to one or more molecules of glutamic acid giving the polyglutamates or folates present in foods and tissues. These molecules are activated into tetrahydrofolates (THF) in the presence of vitamin B_{12} and $NADPH_2$. As co-enzymes of pteroprotein enzymes, THFs participate in the transfer and interconversion of monocarbon groups in histidine and glycine metabolism and in the interconversion of glycine and serine. Protein synthesis starts with the intervention of formylmethionine which is transported by a transfer RNA and whose synthesis requires the presence of THF. This vitamin also acts in the synthesis of purine and pyrimidine bases.

Various signs of deficiency have been observed in several species of fish, but not in the common carp. The most common symptoms relate to a particular form of anaemia – macrocytic anaemia – which is generally associated with anorexia and loss of body mass. Although a concentration of 0.3 $mg.kg^{-1}$ of purified diet is sufficient for maximal survival, 0.6 $mg.kg^{-1}$ is required for maximal growth and 1.1 $mg.kg^{-1}$ for saturation of tissues such as the kidney and liver in young trout. Requirements thus appear to be low, which explains why, in certain cases such as in the carp, they can be met by synthesis occurring in the intestinal flora. American recommendations for salmonids (NRC, 1993) are 1 $mg.kg^{-1}$ of food.

9.3.5 Cobalamines (vitamin B_{12})

The term vitamin B_{12} is applied to several related macromolecules, the cobalamines, of which the best known is cyanobalamin. These consist of a tetrapyrrolic nucleus at the centre of which is a cobalt atom bound to four nitrogen atoms and to a pseudo-nucleotide group. Different radicals can be attached to the Co atom, the main ones being CN, OH, CH_3 and 5′-desoxyadenosine, giving cyanocobalamine, hydroxocobalamine, methylcobalamine and adenosylcobalamine respectively. The last two molecules represent the coenzyme forms which are active in two types of reaction: (a) isomerisation reactions such as those involved in leucine metabolism or the conversion of methylmalonyl-CoA into succinyl-CoA, a metabolite in the Krebs cycle; and (b) transmethylation reactions that require methylcobalamine as a transporter of monocarbon groups such as methyl radicals. The cobalamines play a very important role in the metabolism of folates. Thus, a decrease in input can cause the appearance of symptoms similar to those caused by folic acid deficiency.

Vitamin B_{12}, which only exists in animal tissues and in some micro-organisms, is required in very low quantities and deficiency symptoms only appear after a long period of deprivation. The most characteristic sign is pernicious anaemia (hypochromic microcytic anaemia) which is, however, impossible to detect in fish whose diet contains animal meals. Requirements for this vitamin can be met or partially met by the synthesis carried out by the micro-organisms in the intestinal flora, at least in warm water fish such as tilapia, catfish and carp. Quantitatively, the requirement for vitamin B_{12} is the lowest of all the vitamin requirements; in trout it rises to 7 μg of cyanocobalamine per kg of purified diet. It has not yet been quantified for other species, except for yellowtail where it appears to be higher than in trout (53 $\mu g.kg^{-1}$ of food).

9.3.6 Retinol (vitamin A)

Vitamin A is an alcohol derivative of the carotenoids and is involved mainly in the process of vision, where it plays a role as a coenzyme, and also in other, less well-known hormonal-type mechanisms. This molecule consists of a β-ionone nucleus to which is attached a chain of three isoprenoid units that terminate in an alcohol unit. The molecule resembles half of a β-carotene molecule, the substance from which it is actually synthesised, but transformation occurs in terms of molecules in such a way that, in terms of mass, β-carotene activity is not more than 50% of that of retinol. In fish, much lower values have been found (less than 20%). Vitamin A occurs in three forms, either alcohol (retinol), aldehyde (retinal) or acid (retinoic acid).

There are six isomers as a result of the four double bonds on the lateral chain. The most active form is the 'all E' (historically, 'all trans') form, now obtained synthetically. In fish, mainly in freshwater species, a molecule is found which is only slightly different to true retinol (A_1). Called retinol 2 (A_2), it only differs from the former by an extra double bond on the β-ionone nucleus and it possesses the same activity.

Retinol acts both in the composition of purple retinin or rhodopsin (rod chromoprotein), which is essential for adaptation to darkness, and also in the composition of pigments corresponding to retinal cones or iodopsins, which act in the vision of shape and colour. In all these pigments, the vitamin A molecule is found in the aldehyde form. Perception of light occurs by an isomerisation and return to the initial form, which creates a nervous influx; retinal acts as a coenzyme in this reaction. Vitamin A has also long been known as a 'growth vitamin'. This second role has been less well documented and comes from the fact that retinoic acid, an oxidised derivative of retinol which can no longer operate for vision, is endowed with very important metabolic functions. Cells have retinoic acid receptors. This acid or its derivatives (retinoids) act like steroid hormones by inducing messenger RNA and specific protein synthesis; they stimulate cellular differentiation, particularly in epithelial cells, mucus secretion, immune responses, reproductive functions and protection against cancers.

In fish, deficiency leads to a slowing down of growth. Little is known about specific symptoms, apart from ocular lesions. The resultant mortality is not, however, linked to vision. Excess vitamin is toxic to fish, or leads to fusing of the vertebrae, as found in higher vertebrates. This is only found with very high levels and in exceptional circumstances, since the use of whale or seal liver in fish farm food has ceased. Hypervitaminoses A have, however, been observed in larvae, following excessive supply of cod liver oil and vitamin A. Currently, the recommendations for vitamin A for trout are those established in a Japanese study in 1967, based on different response criteria, including survival, growth, hepatosomatic index and lipid and protein content of the liver. These estimations of 2500 $IU.kg^{-1}$ of food are of the same order of magnitude for the catfish and are similar to those found for pigs and poultry. However, it would be valuable to quantitatively re-evaluate this requirement with the aid of a purified diet.

9.4 TRANSFER AND OXIDATION-REDUCTION COENZYME VITAMINS

The formulae for these vitamins are given in Box 9.2.

9.4.1 Riboflavin (vitamin B_2)

This vitamin plays a key role in oxidation-reduction reactions. Its biologically active derivatives are riboflavin phosphate or flavin mononucleotide (FMN), which is itself converted by linking with an AMP molecule into a second active form, flavin adenine dinucleotide (FAD). FMN and FAD are involved, along with dehydrogenase (anaerobic) and oxidase (aerobic) coenzymes during the oxidation-reduction reactions of catabolism:

- of FAs, and more precisely, during β-oxidation;
- of L or D series AAs, the deamination of which is carried out by L-aminoacid

Box 9.2 Vitamin formulae

Oxidation-reduction vitamins

Riboflavin (B_2)

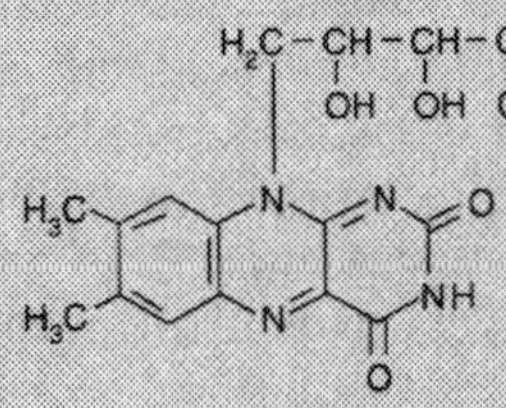

Pantothenic acid

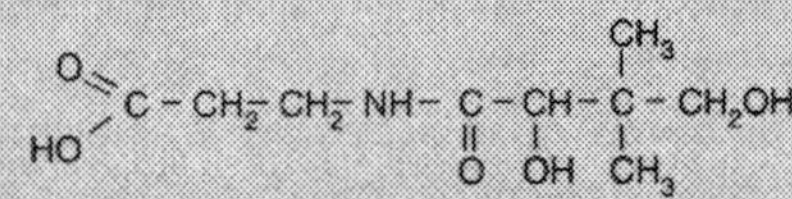

Vitamin K

Niacin (PP)

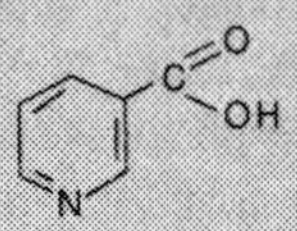

Phylloquinone (K_1)

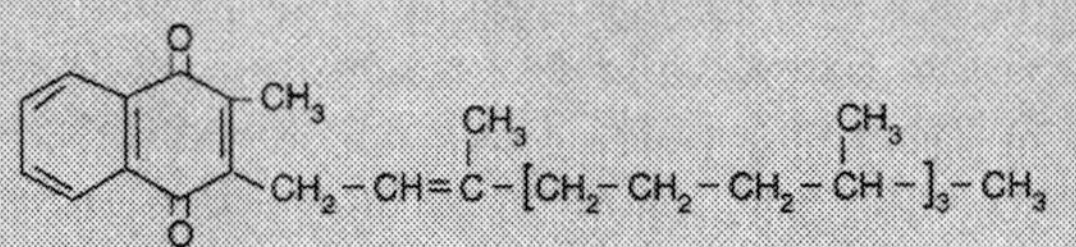

Menadione (K_3)

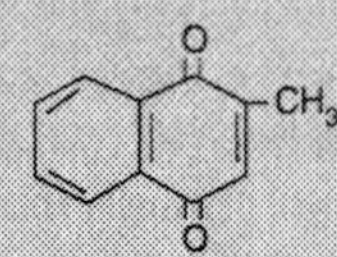

Anti-oxidation vitamins

Vitamin C

Ascorbic acid

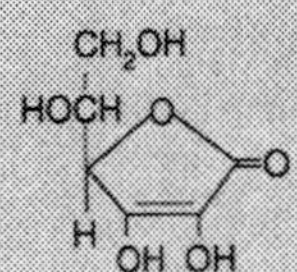

Dehydroascorbic acid

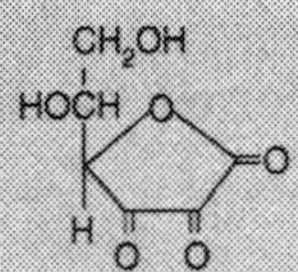

α-tocopherol (E)

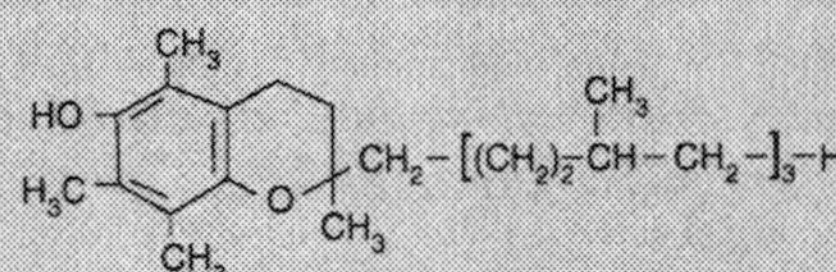

Pro-hormone vitamins

Ergocalciferol (D_2)

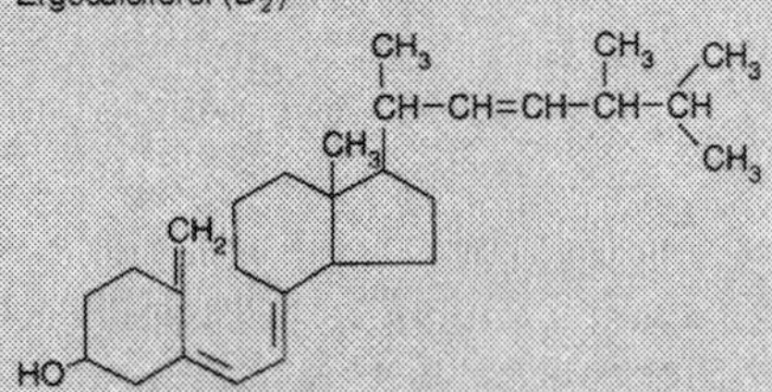

Cholecalciferol (D_3)

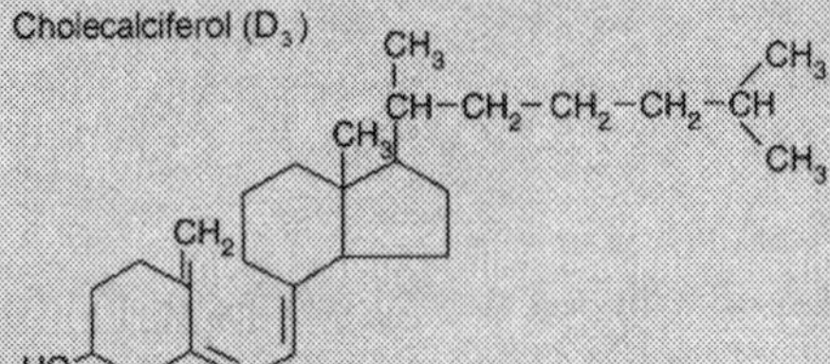

Quasi-vitamins

Choline

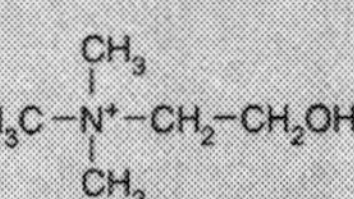

Myo-inositol

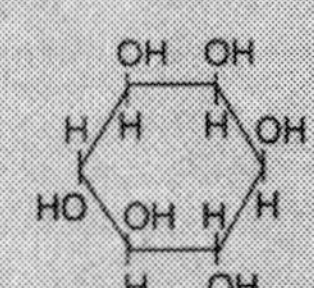

oxidase (low activity) in the presence of FMN, or by D-aminoacid oxidase (DAAO) in the presence of FAD;

- of purine bases such as adenine and guanine, components of nucleic acids, the degradation of which by xanthin oxidase requires FAD;
- of metabolites entering the TCA cycle, where succinate dehydrogenase catalyses one of the four dehydrogenations, with the aid of FAD, while the three other reactions involve another coenzyme system (containing niacin), NAD/NADH, which is itself reoxidised by FAD within the respiratory chain.

Lastly, FAD, which is involved in the glutathion reduction, is indispensable in the maintenance of cellular integrity of red blood cells. Also to be noted is the involvement of riboflavin, in the form of FMN, in the transformation into another vitamin, pyrodoxine, into its biologically active form.

Very many different symptoms of deficiency have been described which vary significantly between species: ocular lesions, various haemorrhages, melanism, etc. Studies on trout fed purified diets have established the required level for maximal growth at around $4\,mg.kg^{-1}$ of diet. Saturation of the tissues by flavins (liver, kidney, spleen) or maximal activity of hepatic DAAO are attained for a concentration of about $6\,mg.kg^{-1}$ of diet. This enzyme has been found to be a better biochemical indicator of vitamin status in trout than erythrocyte glutathione reductase (GR).

9.4.2 Niacin (vitamin PP)

This vitamin is required in much higher quantities than the preceding ones and can be synthesised, albeit inefficiently, by fish, as indicated above. It exists in two vitamer forms, nicotinic acid, pyridine 3-carboxylic acid and its amide or nicotinamide form. The latter is the precursor of two active coenzymes: nicotinamide adenine dinucleotide (NAD) and nicotinamide adenine dinucleotide phosphate (NADP) which are involved in oxidation-reduction reactions. These vitamin forms act as hydrogen transporters in very many (more than 200) carbohydrate, FA and AA metabolism reactions. NAD, a coenzyme of many dehydrogenases, is reduced to $NADH^+\ H^+$ during glycolysis, lipolysis and the Krebs cycle during the production of energy (ATP) via the respiratory chain. NADP is reduced to $NADPH_2$ within the pentose phosphate pathway, becoming a hydrogen donor during very specific energy-consuming reductions, such as fatty acid biosynthesis. It is also the coenzyme for other reductases (GR of red blood cells, dihydrofolate reductase which renders folic acid active) and lastly the hydrolases of steroid and phenylalanine metabolism, etc.

In fish, deficiencies have only rarely been observed. However, requirements determined using growth as the criterion are high: 10, 14 and $28\,mg.kg^{-1}$ of diet for trout, catfish and carp respectively. The scarcity of states of deficiency probably has its origin in the abundance and stability of this vitamin, and also in the possibility for synthesis from tryptophan, which is apparently low but still poorly quantified. Hepatic reserves seem to be the only biochemical criterion used until now to evaluate nutritional state.

9.4.3 Pantothenic acid

Quantitative requirements for pantothenic acid are some of the highest for hydrosoluble vitamins; it occupies a central position in lipid metabolism. It forms part of the composition of coenzyme A (CoA) and acyl-carrier protein (ACP) which function as transporters of acyl radicals.

Acetyl-CoA plays a vital role in cellular metabolism since it is produced by the degradation of simple sugars, FAs and certain AAs and participates in several metabolic pathways, such as the TCA cycle (then the respiratory chain, with production of ATP, CO_2 and H_2O) and takes part in FA biosynthesis. This is achieved by a fatty acid synthetase enzyme complex, at the centre of which is ACP. Acetyl-CoA is involved as an acyl donor and also participates in the formation of malonyl-CoA via a biotin-dependent decarboxylation. A further vitamin, niacin, in the form of $NADPH_2$ is equally necessary.

CoA is also involved in cholesterol biosynthesis, acetylation of various molecules such as choline, aminated sugars which take part in the synthesis of different mucopolysaccharides and chitin in crustaceans. Other derivatives of acetyl-CoA also exist; these are produced from the catabolism of certain AAs or FAs with odd numbers of carbon atoms, such as succinyl-CoA which is necessary for synthesis of the porphyrin nucleus of haemoglobin, in the presence of another vitamin, pyridoxine.

A characteristic symptom of pantothenic acid deficiency (club-shaped gills) has sometimes been observed in several species of fish. Requirements determined by simple use of growth curves and saturation of hepatic reserves lie between 10 and 20 $mg.kg^{-1}$ of purified diet for trout, catfish, yellowtail and sea bream. Some studies carried out on trout and catfish emphasise the importance of determining requirements using criteria which are relevant to the physiological functions of the vitamin.

9.4.4 K vitamins

Vitamin K is necessary for blood coagulation and is represented by a large number of vitamers of plant (phylloquinone: K_1), microbial (menaquinones: K_2) and synthetic (medadione: K_3) origins. These are derivatives of naphtoquinone which has a lateral phytyl chain, like the tocopherols. The most common synthetic vitamer, menadione, does not possess the side chain, but this must be added by the intestinal flora in order for the molecule to become active.

In higher vertebrates, vitamin K plays a very precise role, ensuring the functioning of the very complex mechanism of blood coagulation, allowing the synthesis of four factors which are protein in nature. It acts as a prosthetic group for a carboxylase of glutamic acid which is involved in this process. Its mode of action is therefore very similar to that of the B group vitamins. It also participates in the functioning of proteins which transport calcium. The failure of blood coagulation as a result of vitamin K deficiency is very difficult to demonstrate in fish, except by using either voluntary or involuntary intake of drugs which have antivitamin K properties. Only one case of deficiency has been obtained (in catfish) without the use of antivitamin K. Intestinal synthesis, which is active in mammals, has not been described in fish. It thus appears that vitamin K, like vitamin D, plays a more reduced role than in the higher vertebrates. Studies have shown that excesses of meniadone can become toxic to salmonids. This vitamin remains the only one (along with vitamin A), for which requirements have not been recently evaluated. It is still estimated at 10 $mg.kg^{-1}$ of food following the recommendations of the NRC in 1981.

9.5 ANTI-OXIDANT VITAMINS

The two vitamins included in this category are involved as precedents in many oxidation-reduction reactions, but without having a clear coenzyme role. Their formulae are given in Box 9.2.

9.5.1 Ascorbic and dehydroascorbic acids

The symptoms of vitamin C deficiency in man (scurvy) have been described from time immemorial, while they are unknown for most terrestrial animals. It was only in 1907 that experimental scurvy was obtained by chance in a guinea-pig being fed a cereal-based diet, and its cause identified. Ascorbic acid is not actually a vitamin except for a very small number of species which are lacking in gulonolactone oxidase, an enzyme that allows the conversion of glucose into ascorbic acid.

This molecule has a low molecular mass and consists of a lactone ring with an ene-diol unit, which produces dehydroascorbic acid on oxidation, and two alcohol units. The role of vitamin C in the oxidation-reduction processes can be explained by the reversible interconversion between its reduced form (ascorbic acid) and its oxidised form (dehydroascorbic acid), a reaction catalysed by ascorbate oxidase in the presence of reduced glutathione and NADPH. The two forms are thus equipped with vitamin activity.

Vitamin C acts as a hydrogen donor in many hydroxylation reactions: collagen maturation (hydroxylation of proline and lysine in procollagen), catecholamine synthesis (hydroxylation of phenylalanine and tyrosine followed by those of dopamine and noradrenaline) and carnitine synthesis, which is involved in the β-oxidation of FAs. Other catabolic pathways such as those of phenylalanine and tyrosine, require the presence of vitamin C which protects an enzyme, parahydroxyphenylpyruvate dioxygenase, from inhibition caused by an excess of substrate. In more general terms, this vitamin is involved in many other biochemical processes: neutralisation of free radicals in synergy with vitamin E (blockage of lipid peroxidation, Chapter 7), iron absorption and, more generally, interaction with metallic divalent or trivalent ions; reproduction (vitellogenesis and embryonic development), transformation of cholesterol into bile acids, degradation of exogenous substances (pollutants, drugs) and blockage of carcinogenic nitrosamine synthesis from nitrites, etc. Lastly, at pharmacological doses, vitamin C stimulates immune response reactions.

Because of the extremely labile character of this vitamin, deficiencies have frequently been observed in fish fed artificial diets, which explains the high number of studies carried out on fish. Scurvy-type deficiency symptoms caused by the malformation of collagen maturation, scolioses, lordoses or haemorrhages have been widely described in a large number of freshwater species such as trout, salmon, catfish, carp, tilapia and eel. However, they appear mainly in very young animals and unsystematically. In at least one other species, the turbot, a different pathology is observed, 'renal granulomatous syndrome', which is characteristic of disturbed tyrosine catabolism with its accumulation in the blood and tissues, leading to opacification of the cornea, and appearance of subcutaneous and especially renal nodules; it has only been reproduced experimentally in the turbot but seems to appear in other fish. In general terms, hepatic or renal vitamin C concentration is used for the detection of possible deficiencies. The collagen content of the vertebrae as well as the proline/hydroxyproline ratio are good indicators of deficiency in the trout and catfish, as is tyrosinaemia in turbot.

Several years ago, stable forms, ascorbyl sulphates and especially phosphates, have appeared on the market; many studies have shown that the availability of phosphate forms is always very high, which is not true for the sulphate form. These new forms now allow precise determination of the requirements of fish and crustaceans which have, until now, been poorly understood. For maximal growth and survival, recommended inputs are maintained at 50 $mg.kg^{-1}$ of diet for young trout which are growing fast and whose requirements seem to be ten times lower than was thought a decade ago. Requirements increase during sexual maturation as well as during the embryonic and larval stages. The effect of 'megadoses' on the different responses of the immune system are still being studied.

9.5.2 Tocopherols (vitamin E)

Vitamin E includes an entire series of vitamers belonging to the tocopherol family, which are molecules formed from a double nucleus (tocol) and a lateral chain (phytyl). α-tocopherol is much more active than the other forms. This vitamin has a fundamental role in all living organisms and gives rise to stable free radicals (tocopheryl radicals) which block radical chain reactions. It thus protects numerous molecules against oxidation: vitamin A, carotenoids, co-enzyme Q, and HUFAs. It acts in synergy with the glutathion peroxidase system (and thus selenium) and with vitamin C amongst others, although its liposoluble character tends to spread it throughout all the different parts of the organism.

In fish, one of these roles acquiring particular importance is the protection of n-6 HUFAs and especially n-3 HUFAs, which are very abundant in the phospholipids (Chapter 7). As a result, tocopherol is indispensable to the integrity of cellular membranes and also has a role in the protection of lipids in general against oxidation and, *post mortem*, it favours flesh conservation by slowing down the process of going rancid. This vitamin also possesses other functions, intervening in the mechanisms of plaque aggregation (via prostaglandin synthesis) and haemolysis. Two other major roles should be emphasised: stimulation of immune mechanisms and the maintenance of reproductive functions. The latter appears to be very minor or even non-existent in higher vertebrates but is a real phenomenon in fish and crustaceans; tocopherol thus merits its historical name as the anti-sterility vitamin.

Vitamin E deficiency results firstly in the characteristic symptom of muscular dystrophy, which is a degeneration of this tissue. Also observed are oedemas, exudation of greenish lymph, anaemia, erythrocyte fragility and, sometimes, lack of pigmentation. In broodfish, fertility is reduced. Cases of deficiency are not unusual in fish-rearing units. Requirements are difficult to define because of the interactions between tocopherols and ascorbic acid on the one hand, and HUFAs. It is currently estimated at $50\,mg.kg^{-1}$ of food to obtain maximal survival and growth. The effects of interaction between the two antioxidant vitamins remain to be studied under practical conditions. To a certain extent, synthetic antioxidants play a similar role to the tocopherols, but they are not active *in vivo*, only acting by protecting fragile molecules before absorption. In contrast, α-tocopherol acetate is not active in the intestine before being hydrolysed by esterases in the intestinal wall, a conversion which immediately precedes absorption.

Doses exceeding the requirements contribute to the protection of body lipids from peroxidation during storage and marketing. Excess tocopherol appears to be reasonably harmless, perhaps because this vitamin is easily stored in all lipids in the organism.

9.6 CALCIFEROL (VITAMIN D)

This vitamin, the formula of which is shown in Box 9.2, is the only vitamin that can be assimilated but as a pro-hormone. It is the precursor of 1,25-dihydroxycholecalciferol ($1,25(OH)_2D_3$) which is necessary for the regulation of phospho-calcium metabolism. There are two provitamins of vitamin D: ergocalciferol (D_2), a molecule mainly present in plants, and cholecalciferol (D_3). These two molecules are steroids with three nuclei and a lateral chain attached to an opening on the sterol nucleus; they only differ by a methyl group attached to the lateral chain of the molecule. The relative effectiveness of these two forms is extremely variable from one zoological group to another. In fish, ergocalciferol is three times less active than cholecalciferol. In animals, the latter is derived from cholesterol; however, synthesis requires ultraviolet radiation, since the vitamin character only appears when there is insufficient light – this is often the case in the aquatic environment where rays do not penetrate easily.

$1,25\text{-}(OH)_2D_3$ regulates calcium and even phospho-calcium homeostasis in higher vertebrates, by acting in synergy with parathormone. There are three levels of action: intestinal absorption (the enterocytes have receptors for this hormone), mobilisation of calcium from the bones and reabsorption in the kidneys. In the absence of ultraviolet radiation, vitamin D deficiency leads to decalcification of the skeleton, which can cause rickets and also faults in muscular contraction. Like vitamin A, it possesses other roles that have not yet been elucidated but which involve growth, cellular differentiation and the immune system.

In fish, deficiency can certainly lead to decalcification, but true rickets has never been observed even in the laboratory, which demonstrates the reduced role of vitamin D. This characteristic can be explained by the capacity of the fish to absorb calcium from the water across the gills, a mechanism for which vitamin D does not appear to be required. However, vitamin D regains its role when the fish lives in water with very low calcium levels. Vitamin D deficiency slows growth, leads to tetanus and disturbs the phospho-calcium homeostasis. Requirements determined for trout appear to be extremely high, since they rise to 2400 IU of vitamin $D_3.kg^{-1}$ of purified food, while recommended values for poultry and pigs are only 200 and 220 $IU.kg^{-1}$ of purified food respectively. New research on the metabolic roles of this vitamin are required to explain how such high requirements can be compatible with a reduced role.

9.7 QUASI-VITAMINS AND GROWTH FACTORS

The formulae for these two compounds are given in Box 9.2.

9.7.1 Choline

Choline is the quaternary ammonium compound of trimethylamine. It can be synthesised (it only acts as a quasi-vitamin), at least slowly, from ethanolamine (which itself comes from a common AA, serine) and from methionine, which acts as a methyl donor. In this reaction, the presence of folic acid is necessary for transfer of methyl groups. Choline synthesis occurs by the transformation of ethanolamine which is already attached to phospholipids.

Choline allows the transfer of nervous input by cholinergic fibres (physiological role of acetylcholine). It also forms part of the composition of phosphatidylcholine. The lipotrophic role of choline follows on from the abundance of phospholipids in cell membranes and lipoproteins. By giving rise, by oxidation, to betaine, a methyl group donor, choline provided in the feed also has one last role: it allows methionine sparing.

Choline deficiency demonstrates the lipotrophic role, leading to triacylglycerol accumulation in the liver and other disturbances of lipid metabolism. Other symptoms do not seem to be specific. When there is an excess of methionine or betaine, the deficiency is greatly alleviated, which demonstrates that synthesis is possible. Depending on the species, recommended levels vary considerably: from 400 to 1000 $mg.kg^{-1}$ for catfish and trout respectively. Note that the hygroscopic nature of choline can increase the instability of other vitamins within the same mixture.

9.7.2 Inositol

This molecule, a cyclic hexa-alcohol, can occur in the form of many isomers, only one of which, meso- or myo-inositol, can be utilised by the fish. Inositol has a structural role, being a constituent of a phospholipid, phosphatidylinositol, which can be considered as the active form of this quasi-vitamin. Because of this, inositol is a lipotrophic factor, similar to choline. Phosphotadinylinositol is also involved in the translation of signals from certain metabolic processes, by the generation of 'secondary messengers' (IP_3). It is synthesised easily by almost all higher vertebrates and has long been known as a vitamin in arthropods, including the crustaceans. It is the only 'vitamin' which has an energetic value (it is degraded like a carbohydrate).

In fish, there is low level, insufficient synthesis in the intestine of the carp as well as sufficient synthesis in the intestine and liver of the catfish. The vitamin nature of inositol has not yet been clearly demonstrated in fish. Signs of deficiency result, as for choline, from the lipotrophic function; there is a decrease in phospholipid synthesis, with accumulation of neutral lipids. Recommendations of 300 mg.kg^{-1} of feed for trout have been produced from preliminary studies carried out without knowledge of the concentrations of inositol in non-supplemented diet. For other species, such as carp and yellowtail, values estimated from gain in body mass or hepatic reserves are 300–400 mg.kg^{-1} of diet. Few data are currently available and, as inositol is a very expensive product, it is very important that requirements for this quasi-vitamin are determined.

Inositol is abundant in plants, but in inactive, phytate forms (Chapter 19). A currently-used insecticide, lindane, is an inositol antagonist and thus a true anti-vitamin.

9.7.3 Growth factors

There are many other compounds whose synthesis appears to be insufficiently fast in animals under certain conditions. These have been termed growth factors. Among these are orotic acid (vitamin B_{13}), ubiquinone, pyrroloquinoline quinone, lipoic acid, carnitine, etc. Even some vitamers (para-amino-benzoic acid or PAB, see folic acid), or poorly defined substances such as pangamic acid 'vitamin B_{15}' have been added to this list. It is only rarely that these factors form the subject of studies in fish. Carnitine, the role of which has been indicated above, appears to stimulate growth in sea bass under certain conditions, but it is extremely abundant in animal meals.

9.8 REQUIREMENTS AND THEIR SATISFACTION

As shown above, requirements are precisely known for salmonids, in which they have been established for eight vitamins, using purified diets. Data are more sparse for other species such as carp, eel and catfish. Data for marine fish are almost non-existent. In other cases, nutritionists and food manufacturers are reduced to using the standards established for salmonids (Appendix C). A huge amount of work to determine requirements thus remains to be done; although the indications may be that interspecific differences are not very large, notable exceptions (vitamin D_3) demand confirmation.

Taking into account the problems of leaching, instability and availability, the practical problems of meeting requirements can be approached using the following equations:

$$R_{total} = R_{maintenance} + R_{growth} + R_{reproduction} + I_{other\ nutrients} + I_{environment}$$

$$S = [R_{total} + L_{manufacture} + L_{storage} + L_{leaching} + f_{(rearing\ density)}] \times 1/A$$

where R = requirement　　L = losses
I = interaction　　f( ) = function factor of
S = standard (recommendation)　　A = availability

In practice, very wide margins of security are always used: inputs of vitamin complement are often equivalent to double the requirements and the vitamin content of the raw materials is not taken into account. These precautions thus alleviate the consequences of lack of knowledge.

9.9 CONCLUSION

In summary, the differences between fish and higher vertebrates in terms of vitamin nutrition are minimal. Certainly there exists in fish, and apparently in fish alone, another form of retinol which has been called vitamin A_2, but the efficiency of the usual form is also high. Fish also require two quasi-vitamins: choline, also required by birds, and inositol, which brings them closer to the invertebrates. Two vitamins, calciferol and vitamin K, appear to have a lesser role in fish than in higher vertebrates but, in contrast, vitamins E and C are of greater importance at least from an applied point of view. The determination of quantitative requirements, which are still poorly known for many species, constitutes the line of research which should receive priority.

9.10 BIBLIOGRAPHY

Combs, G. F. Jr. (1992) *The Vitamins – Fundamental Aspects in Nutrition and Health.* Academic Press, San Diego.

Halver, J. E. (1972) The role of ascorbic acid in fish disease and tissue repair. *Jap. Soc. Sci. Fish*, **38**, 79–92.

Halver, J. E. (1989) The vitamins: In: Halver, J. E. (ed.) *Fish Nutrition.* Academic Press, San Diego, 31–109.

Le Grusse, J. and Watier, B. (1993) *Les Vitamines. Données Biochimiques, Nutritionnelles et Cliniques*. Centre d'Etudes et d'Information sur les Vitamines. Produits Roche, Neuilly sur Seine.

NRC (National Research Council) (1993) *Nutrient Requirements of Fish*. National Academy Press, Washington DC.

Woodward, B. (1994) Dietary vitamin requirements of cultured young fish, with emphasis on quantitative estimates for salmonids. *Aquaculture*, **124**, 133–168.

10

Mineral nutrition

As with terrestrial animals, fish require minerals which are the constituents of certain tissues (mainly skeletal structures) or molecules that serve as enzyme co-factors and participate in intra- and extra-cellular ionic balance as well as regulation of endocrine functions. Seven macro-minerals and 15 trace elements are known to have a definite physiological role in most animals. However, in fish and crustaceans, a dietary requirement has been clearly established for only 11 of these (Table 10.1).

It is in the domain of mineral nutrition of aquatic animals that our knowledge remains the most fragmented. Since the functions of macro-minerals and oligo-elements are generally very diverse, ranging from a plastic role (tissue constitution) to a catalytic role (enzymatic activity), there is a relatively large heterogeneity in tissue contents and quantitative requirements. In addition, there is no correlation between tissue concentration and requirements as is found for other nutrients, such as essential amino acids.

10.1 CHARACTERISTICS OF MINERAL NUTRITION

10.1.1 General

Intestinal absorption mechanisms vary depending on the element being considered: these include passive diffusion and active transport, with or without the intervention of specific ligands. Thus, dietary iron is absorbed by linkage to a protein present in the intestinal mucus (transferrin–ferritin couple). However, absorption via the gills appears to be possible. Similar mechanisms exist for other elements (zinc and copper) but the respective roles of the gill and intestinal pathways are very poorly understood. Most importantly, the regulation of the mechanisms in different organs in relation to dietary supply and concentration in the water remains virtually unknown, in both freshwater and marine fish. It is worth recalling that the global absorption

Table 10.1. Macro-minerals and trace elements with confirmed nutritional roles in fish.

Macro-minerals	Trace elements
Calcium (C)	**Iron (Fe)**
Phosphorus (P)	**Zinc (Zn)**
Potassium (K)	**Manganese (Mn)**
Magnesium (Mg)	**Cobalt (Co)**
Sodium (Na)	**Copper (Cu)**
Chlorine (Cl)	**Iodine (I)**
Sulfur (S)*	**Selenium (Se)**
	Fluorine (F)
	Nickel (Ni)
	Vanadium (V)
	Silicon (Si)
	Tin (Sn)
	Chromium (Cr)
	Aluminium (Al)

In bold: elements for which a clear nutritive requirement has been established in fish or crustaceans.
* Element generally considered as organic, in the same way as carbon or nitrogen, because of the very poor bioavailability of inorganic sulfur.

of trace elements almost certainly corresponds to homeostatic regulation, minor deficiencies leading to a much greater efficiency of the absorption mechanisms and slight excesses having the opposite effect (analogy with Figure 9.1). Added to this are the problems of availability linked as much to the animal (physiological stage, species) as to the form of supply and to the numerous interactions that exist between both the elements themselves and the other constituents of the diet. The elimination pathways are also diverse: renal excretion, elimination via the gills, bile, faeces or even through the epidermis. Toxicity resulting from excess of trace elements is frequent, the excess arising not only from the diet but also from the water. In contrast to studies on requirements, there are many studies related to heavy metal toxicity in the aquatic environment.

10.1.2 Interactions between environment, nutritional supply and requirements

In the aquatic environment, management of dietary supply is closely linked to absorption capacity (via the gills, skin and mouth) from the environment itself. As mentioned above, aquatic animals are the only organisms capable of obtaining a certain number of minerals (Na, Cl, K, Ca) and trace elements (mainly Fe, Zn, Mn, Mg, Cu, I, Co) other than from their food supply.

In fresh water, the external environment is very hypotonic in relation to the internal environment. The difference in osmotic pressure leads to a net loss of salts and absorption of water across the skin, with renal excretion of large volumes of very dilute urine (about 100 $ml.kg^{-1}day^{-1}$). In the gills, sodium loss is compensated for by an increased efficiency of the sodium pump. Concentrations of dissolved minerals are often low and often very variable; hence dietary supply plays a major role (Figure 10.1).

In salt water, the external environment is hypertonic in relation to the internal environment. Osmotic pressure leads to dehydration and significant absorption of salts via the teguments occurs. To maintain osmotic balance, fish drink large amounts of water which is itself rich in salts and ions. Excess ions are eliminated via the renal pathway (the urine being very concentrated), via the gills mainly for sodium and chlorine and via the intestine, for divalent ions (Figure 10.1). The possibility of uptake of minerals from the environment is much higher than in the freshwater environment, since the concentrations of dissolved minerals are much higher and almost constant and because the fish are obliged to drink continuously in order to maintain their osmotic equilibrium. Although this assertion appears obvious, it is as yet supported by only a few experimental data.

Under these conditions, the definition of requirements is difficult. In general, it is agreed that 'requirements' are the dietary levels that allow the animal to achieve optimal deposition in its tissues, in the absence of waterborne minerals. Since the latter is often sufficient, it is necessary to refer to diagrammatic dose-response relationships as shown in Figure 10.2.

10.2 MINERALS

The main roles, symptoms of deficiency and sources or minerals are summarised in Table 10.2 and the requirements given in Table 10.3.

10.2.1 Calcium

Calcium is vital to the development and growth of the skeleton (bony structures), maintenance of osmotic equilibrium, muscular activity and functioning of the nervous system. It is also involved in blood coagulation and the regulation of hormonal secretions (calcitonin, prolactin, catecholamines, parathyroid, etc.). A deficiency in calcium in the food rarely causes a decrease in fish growth. Indeed, in teleosts the skeleton is reduced in comparison to terrestrial vertebrates of the same size; in addition, more than 50–60% of the quantitative requirements can be met by the water. In the freshwater environment, the dietary requirement can vary in relation to the concentration of calcium (hardness) in the water. In marine fish, the absorption capacity and consequently the requirement for dietary calcium appear to vary. During very early development, a deficiency can lead to deformation of the skull and vertebrae. However, under certain conditions (prolonged fasting and sexual maturation, two states which are observed simultaneously in salmon during anadromous migration), intense decalcification can be observed, especially in the scales. In species without scales, mobilisation occurs from the skeleton, which demineralises, although true rickets is unknown.

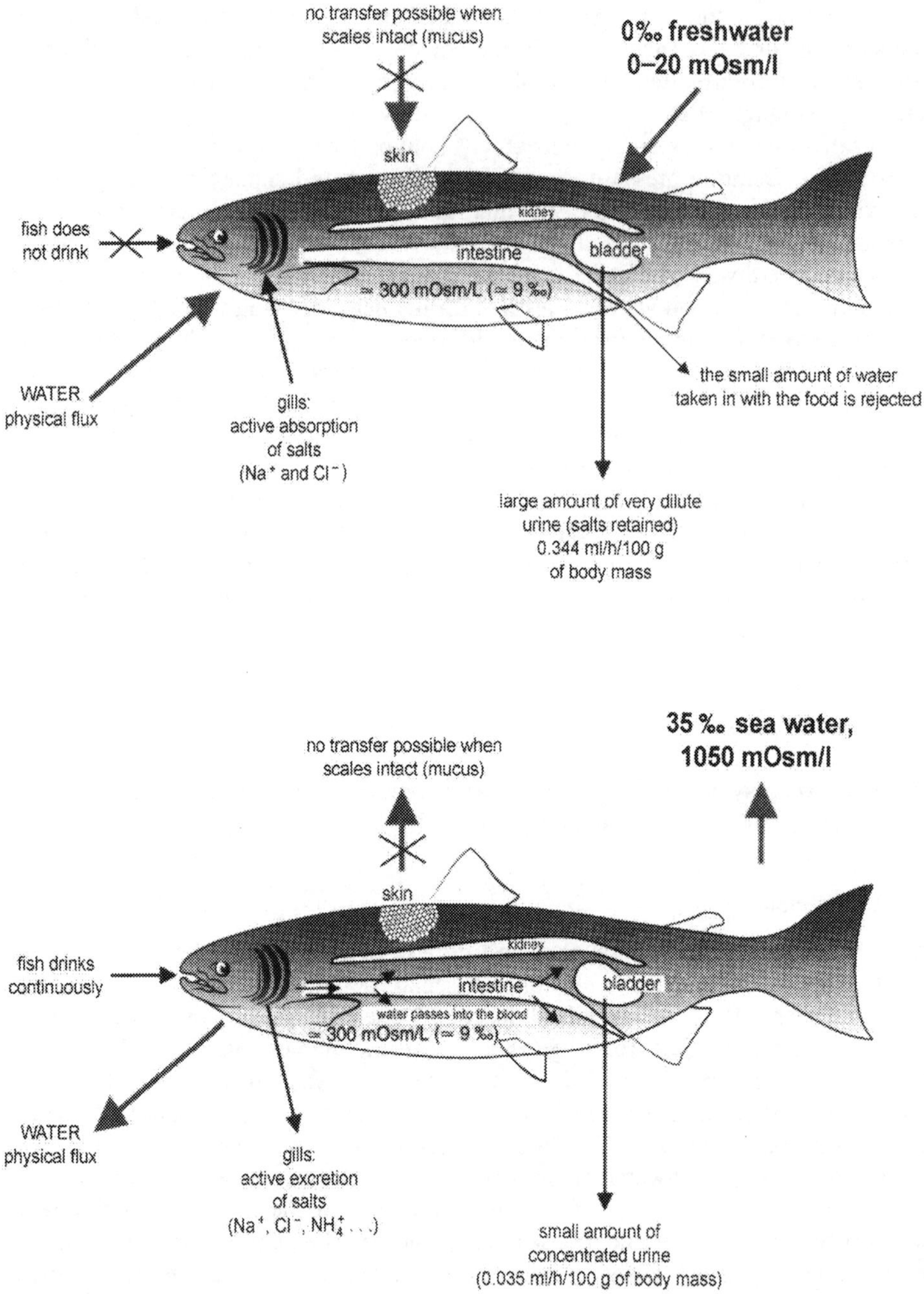

Figure 10.1. Ionic exchanges in the trout, a euryhaline fish, and the environment, in fresh and salt water (from Boeuf, 1987).

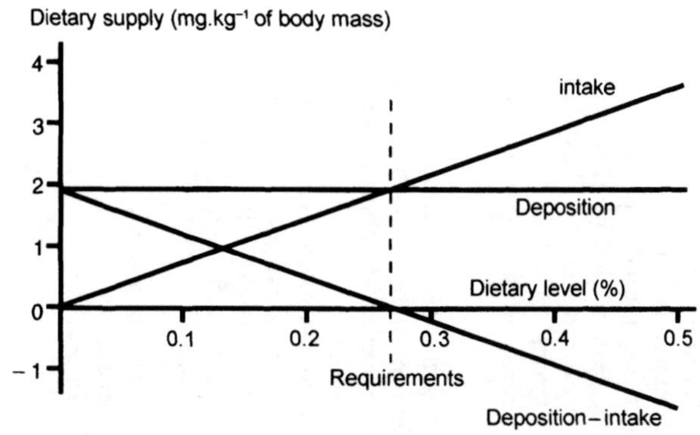

Figure 10.2. Diagrammatic relationship between dietary potassium supply and deposition in the Channel catfish. 'Requirements' correspond to the absence of input from the water (after Wilson and El Naggar, 1992).

The main dietary sources of calcium are ingredients of animal origin (fish meals). In fish, calcium concentrations are highest in the skeleton (more than 30% calcium) and the scales (more than 80%). The calcium in the scales generally constitutes the most labile store. In fish flesh, the concentration can vary from 0.02% to 0.5% wet weight. The interaction of calcium with other nutrients (vitamin D, Mg and Zn) has also been established, although the role of vitamin D in fish appears to be limited (Chapter 9).

10.2.2 Phosphorus

Phosphorus, like calcium with which it is associated, is present in the form of hydroxyapatite in bony tissues. The endoskeleton of fish contains about 15% phosphorus. However, more than a third of the whole body content of this element is in the soft tissues (phospholipids, nucleic acids, energy-rich compounds, etc.). It is a constituent of cells and membranes and is vital for phosphorylation reactions, playing a central role in nitrogen, lipid, carbohydrate and energy metabolism. Fish flesh contains between 0.2 and 0.8% phosphorus.

A deficiency in phosphorus leads to a decrease in skeletal and body growth. The mineralisation of the vertebral column is very dependent on phosphorus supply. Other symptoms of deficiency are an excess of lipid deposition and an increase in the activity of certain enzymes (alkaline phosphatase and enzymes involved in gluconeogenesis).

Fish cannot meet their requirements by absorption across the gills or skin. Although absorption from the environment has been demonstrated in salmonids, it is low and can only meet a small part of the requirements; a dietary supply is vital. Among all the macro-minerals, phosphorus is the one for which dietary supply must cover the greatest part of the requirements. In different species under aquaculture,

Table 10.2. Macro-element sources, roles and symptoms of deficiency in fish.

Macro-mineral	Sources	Roles	Interactions	Deficiency symptoms
Calcium	Water Raw materials of animal origin Inorganic salts	Bone growth Blood coagulation Enzyme co-factor Neurotransmission	Vitamin D P, Zn, Mg	Decrease in growth Decalcification of bones and scales
Phosphorus	Raw materials of animal and plant origin Inorganic salts (mono- or bicalcic phosphates)	Bone growth Energy metabolism Cell and membrane constituent Co-enzyme constituent: $NADP^+$, thiamine pyrophosphate, pyridoxal phosphate, co-enzyme A	Ca, Mg, Mn Niacin Vitamin B_1 Vitamin B_6	Decrease in growth Deformities of skull and vertebrae Demineralisation Increase in lipid deposition Decrease in blood P, haematocrit Increase in alkaline phosphatase in serum Decrease in hepatic glycogen
Magnesium	Water (−) Raw materials of animal and plant origin Inorganic salts	Growth Muscular integrity Respiration, energy metabolism Enzyme co-factors Thyroid metabolism	Proteins Ca, P	Muscular degeneration Deformation of vertebrae Decrease in Mg concentration in tissues Calcinosis, anorexia, convulsions Increase in extracellular fluid volume Decrease in growth Decrease in SOD activity Cataract
Potassium	Water (+) Raw materials of plant origin	Enzyme regulation: $Na^+-K^+-ATPase$ phosphofructokinase Osmoregulation Intracellular ionic balance Muscular contraction Neurotransmission	Na	Decrease in growth Convulsions Tetanus
Sodium	Water (−) Raw materials of animal origin	Extracellular ionic balance Osmoregulation Enzyme regulation	K, Cl	

(−) secondary source, (+) significant source

Table 10.3. Nutritional requirements for macro-minerals (% of dry matter) and trace elements (ppm) in some species.

	%				ppm						
	P	Ca	Mg	K	Zn	Mn	Co	Cu	I	Fe	Se
Atlantic salmon**	0.6					20		6		60	
Pacific salmons	0.6			0.8	R***	R***			0.6		R***
Channel catfish	0.45	0.45*	0.04	0.26	20	2.4		5		30	0.25
Common carp	0.65	0.3	0.05		25	13	0.10	3		150	
Tilapia	0.90	0.65*	0.6		10	12		3.5			
Red drum	0.86				20						
Trout	0.5–0.8	0.03–0.24	0.05		15–20	12		3			0.15–0.4
Eel	0.3		0.14							170	
Red sea bream		0.34								150	

* water containing no Ca
** in freshwater and seawater
*** essential element (required) but requirements not quantified.

phosphorus requirements range between 0.5% and 0.9% of the diet. In contrast to terrestrial animals, for which it is possible to determine precisely the ideal calcium/phosphorus ratio, no such information exists for fish, apart from some data where, under conditions of excessive inputs of phosphorus, a ratio Ca/P of 1 appears to be desirable. This results partly from the variations in water quality, which can affect calcium requirements, and partly from the variability in the availability of phosphorus contained in the dietary ingredients.

One of the problems linked to phosphorus in fish farming is that of optimal supply, which allows requirements to be met while avoiding an excess of phosphorus-rich catabolites. Excess phosphates can lead to eutrophication of natural environments (degradation of water quality). As a result, the nutritional requirements and availability of phosphorus have received more attention than those of other minerals.

Plant sources (with the exception of yeasts) are rich in phytic phosphorus, of which about 40–60% is available to fish. Phytic acid (inositol hexaphosphoric acid) can bind to Zn^{2+}, Fe^{2+} or Ca^{2+} ions, decreasing their intestinal absorption and thus playing the role of 'demineraliser' (Chapter 19). The availability of phytic phosphorus can be improved by adding phytases, for example, of microbial origin to the diet. It also appears that some crustaceans are capable of utilising phytic phosphorus although the enzyme responsible for phytase activity has not been identified. The most soluble inorganic sources (mono- or bicalcium phosphates) are generally much more available than the less soluble salts such as tricalcium phosphate. Equally importantly are differences between agastric species and those with a stomach. The absence of a zone with acid pH impairs the solubility of salts such as bicalcium phosphate; as a result, availability is higher in salmonids than in stomachless species such as the carp (Table B.3, p. 377).

Recognition of the possible effects of an excessive release of phosphorus into the aquatic environment has recently led to the adoption of rigorous standards for the maximal level of phosphorus in fish farm feed. Taking into account the importance of an adequate supply of available phosphorus for growth and metabolism, and given the progress made in improving feed/gain ratios, it would be judicious to use terms of absolute requirements (quantity necessary for unit mass gain) rather than in terms of dietary levels. It is necessary to rely on precise knowledge of nutritional requirements and the availability of phosphorus in the ingredients. Dietary supplies should be sufficient to meet the requirement while avoiding all excess supply in organic or inorganic form. The addition of enzymes such as phytase is only worthwhile if the cost of the enzymes is competitive, in diets which are very rich in ingredients of plant origin.

10.2.3 Magnesium

The concentration of magnesium in fish flesh is about 20–100 mg.100 g^{-1}. This metal is also a major constituent of bony tissues. Its metabolism shows some similarities to calcium. Magnesium is a co-factor which is involved in numerous enzymatic reactions related to phosphorus (pyruvate kinase, phosphokinase, phosphatase, Mg dependent ATPase, etc.) involved in osmoregulation, protein synthesis and growth.

Magnesium deficiency results in symptoms such as decrease in growth, anorexia, skeletal deformity, renal calcinosis and muscular dystrophy. Requirements for magnesium in freshwater fish are relatively low (0.05–0.1%) and can, at least in part, be met by the water which contains 1–3 $mg.l^{-1}$. A correlation with the dietary protein content appears to exist, diets rich in protein requiring a higher magnesium supply.

The availability of magnesium from ingredients of plant origin is quite high. Supplementation with inorganic salts is only necessary in the case of purified diets. The quantity and availability of magnesium in fish meals can vary quite a lot. With inorganic salts ($MgCl_2$), the availability is greater than 75% in salmonids.

10.2.4 Sodium, chlorine and potassium

In their ionic form, these three elements are responsible for the maintenance of extra-cellular (Na^+ and Cl^-) or intra-cellular (K^+) equilibrium. In osmoregulation, the role played by $Na^+ - K^+$ ATPase and the chloride cells is very significant. The action of the chloride ion is also of vital importance in gastric digestion. The aquatic environment is generally rich in Na^+ and Cl^- ions and a 'dietary' requirement for sodium and chlorine has not been demonstrated. However, some beneficial effect of a massive input of sodium chloride (between 4% and 10% of the diet) on physiological adaptation of salmonids during transfer into the sea has been demonstrated. A dietary input of potassium appears to be vital, especially with purified diets. A deficiency leads to anorexia, tetanus and convulsions which can result in massive mortalities. Quantitative requirements established for several freshwater species show quite wide variability (0.3% to 1.2%), probably linked to the potassium concentration in the environment.

10.3 TRACE ELEMENTS

The main roles, symptoms of deficiency and sources of oligo-elements are summarised in Table 10.4, and requirements in Table 10.3.

10.3.1 Iron

Iron is essential for the synthesis of haemoglobin and is involved in numerous enzymatic reactions linked to oxidoreduction (catalase, cytochrome oxidase, peroxidase). An iron deficiency results in microcytic hypochromic anaemia, with a decrease in haematocrit and the concentration of iron in the blood. Fish are capable of absorbing iron dissolved in the water. Quantitative requirements are known for some species of fish (30–150 $mg.kg^{-1}$ diet). An excess of iron can lead to an increase in lipid peroxidation, polyunsaturated fatty acids being particularly sensitive. As a result, in diets rich in fatty materials it is advisable to pay greater

Table 10.4. Oligo-element sources, roles and symptoms of deficiency in fish.

Trace elements	Sources	Roles	Interactions	Deficiency symptoms
Iron	Water (−) Raw materials of animal origin (bloodmeals)! $FeCl_2FeSO_4$, Citrate	Proteins: haemoglobin, myoglobin, cytochromes Respiration, coagulation Enzymes: peroxidases, catalases	Cu, Co, Mn, Zn, Phytic acid, PUFA	Anaemia (with decreased haematocrit) Yellow liver Tissue concentrations Lipid oxidation
Copper*!	Water (+) $CuSo_4$ Primary ingredients of animal origin	Enzymes (cuproenzymes): cytochrome oxidase, SOD, tyrosinase ... Electron and oxygen transport: haemocyanin (crustaceans)	Vitamin C, Zn, Se, Fe	Decrease in growth Cataract Sensitivity to infection
Zinc	Water (−) Primary ingredients of animal origin $ZnSO_4$	Enzymes: dehydrogenases, peptidases, aldolase, dismutase	Phytic acid, Ca, P	Decrease in growth Cataract, nanism External haemorrhages Decreased fecundity
Manganese	Water Primary ingredients of animal origin $MnSO_4$	Enzymatic co-factor	Ca, P, Phytic acid	Decrease in growth Skeletal anomalies Decrease in enzyme activity Decreased fecundity
Selenium	Water Fishmeal Na_2SeO_3	Prevention of lipid autoxidation, SOD co-factor, glutathione peroxidase	PUFA, Vitamin E	Lipid peroxidation Decrease resistance to pathogens Decreased enzyme activity
Iodine	Water (+)	Thyroid hormones		

(−): secondary source, (+): significant source
* Toxicity: > 700 $g.kg^{-1}$ of food and 0.8 − 1 $g.l^{-1}$ in water
! Now prohibited in the European Community

attention to iron input. Crustaceans, which do not have haemoglobin, have much lower requirements than fin fish and are more sensitive to excess amounts.

Raw materials of animal origin, particularly blood meals, are rich in iron, but current EU regulations do not allow the use of this ingredient in fish feeds. In purified diets, supply of iron in the form of sulphate or chloride or, to a lesser extent, citrate can meet nutritional requirements. Among other interactions, vitamin C affects iron availability: ferrous iron is absorbed more efficiently than ferric iron at a neutral pH and vitamin C allows the reduction of ferric salts into ferrous salts in the intestinal lumen.

10.3.2 Copper

Copper also facilitates the absorption of other trace elements such as iron or zinc. Like iron, it is involved in electron transport. In addition, the activity of numerous enzymes (Cu-Zn-superoxide dismutase or SOD, catalase, tyrosinase etc.) depend on the presence of copper. In the plasma, transport is carried out in the form of a metalloprotein, ceruleoplasmin.

A dietary deficiency in copper results in a decrease in SOD and cytochrome oxidase activity and sometimes leads to cataracts. Fish are more tolerant of an excess of Ca in dietary supply than to an increased concentration in the ambient environment. They easily absorb dissolved copper which can thus pose problems with toxicity. A concentration of copper sulphate ($>0.8\,mg.l^{-1}$) in the water can lead to chronic toxicity for many species. However, inputs of up to $600\,mg.kg^{-1}$ of diet do not appear to have harmful effects on trout when they are used over a brief period for disinfection purposes (bactericidal role of copper sulphate).

Requirements for copper are of the order of 3 to $5\,mg.kg^{-1}$ of feed. In crustaceans, copper constitutes the main element of proteins such as haemocyanin or cyanodin, which play a role similar to haemoglobin in oxygen transport; requirements appear to be higher than in fish, but have been poorly quantified.

10.3.3 Zinc

The main role of zinc is as a co-factor in many enzymatic systems involved in the utilisation of almost all nutrients. Thus, the activity of several aldolases, peptidases and phosphatases involved in digestion is zinc-dependent. The importance of zinc, selenium and manganese in protection against auto-oxidation (via SOD) has been clearly demonstrated in salmonids. Zinc deficiency causes cataracts, decrease in growth, dwarfism and lowering in phosphatase and SOD activities. Absorption of dissolved zinc, which occurs mainly across the gills, is low. A dietary input is vital, at least in freshwater fish; requirement levels are between 15 and $30\,mg.kg^{-1}$.

The availability of zinc is variable and depends on numerous other factors of dietary origin. Intestinal absorption of zinc is closely linked to the presence of calcium and phosphorus. The presence of phosphorus in the form of hydroxyapatite (fish meal) or phytates (oil seed meals, cereals and by-products) considerably decreases the availability of zinc. Among the inorganic salts, the availability of sulfate or nitrate is the highest.

10.3.4 Manganese

Manganese acts either as an integral part of enzymes (pyruvate carboxylase, lipase, SOD) or as a co-factor for numerous enzymes involved in nitrogen, lipid and carbohydrate metabolism.

In addition to its effect on SOD activity, a deficiency in manganese results in a decrease in growth rate, dwarfism and skeletal deformities. Absorption from the water is low. Dietary requirements are between 12 and 20 $mg.kg^{-1}$ of dry matter, depending on species and rearing conditions. However, lower values have been noted for the catfish (2.4 $mg.kg^{-1}$). Raw materials of animal origin are quite rich in manganese, although availability varies greatly. Among the inorganic salts, the sulfate or chloride forms appear to be the most bioavailable.

10.3.5 Selenium

The main function of selenium resides in the protection against auto-oxidation of lipid membranes, since it is a major constituent of the enzyme glutathione peroxidase. A synergistic role between vitamin E and selenium in muscle growth in salmonids has also been demonstrated.

A deficiency in selenium results in a decrease in growth and especially a decrease in glutathione peroxidase (GSH) activity in the liver and the plasma. It can also cause an increase in *in vivo* and *post mortem* lipid oxidation reactions. This enzyme catalyses the reduction of hydrogen peroxide or other organic hydro-peroxides (ROOH), allowing a decrease in intracellular peroxide concentration, which has very harmful effects on the membranes (Chapter 7).

Nutritional requirements are between 0.15 and 0.4 $mg.kg^{-1}$ of diet. There is also absorption of dissolved selenium. Its efficiency is all the more important when concentration in the water is low. Fish meals sometimes have high, but very variable, selenium concentrations and it is prudent to incorporate this element in the form of sodium selenite to ensure protection against auto-oxidation, especially in diets which are rich in fat. The toxicity threshold is very low, being 10 $mg.kg^{-1}$ of feed.

10.3.6 Iodine

Iodine is a well-known constituent of the thyroid hormones (thyroxin or T_4 and triiodothyronine or T_3). A major part of iodine input comes from the environment, in both freshwater and marine species of fish.

An iodine deficiency results in thyroid hyperplasia. Although quantitative requirements for iodine have not been determined, an input of about 1 to 5 $mg.kg^{-1}$ of diet appears to be sufficient and necessary for the proper functioning of the hormonal system. Iodine supplements (4.5 $mg.kg^{-1}$ of feed) appear to be effective in protecting against a number of diseases of bacterial origin, including bacterial kidney disease in Atlantic salmon. Fish meals are generally rich in iodine. Supplementation with potassium iodide is essential with purified diets.

10.3.7 Other trace elements

There are few data related to other trace elements. A deficiency in aluminium appears to cause growth disorders in the eel. An input of fluoride (4–5 $mg.kg^{-1}$) is indicated to have a beneficial effect in Atlantic salmon. The nutritional roles of chromium, cobalt, molybdenum, vanadium or silicon are very poorly understood in fish.

10.4 CONCLUSION

From a theoretical point of view, mineral nutrition is one of the most unusual sectors of fish nutrition, as fish can extract minerals from both water and diets. The relative contribution of these two sources remains poorly understood although knowledge of the mechanisms involved should shed light on the possible differences between freshwater and marine species. The nutritional and physiological roles of the different macro-minerals and trace elements are similar in fish and higher animals. Similarly, the data currently available show that in broad terms, the qualitative requirements are comparable in the two groups. Quantitative data are however quite limited, mineral nutrition constituting the least well-known area of fish nutrition.

At first glance and in strictly economic terms, this gap appears to be acceptable since the minerals, with the exception of phosphorus, have only a small effect on the cost of feeding fish. However, the raw materials currently used in fish feeding (fish meals, oils) have very variable mineral contents. Taking into account the significance of these elements in general physiology and metabolism, as well as their interactions with other nutrients, it is essential to develop methodologies for determining food requirements, in relation to the composition of the surrounding water.

10.5 BIBLIOGRAPHY

Boeuf, G. (1987) Bases physiologiques de la salmoniculture: osmorégulation et adaptation à l'eau de mer. *La Pisciculture Française*, **87**, 28–41.

Lall, S. P. (1989) The minerals. In: Halver, J. E. (ed.) *Fish Nutrition*, 2nd edition. Academic Press, San Diego, 219–257.

NRC (National Research Council) (1993) *Nutrient Requirements of Fish*. National Academic Press, Washington DC.

Wilson, R. P. and El Naggar, G. (1992) Potassium requirement of fingerling channel catfish, *Ictalurus punctatus*. *Aquaculture*, **108**, 169–175.

11

Carotenoids and pigmentation

Carotenoids are fat-soluble pigments, the colour range of which extends from yellow to deep red. These substances are very widespread throughout the plant and animal kingdoms and have aroused the interest of scientists since the 1950s. Around 600 naturally occurring carotenoids are currently known. They are not nutrients in the classical sense of the term, but substances which animals cannot synthesise. These compounds are not generally considered as essential but they can play various roles in both invertebrates and vertebrates, although these are still not perfectly understood, of which pigment action is the most obvious. Most studies on carotenoids in aquatic animals involve the salmonids (and to a lesser degree the cyprinids) as well as penaeid shrimps because of the economic importance of pigmentation. The pink coloration of salmon flesh or shrimp carapaces is considered as a criterion of quality. In these cases, the colour is due to certain dietary carotenoid pigments from the food, of which the most important are two keto-carotenoids, canthaxanthin and astaxanthin. Unlike crustaceans, fish cannot synthesise these pigments from simpler carotenoids. In the wild, carotenoids are provided by natural prey, but in intensive aquaculture they are provided in the food. The majority of studies carried out on carotenoids in crustaceans have been aimed at demonstrating the biochemical reactions of inter-conversion of these pigments. It is for this reason that we approach the relationship between carotenoids and pigmentation mainly in the salmonids, directing the reader to studies cited in the references for other aspects of carotenoid metabolism.

11.1 PIGMENTATION EFFICIENCY OF CAROTENOIDS

For fish farmers and feed manufacturers, the most important aspect is the pigmentation efficiency of carotenoids. This is determined by their structure, specific colour, digestibility, metabolic conversion and specific affinity for a particular tissue.

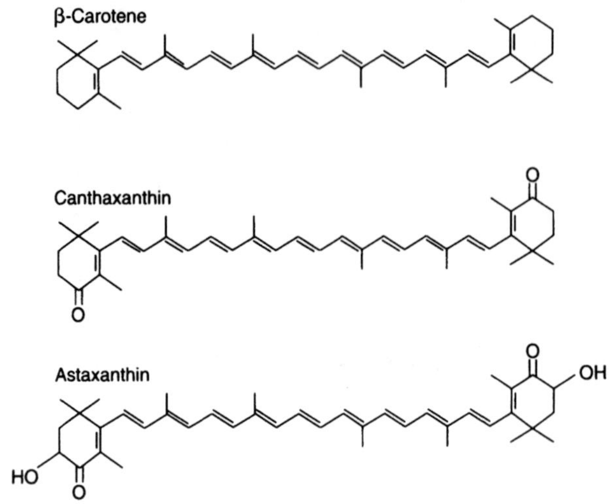

Figure 11.1. Formulae of some carotenoids of aquatic animals.

11.1.1 Structure

Carotenoids are polyenic compounds characterised by a system of conjugated double bonds (Figure 11.1). Their molecule is derived from eight isoprenic chains arranged symmetrically around a central double bond to form a tetraterpene (a terpene is a C_{10} molecule, and so a tetraterpene is a C_{40} molecule). The number of conjugated double bonds varies from 7 to 15 and usually is 11 for fish carotenoids. From this particular chemical structure, there is the possibility for a large number of geometric stereoisomers. In nature, all-E (formerly all-trans) isomers are the most abundant. However, analysis has demonstrated Z-E forms (formerly cis-trans) in significant quantities. The presence of numerous double bonds is also the cause of the instability of these pigments. Carotenoids are easily destroyed and become colourless at high temperatures or light intensities. The most widespread carotenoids among the salmonids are (Figure 11.1):

- β-carotene (β, β-carotene). This carotenoid is one of the most widespread natural substances. Its structure includes two terminal β-ionone units;
- canthaxanthin (β, β-carotene-4-4′-dione). This pigment has two ketone groups in positions 4 and 4′;
- astaxanthin (3,3′-dihydroxy-β, β-carotene 4, 4′-dione). This red pigment has two hydroxyl groups in positions 3 and 3′ and two ketone groups in positions 4 and 4′. It is the main natural pigment in salmonids and many crustaceans. It can be found in both free and esterified forms (mono- or diester).

Other pigments are detected more rarely and generally in small amounts. Some of them also represent important intermediaries in pigment metabolism. It can be assumed that in certain cases their low concentration or even their apparent absence is due to the rapidity of their conversion. Free forms of canthaxanthin and astaxanthin, which are synthesised industrially, are now utilised in intensive aquaculture.

11.1.2 Specific colours

Carotenoid colour is caused by the chromophore which is made up of at least seven conjugated double bonds. These pigments are usually yellow to red. Although β-carotene is orange, canthaxanthin has a longer chromophore and consequently absorbs light at a higher wavelength. It thus appears to be more pink in colour. Astaxanthin has the same chromophore as canthaxanthin. Natural sources of carotenoids always contain a mixture of different pigments, the concentrations of which are not constant. The colour thus depends on the relative proportions of the different carotenoids.

11.1.3 Digestibility or availability

After ingestion, carotenoid pigments can be eliminated intact in the faeces, absorbed or transformed. The products of transformation can in turn be eliminated or reabsorbed by the intestinal mucosa. It should be noted that no transformation of canthaxanthin has been observed in trout.

Since carotenoids are liposoluble compounds, their absorption is linked to that of lipids and their digestibility is influenced by the lipid content of the diet. Thus astaxanthin contained in capelin oil has a higher digestibility (85–90%) than that in shrimp meal (75–80%). Digestibility depends mainly on the form and nature of the carotenoids. Therefore, digestibility of astaxanthin can vary from 10% to 60% and sometimes more depending on its origin, while that of canthaxanthin only reaches 20% to 30%. Astaxanthin is more digestible in its esterified form than its free form. However, astaxanthin dipalmate appears to be poorly utilised by salmonids. The better digestibility of esterified forms is the subject of some controversy: astaxanthin exists in three forms (diester, monoester and free form) of which two are hydrolysed in the digestive tract and faeces analysis only allows the detection of a high proportion of the free form. This leads to the belief that esters are better absorbed, when this could simply be an artefact: as for other nutrients, digestibility, which is only an estimation by difference of what has been absorbed, can be biased by the degradation of carotenoids in the digestive tract.

Another way of studying the availability of carotenoids is by measuring the amounts of pigments contained in fish blood. Canthaxanthin is found in the serum three hours after a test meal. Maximal concentration is attained 24 hours after ingestion. These time periods are reduced for astaxanthin (Figure 11.2a). For the same level of intake, the serum concentration of astaxanthin is higher (around two to three times) than that of canthaxanthin. Carotenoid concentrations in the blood decrease in less than three days after the complete cessation of carotenoid supplementation. The decrease is comparable for astaxanthin and canthaxanthin

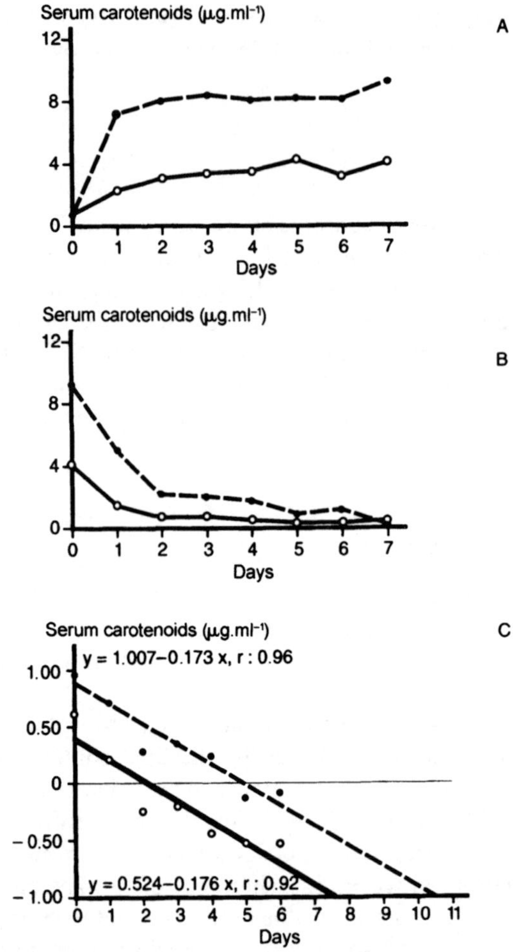

Figure 11.2. A: Pattern of changes in carotenoid pigment content of trout serum; canthaxanthin (——) or astaxanthine (– – –). B: after cessation of pigment intake. C: logarithmic regression lines (Choubert *et al.*, 1994). By kind permission of Elsevier Science.

(Figure 11.2b). However, there is great individual variability, still unexplained, in the capacity for carotenoid absorption in salmonids.

11.1.4 Metabolic conversion

A classification has been proposed based on the aptitude of aquatic animals for metabolising various carotenoid pigments into astaxanthin, their dominant pigment. It consists of three groups:

- animals that can convert lutein or zeaxanthin (but not β-carotene) into astaxanthin (cyprinids);

- animals that can synthesise astaxanthin from β-carotene (almost all crustaceans);
- animals that cannot carry out any conversion (salmonids).

The most important catabolic pathway is the oxidative degradation of the chain of double bonds, such as the biological degradation of β-carotene into vitamin A.

11.1.5 Specific affinity for certain tissues

Specific affinity concerns the aptitude of each carotenoid for being deposited in a particular tissue (skin, muscle, exoskeleton, digestive gland). Some carotenoids, such as the mycoxanthophylls or glycosyled xanthophylls, are not deposited in trout muscle. Others, such as citranaxanthin, are of limited use as pigments for salmonids. However, this characteristic is actually only the result of various processes which end in pigment fixation. It depends on factors such as the availability of carotenoids in the feed ingredients and their digestibility (already mentioned), concentration in the diet, duration of supply and the aptitude of the organism for transforming or depositing pigments in the tissues.

11.2 CRITERIA FOR ASSESSMENT OF PIGMENTATION EFFICIENCY

The pigmentation efficiency of carotenoids is defined as the gain ratio supply. Different criteria are used to assess it: chemical analysis of carotenoid concentration in blood or muscle, measurement of muscle colour and sensory analysis.

11.2.1 Chemical analysis

Carotenoids are unstable in the presence of light, heat, oxygen, peroxides, acids and, in certain cases, to alkalis. Particular attention must be paid to this characteristic during analysis. Measurement of carotenoids is a long process and is carried out after extraction and separation by chromatography (thin layer chromatography or high performance chromatography (HPLC)). The concentration of each pigment is calculated at the wavelength of maximum absorption.

11.2.2 Colorimetric measurement

The definition of a colour involves the determination of three characteristics:

- the hue (or dominant wavelength), allowing basic choice within the colour range;
- the chroma (or excitation purity), which expresses the proportion of the mixture of this colour with white (bright or dull colours). The human eye combines hue and chroma into a unique impression of colour;
- the lightness, which characterises the luminous intensity of the colours (light or dark).

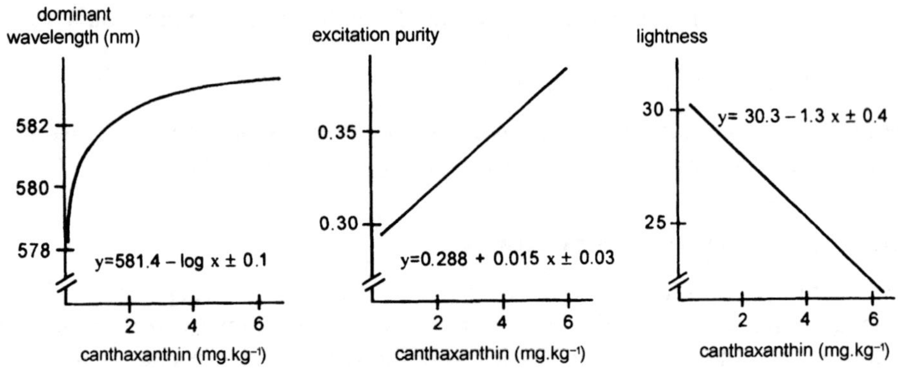

Figure 11.3. Relationship between concentration of canthaxanthine ($mg.kg^{-1}$) and colour characteristics of muscle in the trout (Choubert, 1982).

These physical measurements show (Figure 11.3) a rapid displacement of hue (dominant wavelength) from yellow to orange in relation to the increasing quantity of canthaxanthin in the muscle. The colour of the trout thus tends towards a maximum that cannot be exceeded despite the continued ingestion of pigments. When the concentration of canthaxanthins in the muscle increases, saturation increases, lightness decreases and the colour becomes darker. The characteristics of muscle colour in the trout are affected by the following factors: the hue by the chemical status of the carotenoids; the chroma by the concentration of carotenoids; and the lightness by the physical status of the muscle (crushed or not). It is important to note that, under certain conditions (for example, high concentrations), adding carotenoids to a diet does not necessarily lead to an improvement of colour.

11.2.3 Sensory analysis

Although chemical analysis of carotenoid concentration and the physical measurement of muscle colour are objective and reproducible methods, they remain time-consuming, expensive and complex to utilise. In addition, they only take into account one aspect of pigmentation (concentration or colour for each of the carotenoids studied) and do not allow the part played by each one to be distinguished when different carotenoids are present in the diet.

Subjective methods, which are simpler to put into practice, are also utilised to assess the pigmentation efficiency of carotenoids. These methods are based on the visual comparison of fish flesh, either against each other (rank-test) or in relation to a colour standard. This can be:

- a colour atlas (e.g. the Natural Color System, NCS);
- a colour range, derived from the NCS, but centred more on a segment of colours (e.g. the range established for salmon that have ingested astaxanthin);
- an arbitrary colour scale ranging from 1 (no pigmentation) to x (maximum red), with the knowledge that x can be 8, 10 or even 18, depending on the scale.

These methods are not reproducible and have a certain number of drawbacks, the most significant of which is without doubt linked to the actual principle of these techniques, in knowing how to compare natural pigments with printing inks. The arbitrary scale does not allow the comparison of samples taken at different sites since it is very variable from one producer to another. The rank-test does not allow comparison of samples collected at different times.

11.3 FACTORS AFFECTING PIGMENTATION

The relative proportions of the various carotenoids present in organs or tissues, as well as their different forms, are very variable and depend on numerous parameters. Three main factors influence carotenoid fixation: the diet, the animal and environmental factors.

11.3.1 Dietary factors

11.3.1.1 Carotenoid sources

There are many sources of pigments: raw materials (particular yeasts, algae and krill), industrial derivatives (crab or shrimp offal) and special preparations (shrimp meal). The first two categories have the advantage of existing in large quantities and being relatively cheap. They can also constitute a significant protein source. Utilisation of such carotenoid sources increases fish pigmentation. However, there are problems due to the wide variability in pigment contents and the presence, sometimes in large quantities, of substances that have little nutritional value: calcium carbonate (up to 30% of the by-products of crustaceans) or cell walls which are not very digestible (fibre from algae and yeasts). The technological processes used to eliminate these compounds often have a harmful effect on carotenoid stability. For this reason, attempts have been made to extract pigments from raw materials before they are incorporated into feeds, either as they are or after being dissolved in a vegetable or animal oil which contains no carotenoids. Extracts of crayfish, crab and other crustaceans have been used in this way, as well as those from plants: paprika, chestnut flowers and French marigold (*Tagetes* sp.). The pigmentation obtained using these products is very heterogeneous.

The chemical synthesis of canthaxanthin and then astaxanthin has allowed freedom from using raw materials rich in carotenoids. However, supplementation with synthetic molecules increases the cost of the food by about 15–30%. In addition, it is subject to strict legislation in each country. In the European Union, canthaxanthin and astaxanthin are listed in Appendix I (Directives 88/228/EEC and 87/552/EEC). The recommended dosage levels are 80 $mg.kg^{-1}$ of feed for canthaxanthin and 100 $mg.kg^{-1}$ of feed for astaxanthin.

11.3.1.2 Carotenoid structure

The composition of astaxanthin isomers in the muscle reflects the composition of the diet. In salmonid muscle, astaxanthin does not have the same structure in wild and reared fish. However, the ingestion of optical isomers of astaxanthin, ((3S, 3′S)-, (3R, 3′S)-, (3R, 3′R)-) or of a mixture corresponding to that found in the flesh of wild salmon (1 : 2 : 1) does not lead to the epimerisation of this pigment.

11.3.1.3 Carotenoid stability

The stability of carotenoids incorporated into fish feed is poor. Precautions must therefore be taken, mainly against oxidation, the principal cause of their degradation. For this reason, the protection of synthetic pigments is achieved either by coating them with gelatine or by microdispersion in a carbohydrate carrier substance. The feed can also be conditioned in an inert atmosphere immediately after manufacture; the risks of oxidation are thus greatly reduced. The addition of anti-oxidants (Chapter 18) increases the stability of carotenoids with regard to oxidation.

Feed pelleting (Chapter 21) leads to the loss of a significant fraction of the carotenoid pigments (about 20% for canthaxanthin). The increase in temperature and the abrasion of the coating during passage of the meal through the dyes, rather than the pressure, are responsible for the losses. The damage caused to the pellets hastens the destruction of the carotenoids as a result of the increased surface area exposed to the air. It would also increase losses during storage. Damaging pellets by poor handling and intake of humidity also lead to a decrease in their carotenoid content.

11.3.1.4 Pigment quantity

The influence of the amount of carotenoid pigment ingested by fish on their pigmentation is well established. Fixation of carotenoids in the muscle is increased when the amount of pigments ingested is high (Figure 11.4). However, above a certain level any further increase in pigment intake has no effect. This limit, which is close to 100 mg.kg^{-1} for canthaxanthin, may correspond to a saturation of absorption potential. In addition, fixation of canthaxanthin is less than that of astaxanthin. This difference can be attributed to a better absorption of astaxanthin. In the trout, the change in the slope of the curve, representing the change in fixed canthaxanthin in the entire muscle mass of the fish, only occurs from the moment that the fish has fixed 4–5 mg of canthaxanthin.kg^{-1} of muscle. However, as carotenoids can disappear during storage and final preparation, a concentration of 5–6 mg of carotenoids.kg^{-1} is preferable. For a muscle containing 80 mg of canthaxanthin.kg^{-1} of feed and an average gain ratio, the duration of pigment distribution would correspond to about two to three weeks. However, nowadays the trend is upwards and it is not uncommon to find concentrations of about 8–12 mg of carotenoids.kg^{-1} of muscle, implying higher dietary supplies.

The coefficient of carotenoid retention, defined as the proportion of carotenoids

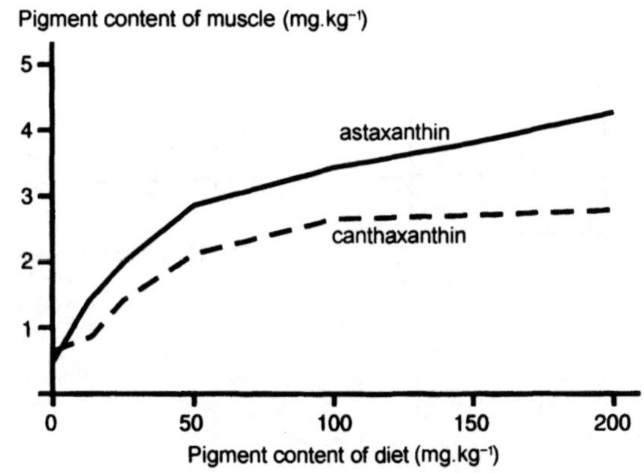

Figure 11.4. Relation between dierary pigment level and muscle carotenoid levels (after Choubert and Storebakken, 1989).

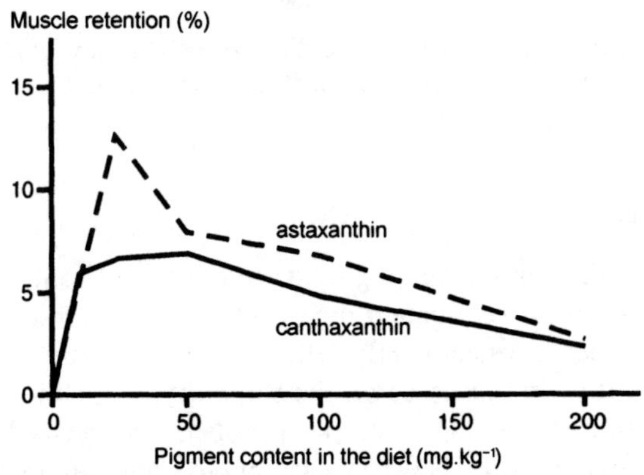

Figure 11.5. Muscle retention of carotenoid pigments in trout in relation to concentration of dietary keto-carotenoids (Choubert and Storebakken, 1989).

ingested which has been retained in the muscle or in the whole animal, decreases as the amount ingested increases (Figure 11.5). For low doses, an increase in retention up to a maximum of 6.9% is observed for canthaxanthin, for a dietary level of 50 mg of canthaxanthin/kg of feed, and for astaxanthin this figure is 12.5% for a dietary level of 25 mg of astaxanthin/kg of food. The average retention coefficient for astaxanthin is 1.3 times higher than that for canthaxanthin, which can be partly explained by the observed differences in digestibility.

If dietary canthaxanthin supply to trout is ceased, the pigmentation of the fish is affected (Figure 11.6). This phenomenon is however more marked in fed fish than in fasted fish. This result suggests the existence of a redistribution of canthaxanthin from one tissue to another, as a result of which the half-life of the muscle tissue is long compared to that of other organs.

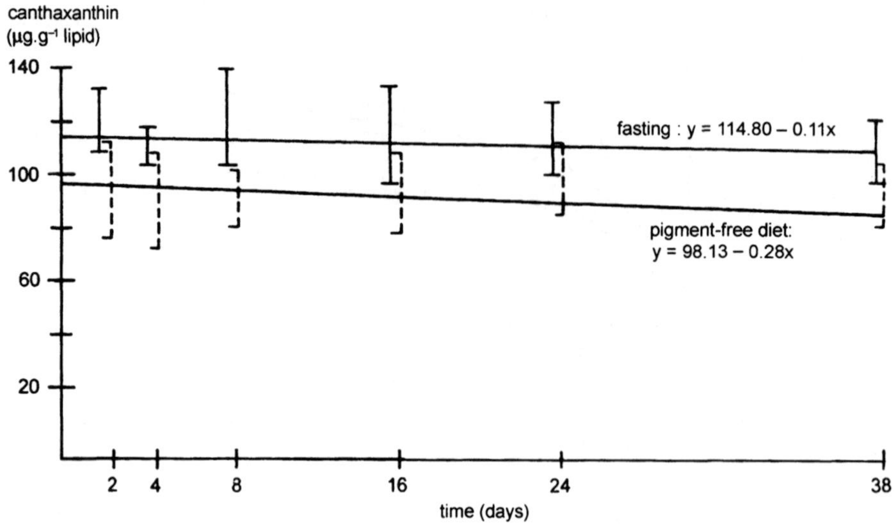

Figure 11.6. Effect of fasting or a pigment-free diet on the canthaxanthin content of trout muscle previously pigmented with canthaxanthin (Choubert, 1985).

11.3.1.5 Dietary lipid

Since carotenoids are liposoluble, adding lipids to the food should increase pigmentation since it tends to improve pigment digestibility. In salmonids, lipids deposited in the muscle or cutaneous tissue partly reflect the nature and amount of dietary lipids. However, direct action of lipids on the solubility and stability of carotenoids probably only constitutes one aspect of the problem. Indirect effects can also be involved, including competition between carotenoids and vitamin A for binding sites on molecules which transport carotenoids in the blood and possibly also across the membranes.

The effect of the lipid source on muscle pigmentation in the trout is still not very clear. In trout, dissolving canthaxanthin in oleic acid rather than in other fatty acids appears to increase its deposition in the muscle. In contrast, the concentration of astaxanthin in the muscle does not appear to be affected by the origin of dietary lipids (corn oil or fish oil).

11.3.2 Animal-related variability

Absorption and fixation of carotenoid pigments can vary considerably depending on the genus, species and even the strain of fish involved. Thus, within a trout population, the coefficient of variation for muscle pigmentation is usually close to 30% in sexually immature animals. This variability can increase greatly during sexual maturation.

11.3.2.1 Body mass and age

Trout weighing less than 150 g fix little canthaxanthin in their muscles, irrespective of dietary supply. The capacity of salmonids to fix carotenoids is linked more to age and physiological state than to body mass or dietary carotenoid content.

11.3.2.2 Tissues

Pigments are fixed in different forms, depending on the tissue being studied (skin or muscle). In salmonid skin, astaxanthin is found in its esterified form, while in the muscle it is in the free form but linked to proteins. According to an initial hypothesis, these forms are produced from esterified forms of astaxanthin contained in the ingredients and are stored in the skin without any modification. In fact, salmonids fix ingested carotenoids as they are, in contrast to some other species of fish, such as the goldfish. However, in a 'perch' (*Cymatogaster aggregata*), astaxanthin after hydrolysis of esterified forms is re-esterified in the intestinal wall and then transported to the skin.

The concentration of carotenoids in salmonid skin is greater than in the muscle (around 10 times higher). This higher concentration can be explained by the role played by carotenoids in phenomena involving light. Although precise functions of carotenoids are still unknown in fish, many observations lead to the conclusion that these pigments play an important role in the protection of cells against the harmful effects of visible light. They might act by modifying or attenuating light provided that the pigments contain a minimum of nine conjugated double bonds in their molecules, which is the case for all carotenoids in trout skin. Carotenoids have been recognised as potential inhibitors of cell damage caused by prolonged exposure to high light intensities in plants, algae and bacteria as well as numerous species of animal including man (photoprotective action).

In crustaceans, carotenoids are found mainly in the form of complexes, of which the carotenoproteins are the best known. These hydrosoluble complexes form the basis of a wide variety of colours, depending on the nature of the prosthetic group.

11.3.2.3 Sexual maturation

During sexual maturation in salmonids, carotenoids in the muscle are mobilised and selectively transferred to the skin in the male and to the ovaries in the female. However, this observation has been made on wild fish, whose feeding activity is reduced or interrupted during anadromous migration. In farmed trout fed until 2–3 weeks before spawning, the process is slightly different. The muscle pigmentation of the trout increases during sexual maturation as a result of dietary carotenoids. While the absolute quantities of canthaxanthin are spread equally between muscle and ovaries, concentrations of pigments are higher in the ovaries than in the muscle. During sexual maturation, carotenoid concentrations in the muscle decrease significantly in both sexes. Four months after spawning, the females again have carotenoid

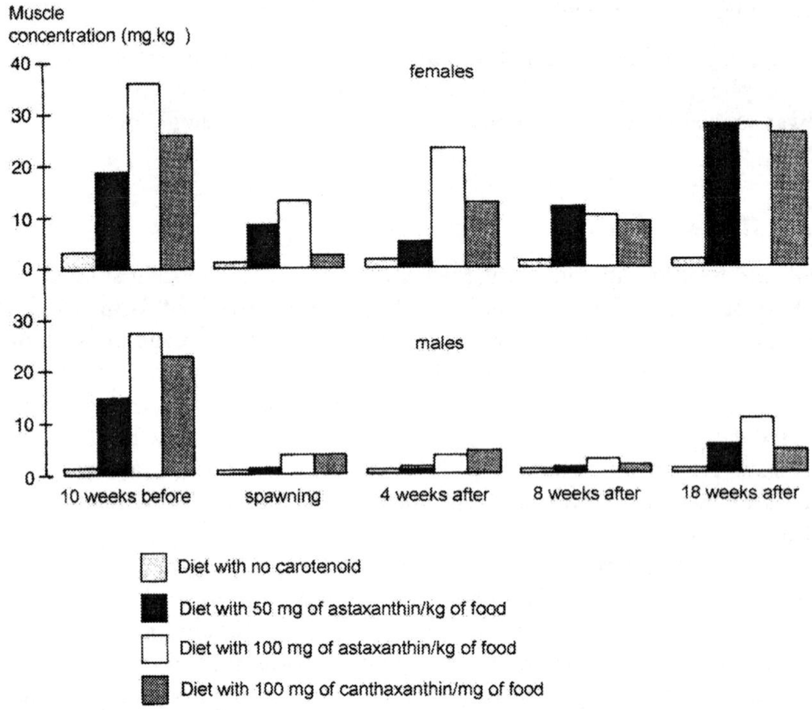

Figure 11.7. Effect of duration of reproductive period on keto-carotenoid content of male and female trout muscle (Choubert and Blanc, 1993).

concentrations in their muscles which are equivalent to those found before reproduction (Figure 11.7), while the males have recovered only one third of their initial concentration of muscle pigments 18 weeks after spermiation. This difference can be explained by the fact that spermiation extends over a longer period than spawning. Therefore, feeding the fish with a diet supplemented with carotenoids just after reproduction has only a limited effect until they have regained their sexually dormant phase.

11.3.2.4 Genetic factors

Salmonids that receive dietary carotenoids show much variability in flesh coloration. There is currently no metabolic explanation for these differences. These colour variations are due to genetic variability between families and *a fortiori* between individuals.

11.3.2.5 Phenomenon peculiar to crustaceans – moulting

The shedding of the old carapace during moulting represents a considerable loss of carotenoid pigments. The pigments present in the new tegument arise without modification from the epidermis to which they are brought from the hepatopancreas by the haemolymph. The definitive and irreversible transformation of the carapace pigments occurs there between stages D_1 and D_2.

11.3.3 Environmental factors

The lives of fish and crustaceans are affected by various environmental factors and therefore it is difficult to separate the effect of each factor. Thus, although water temperature significantly affects the growth and metabolism of fish, it does not appear to have an effect on the pigmentation of trout reared at 7°C or 17°C. In addition, in the same species, salinity does not appear to have any effect on pigmentation. The effect of light has been described above.

11.4 OTHER ROLES OF CAROTENOIDS IN ANIMALS

The pigmentary role is only one of the functions of carotenoids. Another function of these molecules is as precursors of vitamin A. Around 10% of natural carotenoids possess a significant provitamin A activity, notably in the mammals, more in the invertebrates and possibly the lower vertebrates. However, the vitamin A equivalence of any given carotenoid is very variable from one species to another: it never appears to be very high in fish (Chapter 9).

Some carotenoids are involved in the regulation of immune responses, through their protective action against free radicals. These roles have not yet been demonstrated in fish.

The involvement of carotenoids has also been noted in physiological phenomena such as growth and reproduction. In growth, the mode of action of this chemical group of pigments is not known, but it appears that it is linked to their provitamin A character. In the reproduction of fish (as in mammals), the involvement of carotenoids *per se* has been indicated on several occasions: during oocyte maturation, the fertilisation of ovules, egg respiration and embryo development, but this has still not been clearly demonstrated.

11.5 CONCLUSION

In conclusion, salmonid pigmentation depends mainly on the retention of carotenoid pigments. In order to improve retention efficiency, factors such as absorption, fixation and metabolism of these pigments must be taken into account. These factors are numerous, depending on the species and vary during the life cycle. Until now, studies have only been carried out on the pigmentation aspect and the interrelationships between carotenoids and vitamin A. Research should now be aimed at a better understanding of the other aspects of the nutritional role of carotenoids since they play a role in reproduction, growth and, to a lesser extent, in resistance against diseases.

11.6 BIBLIOGRAPHY

Bauernfeind, J. C. (1981) *Carotenoids as Colorants and Vitamin A Precursors*. Academic Press, New York.

Choubert, G. (1982) Method for colour assessment of canthaxanthin pigmented rainbow trout (*Salmo gairdneri*). *Sci. Aliments*, **2**, 451–463.

Choubert, G. (1985) Effects of starvation and feeding on canthaxanthin depletion in the muscle of rainbow trout (*Salmo gairdneri* Rich.). *Aquaculture*, **46**, 293–298.

Choubert, G. and Blanc, J. M. (1993) Muscle pigmentation changes during and after spawning in male and female rainbow trout, *Oncorhynchus mykiss*, fed dietary carotenoids. *Aquat. Living Resource*, **6**, 163–168.

Choubert, G., Gomez, R. and Milicua, J.-C. G. (1994) Response of serum carotenoid levels to dietary astaxanthin and canthaxanthin in immature rainbow trout *Oncorhynchus mykiss*. *Comp. Biochem. Physiol.*, **109A**, 1001–1006.

Choubert, G. and Storebakken, T. (1989) Dose response to astaxanthin and canthaxanthin pigmentation of rainbow trout fed various carotenoid concentrations. *Aquaculture*, **81**, 69–77.

Goodwin, T. W. (1952) *The Comparative Biochemistry of the Carotenoids*. Chapman and Hall, London.

Isler, O. (1971) *Carotenoids*. Birkhaüser Verlag, Basel.

Lenel, R., Nègre-Sadargues, G. and Castillo, R. (1978) Les pigments caroténoïdes chez les crustacés. *Arch. Zool. Exp. Géné.*, **119**, 297–334.

Part III

Fish and Crustacean Larvae

12

Ontogeny, development and digestive physiology of fish larvae

In fish, the term larva is used to describe the youngest individuals, from the moment when they are able to ingest exogenous food to when they take the adult form (Figure 12.1). The larval stage follows the embryonic stage and precedes the juvenile stage and it corresponds to a period of continuous morphological and physiological transformation in the animal. The larva shares certain characteristics with the embryo, such as very small size, very high specific growth rate and incomplete differentiation of certain organs. However, it is already capable of capturing and digesting prey.

12.1 PRELARVAL DEVELOPMENT

Embryonic development merits consideration since it affects certain important characteristics of the larvae and, in particular, size at first feeding.

12.1.1 Utilisation of vitelline reserves

Embryonic development occurs first within the confines of the egg chorion, and after hatching in the external environment. The embryo is then termed 'sac fry', since the vitelline vesicle forms a clearly-visible ventral protrusion, or 'free embryo', since it has a certain capacity for autonomous displacement although its swimming performance is still very poor.

From a nutritional point of view, the embryo meets its growth and maintenance expenditure by using the vitelline reserves of maternal origin present in the egg at the moment of fertilisation. The question concerning the existence of other food sources has been posed. The hypothesis of a trophic supply by absorption of dissolved organic matter has been put forward to explain the growth of leptocephalus larvae found in some species such as the eels. These very particular larvae are very active

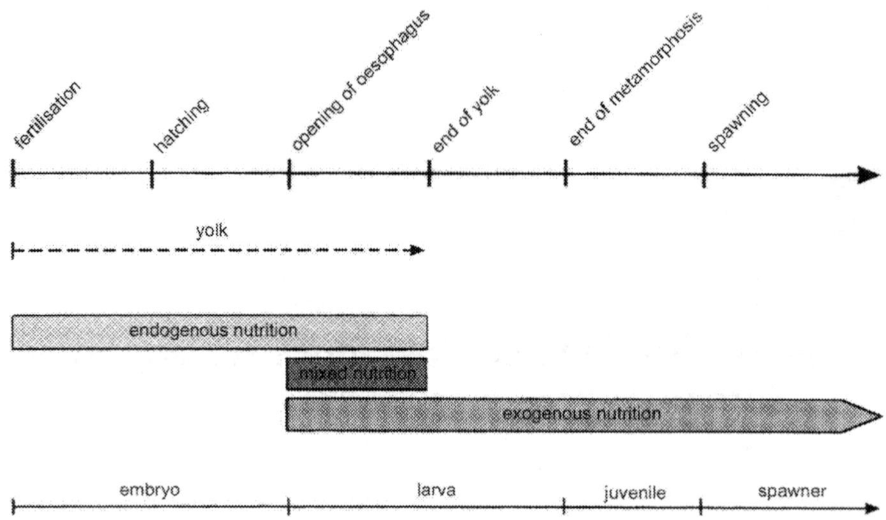

Figure 12.1. Developmental stages and modes of nutrition during the life of fish.

during their migratory phase and their length increases to reach 100 times the diameter of the egg without any predatory behaviour having been observed. The capacity for absorbing amino acids (AA) via the trans-epithelial pathway has been demonstrated in embryos and larvae of carp and white fish using bathing experiments: protein synthesis occurs, using marked AA which are dissolved in the water. However, the possibility of boosting larvae using nutrient baths has hardly been explored and it is considered that under normal rearing conditions supply other than that from vitelline reserves is negligible.

Initial reserves are by definition limited and, taking into account the small size of the eggs of most species, they only allow the formation of a small organism. In addition, large differences exist between fish species (Table 12.1). The turbot embryo initially has at its disposal one joule of reserves, eight times less than the carp and 4800 times less than chinook salmon. In unfed animals, the respective masses of yolk dry matter and the actual organism (embryo or larva) change in a similar way in different species (Figure 12.2). Only the parameters of the equations of the curves appear to differ depending on species, the initial quantity of yolk and temperature.

Hatching occurs during the quasi-exponential growth period following fertilisation. Embryonic mass at hatching is not constant for a given species and depends on the incubation temperature. Opening of the oesophagus occurs after the inflexion point of the body mass curve of the embryo and before maximal mass of the starved larva is reached. At this stage, yolk still remains, the utilisation of which meets maintenance requirements. Thereafter, biomass decreases and the yolk finally disappears. The composition of dry matter in the embryo and in the yolk varies little during the course of resorption and curves similar to those for body mass describe the variations in energetic content of the embryo and yolk respectively. In addition,

Table 12.1. Initial quantity of yolk (mg of dry matter) and maximal mass of fasted larvae (mg of dry matter of whole individual) in some fish.

Species	Initial amount of yolk	Maximal mass of fasted larvae
Chinook salmon[1]	164	112 (6°C)
		104 (8°C)
		98 (10°C)
		80 (12°C)
Brown trout[2]	38	19
Halibut[3]	1.4	0.75
African catfish[4,5]	0.4	0.30
Carp[6]	0.23	0.12
Turbot[7]	0.04	0.02

[1] Heming, 1981. [2] Escaffre and Bergot, 1990. [3] Finn *et al.*, 1995. [4] Kamler *et al.*, 1994. [5] Verreth, 1994. [6] Kamler, 1976. [7] Quantz, 1985.

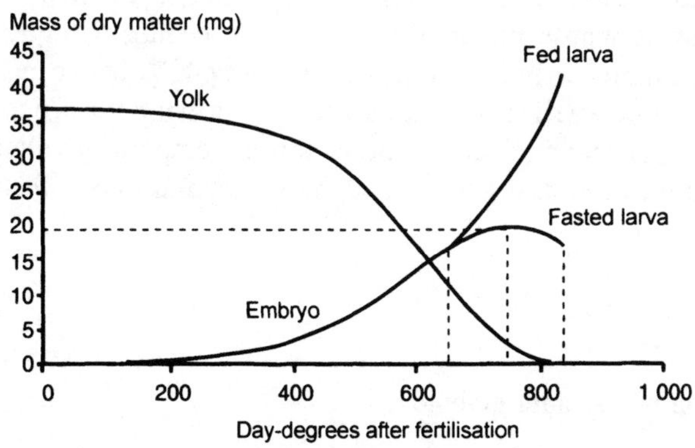

Figure 12.2. Change in mass of dry matter (mg) of yolk and body in the brown trout (after Escaffre and Bergot, 1990).

for a given temperature, the oxygen consumption of the embryo or larva per unit time appears to be directly proportional to the body mass, and not to the body mass raised to the power 0.7–0.8, as in the juveniles. It therefore shows the same variations as the embryonic mass. After hatching, oxygen consumption can be increased by external factors such as illumination, which increases the movement and swimming expenditure of the larvae.

In species that differ greatly for their initial quantities of yolk and incubation temperatures, the maximal dry matter of the unfed larvae represents between 50% and 70% of the initial amount of yolk. This proportion is high and relatively constant and allows the maximal size of unfed larvae and those capable of feeding (which is less than, but close to the former), to be predicted from the initial amount of yolk. In trout, the size of alevins produced by a female and compared at a given stage of development is directly proportional to egg size. Hatching occurs at the same time, independently of egg size, but the maximal mass during fasting is reached slightly later by the alevins produced by the largest eggs. In an organ such as the liver, large alevins have the same size of cells but twice as many as an alevin half the size. If feeding is delayed, large alevins reach the 50% mortality stage later than small ones.

Many rearing difficulties are linked to the small size of the larvae and so it would appear advantageous to select eggs containing a lot of yolk. When the perivitelline space is small, the volume of yolk corresponds approximately to that of the egg, whose size can serve as a selection criterion. For a given species, egg size varies from female to female. In general, eggs produced during the first spawning (first year of spawning in species with an annual cycle, or first spawning of the season in batch spawners), are smaller than in following spawnings. Egg size also tends to decrease in very old females and in the last spawnings of the season.

For a given initial quantity of yolk, it is also worthwhile optimising the transformation yield of yolk into embryo, i.e. the ratio between the maximal mass of unfed embryo and the initial yolk mass. This ratio depends on external factors and in particular on temperature. Chinook salmon eggs incubated at 6°C, 8°C, 10°C or 12°C produce alevins whose maximal mass at first feeding is higher for the lower temperatures, the difference in alevin masses reaching 40% between 6° and 12°C. In general terms, for each species there seems to exist a temperature at which growth rate of the embryo is maximal and another lower, temperature that optimises the yield from yolk utilisation and gives the highest maximal mass of unfed larvae.

12.1.2 Utilisation of the lipid globule

In addition to the yolk, which consists mainly of proteins, the vitelline reserves of certain species (sea bass, sea bream, turbot) contain a lipid globule. This globule is absent in other species such as carp and halibut. In the white fish (coregonids), many small lipid globules fuse progressively during incubation to form a single large globule in the larva.

Resorption of the lipid globule is slower than that of the yolk and a significant proportion of the globule often remains when the yolk is exhausted. The globule is made up of apolar lipids (triacylglycerols and also waxes in some marine fishes). It contains no proteins so cannot be used for the synthesis of new tissues in unfed fish that have exhausted their yolk. In contrast, it constitutes an energy reserve that can meet the energy expenditure of swimming and possibly limit the oxidative catabolism of proteins. In white fish, it is noted that at the moment larval mortality starts, some of them still contain a lipid globule which has not been utilised. Some authors have suggested that the lipid globule, which is less dense than other body parts, plays a physical role as a float, allowing swimming equilibrium in larvae that do not yet possess a swim bladder.

12.2 TRANSITION FROM ENDOGENOUS TO EXOGENOUS FEEDING

12.2.1 Differentiation of the digestive tract

The end of the embryonic stage is marked by the opening of the oesophagus, which occurs after the anus opens. Different studies, in particular in the turbot and African catfish, show that at this moment the absorptive epithelial cells of the intestine in the larvae appear similar to the enterocytes of older animals. From the time of first feed intake, it is possible to distinguish an anterior part where the enterocytes show lipid absorption activity, and a posterior part where there is absorption of protein macromolecules. In the carp, a distal segment, distinct from the latter and which does not absorb protein macromolecules, is also present. This distinction into specialised segments according to the histological type of the enterocytes is exactly what can be found in larger-sized fish (Chapter 4).

The ultrastructure of hepatic and pancreatic cells is also identical to that found in the juveniles. The presence of pancreatic enzymes can be detected in the pancreatic acinar cells even before the oesophagus opens. All the histological data indicate the existence of functional structures allowing the digestion and absorption of lipids in the anterior intestine. Trypsin activity is discernible in the pancreas and in the intestinal lumen, which suggests a capacity for protein degradation and absorption of the products of degradation in the anterior intestine. In addition, absorption by pinocytosis and the intracellular degradation of protein macromolecules by cytosolic enzymes can be observed in the posterior intestine. The capacity for gelatinised starch utilisation has also been demonstrated in the carp at the digestive (presence of amylase activity) and metabolic (induction of liver hypertrophy by addition of starch to the diet) levels.

The significant differences observed between the digestive system of larvae and that of juveniles, with the exception of the absence of a stomach, are neither cytological nor functional but concern general morphology. In larvae, the intestine is very short in relation to the body. It often forms a simple, straight tube, with none of the loops or evaginations (pyloric caeca) which can be present in the adults. The musculosa is thin. The mucosa is not formed into the folds seen in the juveniles. Initially the liver and pancreas each consist of a single lobe, while in the juveniles of several species – and in particular those of cyprinids – these two organs are intimately linked together along the intestinal loops. In addition, the pancreas of the youngest larvae has no adipose cells among the acinar cells.

12.2.2 Effect of feeding delay

The yolk which remains at the moment when the oesophagus opens allows the larvae to grow a little, even when food is absent. The specific growth rate of larvae at the point of first feeding is reduced and becomes negative after the stage of maximal mass. Access to exogenous food allows the larvae to maintain a high specific growth rate which corresponds, at least temporarily, to exponential growth. At a given age, trout alevins fed precociously have higher mass than those fed late, even though the latter, once they are fed, attain the same specific growth rate as the

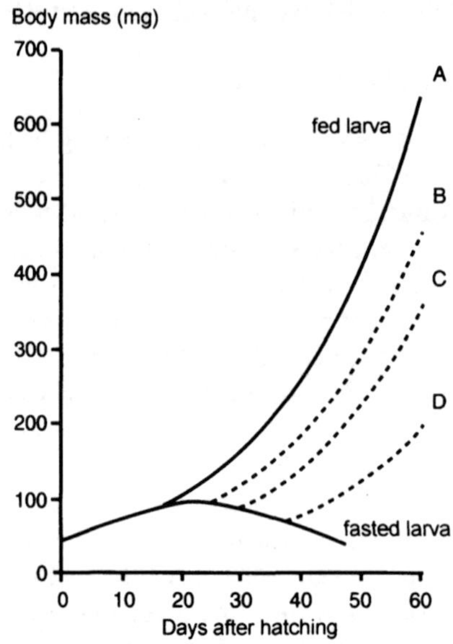

Figure 12.3. Change in body mass of trout fed from opening of the oesophagus (curve A), late (curves B, C and D) or fasted (after Escaffre and Bergot, 1985).

former (Figure 12.3). However, after a certain degree of fasting, called the 'point of no return', the weakened animal becomes incapable of feeding itself and will die, even if food is available.

The smallest larvae are most sensitive to starvation. Large larvae which develop in cold water spend several days (and even several weeks in the case of salmonids with large eggs), finding and learning to catch prey before suffering serious effects of starvation. In contrast, very small larvae living at high temperatures are in a difficult situation. The time scale of their development, expressed in number of days or illuminated periods, is short and they only have a narrow time slot in which to feed under good conditions, especially if their food is restricted to the light phase of the nycthemeral cycle.

The intestine and the liver are the two organs that are rapidly affected by lack of food. Biochemical (RNA/DNA ratio, fatty acid composition, level of enzyme activity) or histological (height of intestinal epithelium, volume of hepatocytes, size of hepatocyte nuclei) criteria are used to assess the nutritional state of larvae, i.e. their degree of starvation or adequacy of their food.

12.3 LARVAL DEVELOPMENT

12.3.1 Features of growth and development

Well-fed larvae have much higher specific growth rates than those of larger fish. These rates depend on temperature, for example they can exceed 60% per day in certain African catfish reared at 27–29°C. At 24°C, carp fed with artemia can reach a daily growth rate of 40–50% during the first days of feeding. Using artificial diets this rate can remain close to 20–25% for several weeks, up to a mass of 1 g, while it is only of the order of a few per cent for animals greater than 1 kg. Larvae conform to a general rule: the specific growth rate decreases inescapably as the size of the animal increases.

Increase in larval mass is accompanied by significant modifications in their morphology, anatomy and physiology. In the carp, as in many species, the larval body becomes proportionately less elongated and more solid. Depth and thickness of the body increase faster than length, until the proportions defining the juvenile are reached. The coefficient b of the relationship:

$$\log(\text{mass}) = \text{a} + \text{b}\log(\text{length})$$

is between 4 and 5 for larvae while it is close to 3 in fish larger than 1 g. The external segments of the body (snout, caudal area) grow faster than the central part. The fins become differentiated during the larval phase. These modifications have been linked to the change in mode of swimming and interpreted as adaptations to the laws of hydrodynamics which govern the environment. The forces of viscosity are significant for small larvae which move slowly, while forces of inertia affect larger fish as they swim more rapidly. Growth allometries (corresponding to different values of the coefficient b in the equation given above) have been observed for various larval tissues and organs. The nervous system and eye tissue, which are relatively well developed at the start, increase more slowly in volume than muscular tissue. Liver growth is particularly fast. The respiratory exchange surface area increases greatly during larval development, with the differentiation of gill lamellae and the decline in initial cutaneous respiration (disappearance of embryonic superficial red muscle). The surface area of the intestinal mucosa increases as a result of the elongation of the intestinal length relative to fish length and of the development of folds in the luminal walls. Differentiation of the stomach in those species where it is present often occurs late and indicates the end of the larval period. However, the start of the juvenile phase is not strictly defined by a unique event, comparable to the opening of the oesophagus for the larval phase. In addition to general shape, criteria such as pigmentation, fin differentiation and presence of scales are used.

In certain species, such as the flatfish (Table 12.2), the difference between the appearance of the symmetrical larva at first feeding and the asymmetrical juvenile is spectacular and the term metamorphosis is used to describe the transformation from one to another (Figure 12.4). For other species, in which the eggs contain a lot of yolk, the animal already possesses juvenile characteristics at the start of the exotrophic phase and it is considered that the larval stage does not exist. Salmon and trout represent an intermediate category between the two previous ones. When the first meal is taken, their 'larvae' possess a functional stomach but can still be distinguished from juveniles by their morphology. They are currently called alevins rather than larvae.

Table 12.2. Development of turbot larvae. After Segner *et al.*, 1994.

Larva length (mm)	Event
4	opening of oesophagus
5	disappearance of yolk
5.5	development of swim bladder
6	development of fin rays
8	curvature of the notochord
18	start of asymmetry
26	resorption of swim bladder and end of eye migration

Larva length (mm)	Event
4	opening of oesophagus
5	disappearance of yolk
5.5	development of swim bladder
6	development of fin rays
8	curvature of notochord
18	start of asymmetry
26	resorption of swim bladder and end of eye migration

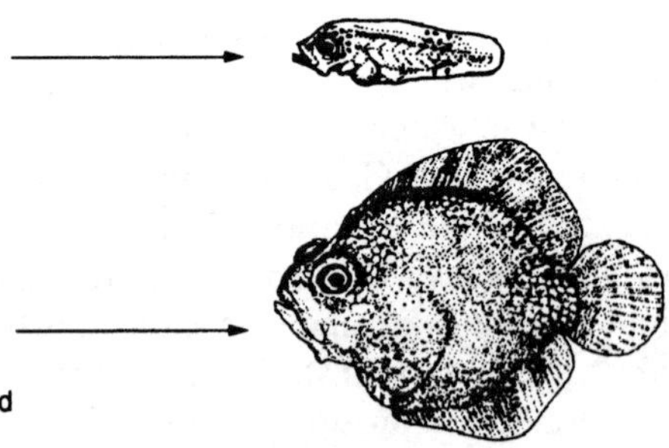

Figure 12.4. Development of turbot larvae (after Segner *et al.*, 1994).

12.3.2 Enzyme systems

12.3.2.1 Variation in enzyme activity during development

In sea bass larvae, a large increase in the activity of pancreatic enzymes such as amylase, trypsin and chymotrypsin is observed at the time when the mouth first opens, independent of food intake. The intestinal enzymes of the cytosol (e.g. leucine-alanine peptidase) and of the brush border (such as γ-glutamyl-transpeptidase, maltase and alkaline phosphatase) can also be detected before the mouth opens, although at low levels.

In studies of sea bass larvae, all the enzymes except pepsin show in diagramatic form the same specific activity profile (enzyme activity in relation to quantity of proteins) during the course of development (Figure 12.5): a rapid increase until day 12, maintenance of activity at a high level, then, towards day 23, a fall in

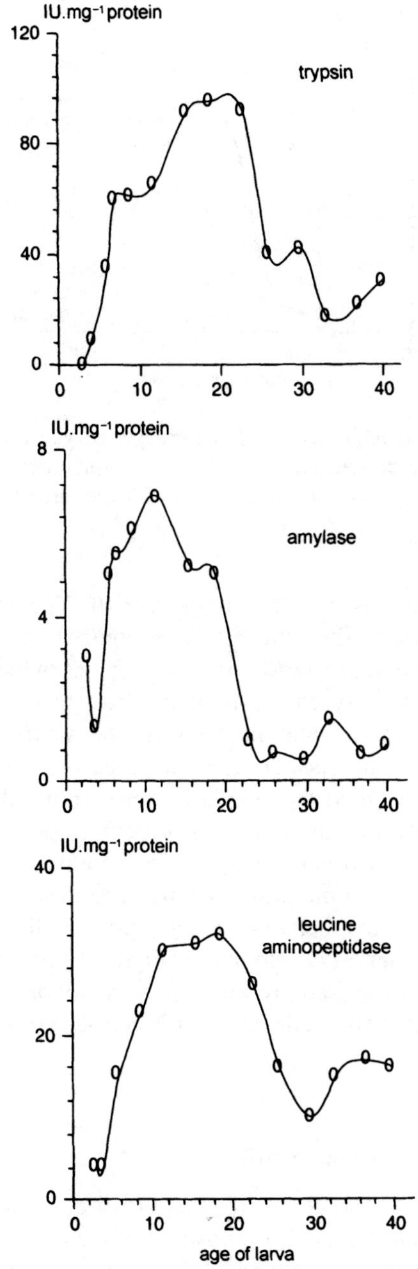

Figure 12.5. Change in specific activity of some digestive enzymes in marine fish larvae during development. IU: international unit of measure for enzyme activity; mIU: milli-unit.

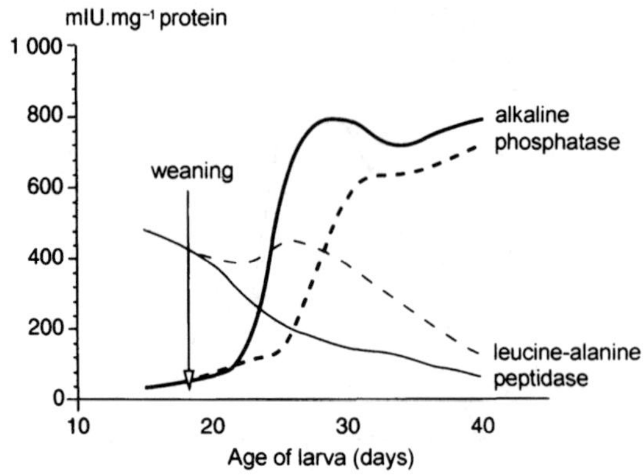

Figure 12.6. Variation in specific activity of enterocyte enzymes during development of fish larvae: increase in a brush border enzyme, alkaline phosphatase, and decrease in a cytosolic enzyme, leucine-alanine peptidase in animals fed with live prey (——) and animals weaned with a compound diet (– – –).

activity which then remains constant until day 40. The fall in activity does not correspond to a decrease in the total digestive capacity of the larva. New enzymes appear and the relative importance of the enzymes which are initially present decreases (the specific activity corresponds in effect to the activity of the enzyme being studied in relation to total proteins in the sample). Pepsin can only be measured after day 25. The specific activities measured in a purified fraction of enterocyte brush border increase regularly throughout development, indicating that the intestinal mucosa increases its digestive potential as the microvilli multiply. The presence of cytosolic enzymes in the enterocytes has been shown by immunohistochemistry and some have recently been measured. The specific activity of these enzymes in the cytosol changes in the opposite direction to that of those in the brush border during larval development (Figure 12.6). Digestion in the intestine in young larvae is characterised by an important involvement of the enterocyte cytosol, while in older individuals the brush border enzymes of the enterocytes become preponderant.

12.3.2.2 Modulation of enzyme activity

From the start of feeding, fish larvae are capable of modulating their enzyme activities in response to the biochemical composition of the food. Thus, it has been shown in sea bass larvae that a diet containing 12% starch results in a very large increase in amylase activity, even during the phase of transition from yolk-sac to exogenous feeding. Similarly, trypsin activity increases in response to the presence of free AAs in the diet. An increase in alkaline phosphotase is also observed when the diet contains phosphorylated molecules, such as phosphoproteins.

Other stimuli affect the secretion of pancreatic enzymes. In herring larvae, trypsin secretion has been demonstrated in response to a mechanical stimulus such as distension of the digestive tract.

For a long time the setbacks encountered during experiments substituting manufactured food for live prey in marine fish larvae have been explained by the following hypothesis: the young larvae synthesise too few enzymes to be able to digest the manufactured food and the live prey provide, directly or indirectly, an enzymatic complement which is needed to their digestion by the larvae. Recent data, which demonstrate the large capacity of the larvae for synthesising digestive enzymes, invalidate this hypothesis. In addition, the direct supply of enzymes by live prey appears to be negligible – the digestive tract of a 20-day-old sea bass larva can only hold five artemias at one time, which corresponds to 2% of the trypsin activity measured in the larva. Lastly, the supplementation of enzymes in the diet does not give very conclusive results. A positive result has been indicated for sea bream larvae fed with manufactured diets enriched with pancreatic enzymes, but such a result appears to be the exception.

Elsewhere, it has been shown that the presence of pepsin after the formation of the stomach is not determinant in explaining the good utilisation of manufactured food in fish larvae. Carp larvae (agastric species) can be fed on manufactured food from the moment the mouth opens.

12.3.2.3 Effect of weaning with a classically formulated diet

In all fish species, the change from a 'live prey' feeding sequence to a manufactured feed causes a more significant delay in growth, the earlier weaning is carried out. In contrast larval survival is little affected, except if weaning is very early, for example before 15 days in the sea bass.

A sudden increase in specific activity of all enzymes is observed during the first week after weaning. This increase results from a mobilisation of the larvae's enzyme capacities to best utilise the food. Thereafter there is a return to values identical to those found in larvae fed with live prey. However, this return to normal is not observed in sea bream larvae which are weaned before day 20. Although the larvae ingest the microparticles of food effectively, their enzymatic profiles remain close to those observed during malnutrition or starvation.

This modification of enzyme profiles can be explained by the fact that the development of digestive functions is still incomplete in the larvae. In most species, the appearance of certain digestive functions occurs during larval development following a genetically-programmed sequence, and it is only after three weeks that the animal has acquired adult-type digestion. In the sea bass, weaning at 10 days affects the secretion of pancreatic enzymes, although they are synthesised efficiently. Similarly, in the intestine, precocious weaning disturbs the appearance of brush border enzymes, while maintaining high levels of cytosolic enzymes. The animal thus retains larva-type digestion and the appearance of adult-type digestion is delayed or nullified. Weaning with a food which is poorly adapted for young larvae (before three weeks) can thus compromise the entire subsequent development of the animal (Figure 12.6).

12.3.2.4 Adaptation of food to digestive specifications of larvae

A criterion for the good adaptation of a feed for larvae is that there is no change to the normal sequence of the appearance of the different digestive functions. Recent experiments show that the addition of 10% free amino acids to the manufactured feed results in a noticeable increase in the secretion of pancreatic enzymes. In addition, incorporation into the food of 12% protein hydrolysate limits the delay of the appearance of intestinal brush border enzymes. The modification of part of the dietary protein supply of the diet thus allows a better maturation of the digestive functions of the larvae, resulting in gains in body mass (60%) and survival (30%) in comparison with larvae weaned using a diet containing nature protein. It is also possible that the form of lipid and carbohydrate supplies influences maturation of digestive functions.

The results obtained during the past few years show that the larvae possess sufficient digestive capacity to utilise a manufactured feed as long as the formula takes into account the digestive characteristics of the organism. This point of view is confirmed by the recent demonstration that it is possible to feed manufactured feed to sea bass larvae as their first meal, without any live prey supply.

12.4 QUALITY OF JUVENILES

In larval rearing units, the unit taken into account is often a 'batch', usually consisting of a large population of the order of 10^4 to 10^6 individuals sharing the same tank and the same food over a period of several weeks. One of the goals of juvenile production is to obtain homogeneous cohorts containing no abnormal fish.

12.4.1 Homogeneity of cohorts

For larvae produced from the same spawning batch and incubated under the same conditions, the initial variability in mass is low and the coefficient of variation is usually less than 15%. This coefficient increases during rearing to a value of about 30% in both trout and carp. Poor-quality feed results in an increase in the coefficient of variation. The coefficient of asymmetry of mass is usually nil or slightly negative at the start. It can decrease at the start of feeding if some of the larvae do not grow. Thereafter it increases with time, partly due to the mortality of the smallest larvae and partly to the appearance of a small number of animals with much higher growth rates than the population average. The appearance of these animals is often associated with cannibalism. This abnormal behaviour results from poor rearing conditions and in particular from a lack of food or poor adaptation to the feed.

12.4.2 Developmental anomalies

Numerous developmental anomalies can be found in rearing units. They are not observed in the wild (which does not mean that they do not exist), except in very polluted environments. The most serious anomalies disappear from the cohort by the death of the affected individuals. In contrast, non-lethal serious anomalies can remain in protected rearing units in the absence of predation other than cannibalism.

Three factors have been identified as the cause of the anomalies. The first is physico-chemical. Thus, the non-inflation of the swim bladder in sea bream and sea bass larvae can be prevented by cleaning the surface water of the tanks in order to eliminate the only film. This absence of inflation leads to lordosis in the Perciformes. It is known that temperatures too far removed from those preferred by a species, especially at the embryonic stage, can lead to irreversible deformities in the survivors under these extreme conditions.

Nutritional deficiencies constitute a second source of abnormalities. Starvation can affect not only the digestive apparatus but also external morphology, for example a flattening of the head and exophtalmy. Pigmentary anomalies in flatfish appear to be due mainly to poor nutritional quality of the artemia, although the nutrients involved have not been clearly identified. A deficiency in phospholipids can cause scoliosis in the ayu. Vitamin C deficiency in carp larvae causes gill deformities and erosion of the caudal fin. An excessive input of vitamin A causes compression of the vertebrae in the Japanese flounder.

Lastly, microbial infections can be at the origin of larval malformations. This is the case for a virus that causes epidermal hyperplasia in the Japanese flounder. Pathogenic vibrions can also cause intestinal and muscular necroses. Other bacteria such as *Flexibacter ovolyticus* can destroy the egg's chorion, thus leading to premature hatching of malformed embryos.

12.5 CONCLUSION

The study of fish ontogeny during the larval period has long been the preoccupation of biologists alone. The physiological approach recently undertaken has opened a new perspective on nutritional problems treated previously by empirical means. Recently acquired data should facilitate the task of nutritionists in the development of diets adapted for fish during these stages.

12.6 BIBLIOGRAPHY

Alami-Durante, H. (1990) Growth of organs and tissues in carp (*Cyprinus carpio* L.) larvae. *Growth, Development and Aging*, **54**, 109–116.

Escaffre, A.-M. and Bergot, P. (1984) Utilization of the yolk in rainbow trout alevins (*Salmo gairdneri* Richardson): effect of egg size. *Reprod. Nutr. Dévelop.*, **24**, 449–460.

Escaffre, A.-M. and Bergot, P. (1985) Effet d'une alimentation précose ou retardée sur la croissance d'alevens de truites arc-en-soleil (*Salmogairdneri*) issus d'oeufs de taille différante. *Bull. Fr. Pêche. Piscic.*, **296**, 17–28.

Escaffre, A.-M. and Bergot, P. (1990) Résorption du vitellus chez les alevins de truite commune (*Salmo trutta* L.) gardés à jeun ou nourris. *Bull. Fr. Pêche Piscic.*, **317**, 58–67.

Escaffre, A.-M., Bergot, P. and Szlaminska, M. (1989) Site of absorption of protein macromolecules in the intestine of carp (*Cyprinus carpio* L.) larvae and rainbow trout (*Salmo gairdneri* Rich.) fry. *Pol. Arch. Hydrobiol.*, **36**, 263–271.

Finn, R. N., Ronnestad, I. and Fynh, H. J. (1995) Respiration, nitrogen and energy metabolism of developing yolk-sac larvae of Atlantic halibut (*Hippoglossus Hippoglossus* L.). *Comp. Biochem. Physiol.*, **IIIA**, 647–671.

Gatesoupe, F. J. (1994) Development of fish larvae and rearing conditions in hatcheries. In: Lahlou, B. and Vitiello, P. (eds.) *Coastal and Estuarine Studies. Aquaculture: Fundamental and Applied Research.* American Geophysical Union, Washington, DC, 159–172.

Heming, T. A. (1982) Effect of temperature on utilization of yolk by chinook salmon (*Oncorhynchus tshawytscha*) eggs and alevins. *Can. J. Fish Aquat. Sci.*, **39**, 184–190.

Kamler, E. (1976) Variability of respiration and body composition during early developmental stages of carp. *Pol. Arch. Hydrobiol.*, **23**, 431–485.

Kamler, E. (1992) *Early Life History of Fish: An Energetic Approach.* Chapman and Hall, London.

Kamler, E., Szlaminska, M., Kuczynski, M., Hamackova, J., Kouril, J. and Dabrowski, R. (1994) Temperature-induced changes of early development and yolk utilization in the African catfish *Clarias gariepinus. J. Fish Biol.*, **44**, 311–326.

Osse, J. W. M. (1990) Form changes in fish larvae in relation to changing demand of function. *Neth. J. Zool.*, **40**, 362–385.

Quantz, G. (1985) Use of endogenous energy sources by larval turbot *Scophthalmus maximus. Trans. Am. Fish Soc.*, **114**, 558–563.

Segner, J., Storch, V., Reinecke, M., Kloas, W. and Hanke, W. (1994) The development of functional digestive and metabolic organs in turbot, *Scophthalmus maximus. Mar. Biol.*, **119**, 471–486.

Verreth, J. (1994) Nutrition and related ontogenetic aspects in larvae of the African catfish, *Clarias lazera.* Thesis, University of Wageningen.

Zambonino Infante, J. L. and Cahu, C. (1994) Development and response to a diet change of some digestive enzymes in sea bass (*Dicentrarchus labrax*) larvae. *Fish Physiol. Biochem.*, **12**, 399–408.

13

Feeding fish larvae with live prey

As a result of their ontogeny, fish larvae must feed themselves actively before their digestive system is fully developed (Chapter 12). At the current state of knowledge, for most marine species the inert foods that have been developed are not suitable for the first phases of larval development. It is therefore necessary to feed these larvae with live prey from the point of first feeding (larvae of 0.2 to 0.4 mg, depending on species) up to the post-larval stage (more than 50 mg) when the animals can adapt to inert food (called the weaning period). This phase is short (one to two months), but growth is rapid. The dimensions of the larval mouth determine the size of particles that can be used: usually about 200 µm when the mouth opens up to a few millimetres at the end of rearing.

In the wild, the larvae feed on plankton. The *a priori* simplest solution is to reproduce a more-or-less closed and controlled environment to encourage the development of first phytoplankton and then zooplankton before introducing the larvae. This extensive type rearing can be applied to low densities of organisms. Traditional aquaculture of freshwater fish owes its success to the development of this method. Its application in the lagoon system is very limited and the expansion of marine fish production rests principally on intensive rearing methods for larvae.

In intensive rearing, it has been necessary to create an artificial food chain by producing unicellular algae to feed filter feeding animals, themselves destined to be prey for fish larvae. Many studies have been necessary to manage this succession of production; prey nutrition has been the subject of studies in the same way as larval nutrition. Production costs remain high and it is these, and not the intrinsic quality of the live prey, which has guided the choice of the few selected species. The nutritional quality of the prey has been studied and attempts have been made to improve it. In addition, these intensive rearing units are operated in an environment containing a lot of organic matter which favours bacterial proliferation. These bacteria constitute an additional risk and the quality of the bacterial environment of the prey items must be considered at the same time as their food value.

This chapter describes the few species which have been retained, production techniques, the food value of these products and the methods used for their improvement, as well as quantitative aspects of feeding. Lastly, the bacteria associated with these rearing units will be considered, as well as their effects on larvae and methods for controlling them.

13.1 LIVE PREY USED IN MARINE AQUACULTURE

The first level in the food chain is made up of unicellular organisms which can be very different from the natural phytoplankton, for example, yeasts or bacteria. They can be used live, lyophilised or spray-dried, directly or to feed prey-animals (second level). Like those at the first level, the organisms at the second level do not necessarily belong to the marine zooplankton. It generally includes filter feeders that can easily feed on inert microparticles.

13.1.1 Unicellular organisms

The great value of these organisms resides in their high concentrations (50–80% of dry matter) of good-quality proteins. These prey are also rich in nucleic acids and vitamins. Some are produced industrially, which allows substantial economies to be made in the cost of large-scale production of the food chain.

13.1.1.1 Yeasts

There are a large number of commercial sources of live or spray-dried yeasts which can be used as food for filter-feeding prey. Live baker's yeast (*Saccharomyces cerevisiae*) is the most widely-used food for rotifer production. In Japan, this yeast is cultivated specially in an environment containing cuttlefish liver oil, in order to enrich it with essential fatty acids (EFA) for marine fish. Dried yeasts, in spray-dried form, constitute good protein sources for growing on of artemia. Marine yeasts have also been proposed, but their utilisation has not been developed: their EFA content is higher than that of baker's yeast, but is still insufficient to meet the requirements of marine fish larvae.

13.1.1.2 Algae

Freshwater chlorella is currently the major alga produced industrially (for human consumption) which can be used in significant amounts in aquaculture, either in live or spray-dried form. It is expensive, and most hatcheries cultivate unicellular marine algae specially, since they remain necessary to the maintenance of stocks of rotifers and for starting off mass production. Many species are used, including *Pavlova lutheri* (4 µm) and *Platymonas suecica* (10 µm). Algal culture is relatively costly, both in terms of capital installations and labour. It requires work to be carried out in a sterile environment, for the maintenance of monospecific stocks (not necessarily bacteria-free). Light and temperature must also be controlled, as well as enriching the air with CO_2 and providing minerals and certain vitamins. Algae vary widely in biochemical composition in relation to growth conditions, but strictly controlled culture allows their chemical composition and thus their nutritional value to be regulated to a certain degree. When climatic conditions are favourable, open-air culture in large ponds allows mass production at a reduced cost. However, this more risky method of production is not always suitable for the requirements of a hatchery in which live food must be provided every day.

13.1.2 Live animal prey

Animal prey used to feed fish larvae are always invertebrates. They are produced in hatcheries, either from maintained populations (rotifers, copepods) or from artemia cysts (also called 'resting eggs') which are sold after collection from salt marshes in which they encyst. Alternative sources can be harvested from natural resources (zooplankton collection, fish eggs or mollusc larvae).

13.1.2.1 Copepods

Copepods from the marine zooplankton represent the majority of prey consumed in the wild; they are thus used as a reference food, with a recognised effectiveness for larval rearing. The range of sizes amply covers all the requirements throughout development, since the first nauplius is 80–200 µm in length and the adults are 1–5 mm. Unfortunately, few species are suitable for mass production, since the production cycle is long (20 days to obtain adults) and maximal densities are low, so high volumes are required. Such rearing units constitute a heavy burden for hatcheries. When harvest of copepods from the wild is possible, it constitutes a very high-value secondary food source for the hatchery. Harpacticoid copepods, for example *Tigriopus japonicus*, are benthic detritivores. Their reproductive cycle is just as slow as the planktonic forms, but they can be fed at reduced cost and withstand higher concentrations. In a sufficiently warm climate they can be cultured in open-air ponds.

13.1.2.2 Rotifers

In traditional Japanese aquaculture, these animals were first considered a pest since their extremely rapid proliferation causes deterioration of environmental quality of eel-rearing ponds. Subsequently, this exceptional productivity has been put to good use in developing aquaculture of marine fish. A single type of rotifer belonging to the genus *Brachionus* was chosen because of its euryhaline characteristics. Its size, 80–300 µm (Table 13.1) is suitable for first-feeding many fish species. This rotifer was long considered as a single species, *B. plicatilis*. In fact there are two, clearly different size types that have now been allocated to two distinct species: *B. plicatilis* (large) and *B. rotundiformis* (small). The most important strains for aquaculture are those that can withstand large fluctuations in their environment without a sexual reproductive cycle being triggered. The high productivity of the species relies on parthogenetic multiplication. At 20°C, a female can reproduce itself two days after

Table 13.1. Mean mass and size of prey used in intensive larvae production.

Prey	Mean mass (μm)	Length (μm)	Width (μm)
rotifers			
S type	0.10	166	122
L type	0.45	262	177
artemias			
at hatching	1.9	460	164
enriched 24 h	2.0	617	164
fed 48 h	3.4	895	181

hatching, and lay about 20 eggs in eight days. The reproductive rate depends in particular on the concentration of individuals and their feeding. Under the best conditions, it can exceed a doubling in population per day. Almost all nutritional requirements of brachionus are unknown, apart from that they require vitamin B_{12}.

Source stocks are kept in small volumes where a population of females is maintained at low concentration by the addition of small amounts of algae twice weekly. Production is carried out by stabilised continuous cultures from which samples are taken daily; in hatcheries, this volume varies from several hundreds of litres to a few m^3. They are started off by inoculation into a high concentration algal culture (several million cells.ml^{-1}). When the concentration of rotifers exceeds 100 individuals.ml^{-1}, a larger volume can be inoculated. The most practical food for production purposes is fresh baker's yeast, distributed at a rate of about 200 mg (60 mg of dry matter)/l of culture/day. This food supply allows the maintenance of a population greater than 200 individuals.ml^{-1}, with daily collection of a quarter of the culture volume. The use of fish oil in emulsion with the yeast improves rotifer quality (Figure 13.1). However, this food is not always sufficient in the long term and thus input of algae is necessary. When algal culture is possible in outside ponds, direct 'seeding' can be carried out and so larger, though less well controlled, volumes become available.

Recently, various methods were proposed for cultivation of rotifers at high density (several thousands individuals per ml). The main factor limiting population growth and density is the degradation of water quality induced by high food level. This constraint can be overcome either by automatic monitoring of oxygen level and pH of the medium in short batch cultures, or by a recirculation system in semi-continuous culture in order to maintain an acceptable water quality with high daily food amount.

13.1.2.3 Artemia

The genus *Artemia* (branchiopod crustacean), previously reduced to a single species, *A. salina*, has been divided into several species using chromosome characteristics. These species are utilised with distinction in aquaculture and usually the one generic

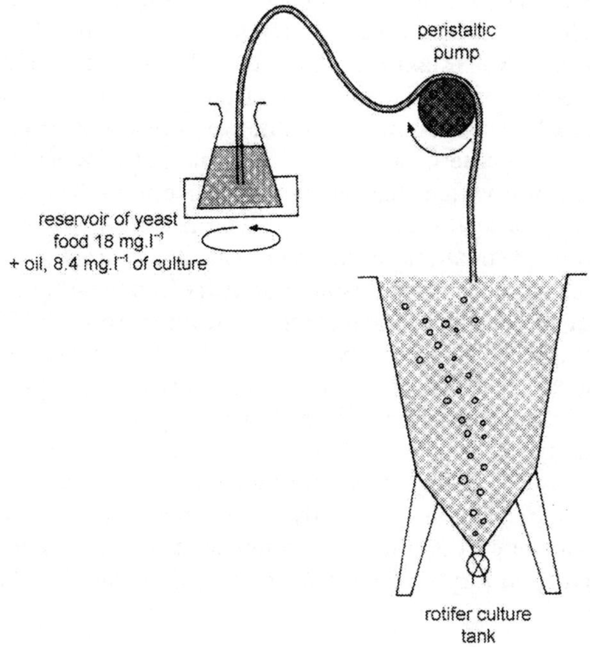

Figure 13.1. Diagram of rotifer feeding system using a semi-continuous technique ($\frac{1}{4}$ of the volume turned over per day) and fed on fresh yeast and oil emulsion.

name is used. The animal is adapted to hypersaline environments and produces large numbers of cysts in many salt marshes and other extended very salty areas of water where they thrive. The dormant cysts are easy to collect and are kept in a dry state under vacuum and sold. They are very easy to use in the hatchery: they can be reactivated by hydration and oxygenation to obtain hatching of larvae at the nauplius stage in 24 h at 28°C (Table 13.1). These nauplii should be released from the shells (chorion) of the resting eggs before use. This is carried out by decanting or sieving in water, or by a technique called 'decapsulation' which consists of 'dissolving' the chorions in hypochlorite solution (2% sodium hypochlorite for 7–10 min) before incubating the eggs. The nauplius larvae can be used immediately as prey. Rearing for several days allows larger prey to be obtained, but this becomes less and less practical and more constraining. The development of early weaning (with manufactured feed) also reduces the necessity of pre-grown artemias.

13.2 NUTRITIONAL REQUIREMENTS OF LARVAE

Feeding larvae using live prey does not easily lend itself to the study of requirements since it is difficult to obtain experimental live food with well-controlled composition and to vary one factor precisely, except in the case of fatty acids (FA).

The requirements of marine fish for long chain, polyunsaturated FA (HUFAs), in particular EPA (C20 : 5, n-3) and especially DHA (C22 : 6, n-3) (Chapter 7, Box 7.1 for terminology) is now well established. Since the HUFA content of prey can be modified, the positive effects of enrichment on larval survival, growth and resistance to stress have been clearly demonstrated. Quantitative evaluations of requirements have been undertaken using criteria such as the total amount of series n-3 C20 and C22 FA. This requirement is variable depending on the species of fish but it appears to be higher in larvae than in juveniles (1.3% in contrast to 0.8% of dry matter respectively in the turbot). Estimations of this requirement also differ widely between authors. Thus, in the gilthead sea bream, estimations vary from 0.56% to 3% of prey dry matter and appear to depend on the methods used for these studies. Moreover, DHA is more efficient than EPA, especially for survival and resistance to stress. It is also implicated in the laying down of cutaneous pigmentation in flatfish. Requirements for n-6 series FA have not yet been defined, but a requirement for arachidonic acid must exist for the synthesis of prostaglandins.

There are no precise data for other nutritional factors, at least for marine species. Vitamin C requirements are currently being studied. It is for this reason that the nutritional value of prey is still mainly estimated from juvenile requirements or from data obtained from larvae of freshwater species, in which feeding of inert food is possible.

13.3 NUTRITIONAL QUALITY OF PREY

The quality of live prey is naturally rather variable since these organisms react to environmental and feeding factors during their production right up to the moment at which they are ingested by the larvae. These factors are more-or-less controllable and variations in biochemical composition of prey are observed between hatcheries and even within one hatchery. During nutritional studies on larvae, it is necessary to verify the composition of the prey as well as their variability and possibly to modify it. The nutritional quality of artemias or rotifers can be altered just before they are distributed to larvae via culture or using what are called enrichment methods.

13.3.1 Culture methods

Using a relatively balanced diet, a culture of at least two days is necessary to modify FA composition and obtain growth in artemia. This requires rearing at low densities (10 to 20 artemias.ml^{-1}). Feeding methods only result in lipid contents of 10% to 20%.

13.3.2 Enrichment methods

Enrichment methods were initially achieved while aiming to add nutritive value to the prey itself in terms of its digestive content. Biocapsules were thus obtained to which nutrients or pharmaceutical products could be added. The actual input from the digestive content is ephemeral and debatable, but these are undeniably efficient for lipid supply. Live prey have a remarkable capacity for incorporating lipids.

These enrichments are carried out either using nutrient baths or feed rich in lipids, in which an oil emulsion provides high doses of EFAs. Various commercial and experimental products have been developed for this purpose. Short-term enrichments ($\frac{1}{2}$ hour) have already proved effective for 'pregrown' artemia and also rotifers. After hatching, artemia nauplii require longer duration enrichments (24 hours) carried out at high concentrations (300 artemia.ml^{-1}). These periods include the duration of predevelopment until the mouth opens and the true enrichment period. Enrichment over a 24-hour period using an emulsion rich in EFAs is more widely used than culture. It leads to high levels of lipids (20–30% of dry matter in the prey) while providing FAs which add onto pre-existing lipids. Incorporation of EFAs by rotifers is quasi maximal in six hours and enrichment over 24 hours using marine algae is also currently used in certain hatcheries.

13.3.3 Energy

The concept of energy is difficult to take into consideration in larval rearing units where it is technically impossible to measure precisely food intake and digestibility. Mean mass, chemical composition and thus available energy in the prey vary widely in relation to their physiological state. The important measurement in nutritional terms is the energy value of the prey at the moment when they are consumed by the larvae, a different value from that which can be estimated at the time of distribution. While they are present in the rearing tanks, the prey utilise their energetic reserves and lose part of their nutritive value. The aim must therefore be to limit these losses by adjusting distribution to the quantitative requirements of the larvae while attempting to minimise them.

13.3.4 Lipid and essential fatty acid composition

Comparison of FA composition of various organisms used as live prey shows large differences (Table 13.2), depending on organism type and feeding methods (using production or enrichment).

Series n-3 EFA (EFA n-3) content is generally high in copepods, which seem to be able to concentrate them, particularly DHA. In comparison, EFA content as well as the DHA/EPA ratio are low in rotifers and artemias.

Some authors consider that marine copepod HUFAs come directly from the phytoplankton which they consume. However, great capacities for elongating and desaturating n-3 FAs have been demonstrated in certain species of copepod, while these capacities are very low in rotifers and artemias.

Rotifers fed on baker's yeast alone are highly deficient in HUFAs and their nutritional value is insufficient. In this case, enrichment methods are necessary. An oil emulsion rich in n-3 EFAs can also be added to the yeast used to feed the rotifers. In addition, commercial feeds exist that allow cultivation of rotifers with good EFA

Table 13.2. Lipid composition (% dry matter) and main polyunsaturated fatty acids of live prey.

		Rotifers (*Brachionus plicatilis*)				*Artemia* sp.		
		Fed with						
	Copepods (*Tisbe* sp.)	Yeast	Algae (*Platymonas suecica*)	Yeast + super Selco®	Enriched with Super Selco®	Nauplius at hatching	Enriched 24 h with Super Selco®	Fed 48 h with yeast and cod liver oil
Lipids (% D.M.)	5.1	11.8	8.5	8.5	18	18.1	21	18.9
Fatty acids (% D.M.)	2.4	8	5.5	6.8	15	14.4	15.1	14.7
HUFAs (% D.M.)	1.1	0.2	0.7	1.4	5.7	0.5	3.3	2.5
FAs (% total FAs)								
Total saturated	20.8	20.6	23.4	8.7	9.6	17.7	13.6	15.2
Total n-7	11.2	15.8	5.9	26.6	11.6	13.4	11	17.8
Total n-9	8.5	26.8	9.4	28.8	19.8	21.9	18.4	29.3
Total n-6	5.4	22.6	29	9.9	9.3	9.8	7.6	5
of which 18 : 2 n-6	1.8	18.6	18.6	7.6	7.2	8.3	5.4	3.6
20 : 4 n-6	2.7	1.1	5.1	1.1	1.2	1.4	1.3	1.1
Total n-3	44.9	6.7	25.3	24.4	41.7	29.4	42.1	24.9
of which 18 : 3 n-3	0.4	2.9	7.2	1.7	2.2	22.3	17.8	6
20 : 5 n-3	14.7	1.1	6.2	10.3	16	2.9	12.7	11.7
22 : 5 n-3	tr	0.7	1.9	3.1	3.4	–	1.1	0.6
22 : 6 n-3	29.5	0.2	0.1	6.6	16.2	–	7.2	4.5

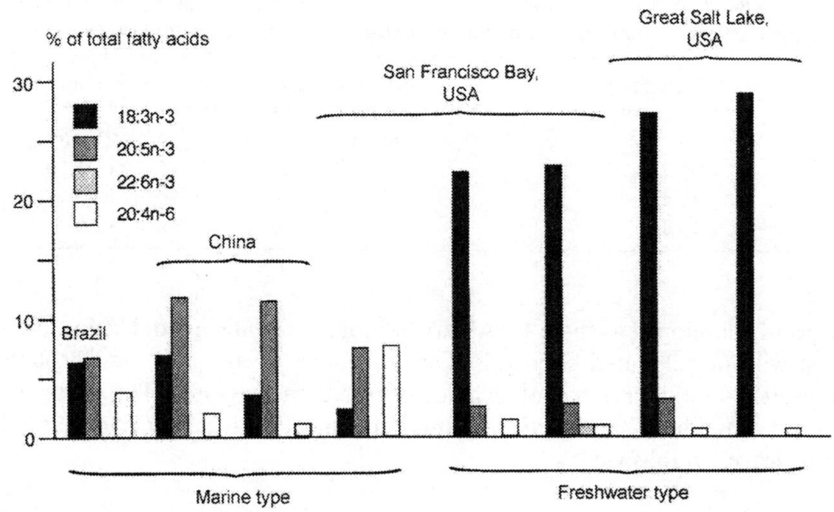

Figure 13.2. Main polyunsaturated fatty acids from artemia samples from different geographical origins.

contents. If marine algae are used, the species and culture conditions determine their FA content and that of the rotifers which they are feeding.

The composition of artemias at hatching varies depending on the batch (location and date of harvest); in effect, it depends on the natural food of the animals. Two types can be defined in relation to their FA composition (Figure 13.2): the 'marine' type which is relatively rich in EPA (required by larvae or marine animals) and the 'freshwater' type, which is rich in linolenic acid 18 : 3 n-3 (which freshwater fish larvae can use as a precursor for EPA or DHA). In practical terms, successful rearing of some marine fish larvae can be achieved using marine type nauplii. Nevertheless, the latter have, even with the best batches of cysts, a lower HUFA content than the estimated requirements of the demanding marine species. DHA is generally absent or present only in very low amounts in the newly-hatched nauplius, while this FA is the one most efficient in the larvae. As a result, supplies must be made by feeding or prey enrichment.

13.3.5 Protein composition

Live prey have protein contents higher than 60% of dry matter, a value which can be decreased by enrichment with high quantities of lipids. In rotifers, the growth rate of the culture positively influences the amount of protein per individual. The quality of proteins in live prey is considered to be good and intrinsically varies little. The amino acid compositions of different species used as live prey are broadly comparable and meet the requirements for essential amino acids estimated for juvenile fish.

The proportion of free amino acids (FAA) present in zooplankton organisms has sometimes been considered as a quality factor for larvae in which the digestive

Table 13.3. Oligo-elements in different live prey: rotifers produced using algae or yeasts and newly hatched artemia nauplii (mean of different stocks in $\mu g.g^{-1}$ of dry matter).

	Rotifers/algae	Rotifers/yeast	Artemia nauplii
Fe	216 ± 52	135 ± 68	265 ± 94
Zn	38 ± 2	98 ± 28	218 ± 56
Mn	19 ± 2	17 ± 8	15 ± 5
Cu	14 ± 3	18 ± 8	12 ± 7

tract is poorly developed as these FAAs can be more easily absorbed. FAA content is higher in wild or cultivated copepods than in rotifers or artemias, and protein or mixed (protein + lipid) enrichment techniques have been suggested. They allow FAA content to be modified, but the effect of this enrichment on rearing performance has not been clearly established.

13.3.6 Minerals

Relatively few data exist about the influence of minerals in prey on larval feeding. In sea water, mineral supply from the water is often considered to be sufficient. However, in intensive culture it is limited and an overall mineral enrichment benefit has been discovered in this type of culture. Some data exist for the macro-mineral and trace elements compositions of artemias, showing values *a priori* correct, but variable depending on the stock. Rotifer cultures often depend on algal ones, themselves produced by intensive culture which provides minerals (Table 13.3). The medium used for algal production vary from one hatchery to the next, with some providing oligo-elements and some not and even adding chelating agents, which limit the incorporation of trace elements. These factors affect the mineral composition of the rotifers.

13.3.7 Vitamins

Little information exists about the vitamin content of prey animals. Rotifers require vitamins that are mainly provided by the bacteria associated with their rearing. Culture conditions affect this synthesis by associated bacteria, with low oxygen levels being favourable. It is possible to enrich the prey with liposoluble vitamins via their food, while direct input of hydrosoluble vitamins seems to be ineffective. Thus, vitamin C can be provided in the form of ascorbyl palmitate, while hydro-soluble forms have no enriching effect.

13.3.8 Pollutants and toxins

Different studies have shown that artemia nauplii used in aquaculture contain various organochlorides from pesticides used around the cyst production areas. It is suspected that these compounds influence artemia quality. In general, tests comparing effectiveness of artemia stocks in larval rearing performance mainly show correlations with other characteristics such as EFA composition.

13.4 PRACTICAL FEEDING

13.4.1 Feeding behaviour of larvae

Behaving as a predator, the larva expends energy to catch its prey and each attempt which does not result in a capture increases this expenditure. This behaviour has been more specifically studied with the aim of optimising feeding conditions for larvae. Environmental factors in the rearing tanks (colour of walls, lighting, water flow) influence the success rate of captures and hence rearing success. Methods of prey distribution in the cultures have important consequences for prey quality at the moment of capture, in particular on their energy value as well as the level of effective feeding. The type, density and size of prey influence the ratio between the energy retained by larvae and the energy expended for their capture.

13.4.2 Prey size and type

In practical terms, the feeding sequence consists of a succession of prey types that provide a balance between food size and mouth size during development. Early introduction of artemia nauplii (larger than rotifers) from the time at which the larvae can ingest them allows larval growth to be accelerated. Similarly, final utilisation of pregrown and enriched artemias allows an adaptation of artemia size to those of the larvae. A study of carp larvae fed using different strains of artemia has shown that growth results are in proportion to the size and mass of the nauplii of the utilised strains (it is known that for this fish species, differences in FA composition of artemia stocks do not appear to be important).

13.4.3 Quantitative aspects

In intensive rearing methods, great care must be taken with the distribution of live food. The daily requirements of the larvae must be met completely and exactly. However, the number of larvae at any given moment cannot be known precisely because of the relatively high mortality rates at the start of rearing and of the impossibility of correctly counting mortalities when the larvae are small. In addition, growth is very rapid and the amount of food distributed must increase accordingly (Figure 13.3).

Meeting daily quantitative requirements must reconcile two contradictory principles:

- ensuring optimal prey density to favour their capture by larvae;
- avoiding prey remaining for a long time in the tank before being caught and thus losing some nutritive value.

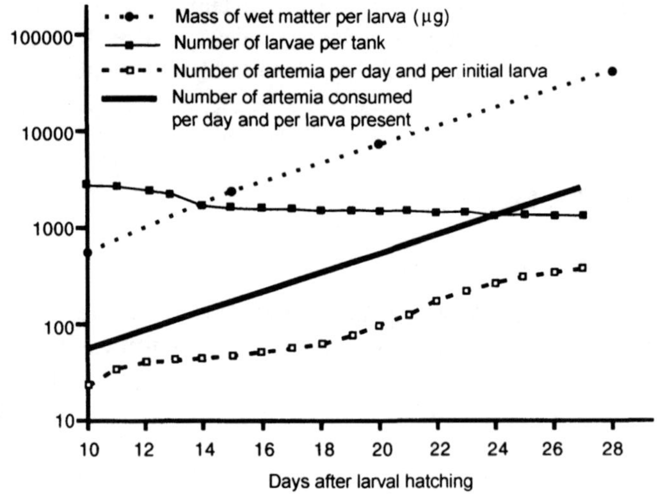

Figure 13.3. Quantity of enriched artemia consumed by turbot larvae (logarithmic scale) in a rearing unit.

The distribution method for prey depends on the rearing techniques, whether they are in green or clear water:

- in green water (with phytoplanktonic algae present), prey density is maintained approximately constant by the introduction of new prey which compensate for those ingested by the larvae;
- in clear water, the method widely used is rather empirical. It consists of quantifying the prey remaining in the rearing tank and then adjusting the subsequent distribution in relation to these remainders and to a pre-established plan. The objective is to obtain a minimum of unused prey remaining in the rearing unit on the following day.

13.4.4 Management of prey quality in the larval tank

In clear water rearing units, the prey remain unfed in the larval tank. Different techniques can be used to minimise loss in food value and allow the best possible larval growth, either by distribution of several meals per day or by continuous distribution of prey using a pump system from a prey reservoir (Figure 13.4). Prey which are about to be distributed are held in an enriching environment or chilled (for artemia nauplii). For continuous distribution, excess prey are evacuated in the outflow from the rearing tank, which is fitted with a filter that allows prey to pass through while retaining the larvae.

The 'green water' technique allows the maintenance of the nutritional value of the prey at a relatively satisfactory level, since they are able to feed themselves in the tank. However, the body composition of the prey develops under the influence of the algae consumed. This phenomenon is effective for rotifers or copepods, but for

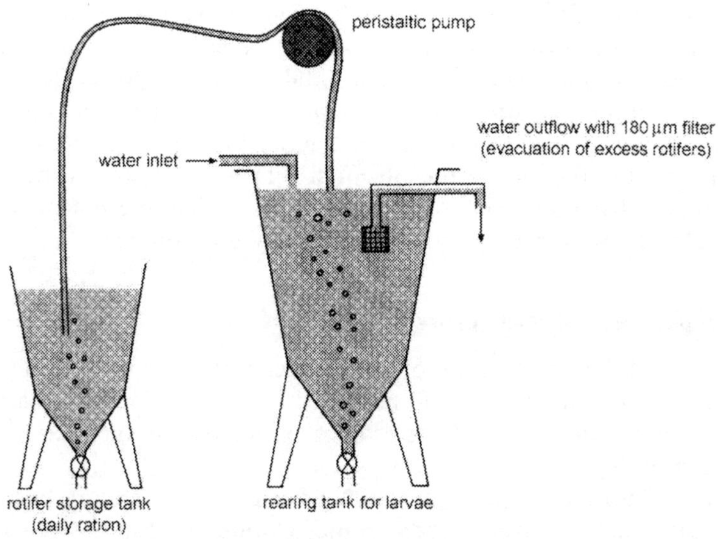

Figure 13.4. Diagram of continuous distribution method of prey (in this case, rotifers) to larvae.

artemia nauplii quality maintenance is ensured by continuous or fractionated distribution.

13.5 BACTERIAL QUALITY OF LIVE PREY

13.5.1 Significance of bacteria in the food chain

Living organisms present in an artificial food chain are not all introduced voluntarily. Bacteria inevitably develop within it unless germ-free conditions are used, delicate methods which are generally reserved for experimental work. The participation of the intestinal microflora in fish feeding has not been studied in this book, since the scale of this flora varies from 10^3 to 10^8 cells.g^{-1} of digestive content, depending on diet and climatic zone of the species being considered. It is considered that a minimum of 10^7 cells.g^{-1} is required for the concentration of bacterial enzymes to be sufficient to have a significant effect on the digestive process. This level is usually attained in rotifer cultures where between 10^7 and 10^8 cells.ml^{-1} of environment and about 10^9 cells.g^{-1} of rotifer are found. Healthy turbot larvae fed with these rotifers contain, on day 8, between 10^7 and 10^8 cells.g^{-1} of larva. It is thus possible that bacteria play a role in larval nutrition. This is certainly the case for the feeding of filter feeding predators. For example, *Pseudomonas* is a good feed for rotifers; in particular, it constitutes a source of vitamin B_{12}.

13.5.2 Ambiguous role of vibrions

Vibrions generally dominate the flora associated with marine fish, although the genus *Pseudomonas* can also be present in large numbers. They also dominate the flora associated with filter feeding preys. It might be assumed that most of these strains are inoffensive or even useful. However, the pathogenic character of some strains has been demonstrated by experimental infection. This is the case for an opportunistic 'vibrion' (a strain of *Vibrio splendidus*), which can reach levels of more than 10^9 cells.g^{-1} of turbot larvae, leading to massive mortality of turbot about 30 hours after the start of the infection. If infection is not caused experimentally, it appears that the strains only expresses its pathogenic potential when the larvae are already weakened by unfavourable rearing conditions.

13.5.3 Prophylaxis and food quality

In order to avoid the proliferation of these undesirable vibrions, a first precaution can be taken. The rearing of larvae produced by a spawning of dubious quality should be avoided and the standards established for rearing the species should be followed rigorously.

A second precaution consists of limiting bacterial proliferation in its entirety. Particular care must be taken to keep rotifer culture as 'clean' as possible, while maintaining it at artificially high levels (300–400 individuals.ml^{-1}) and transferring it regularly into a completely clean tank with a new medium.

Enrichment of prey EFA content should not be carried out by using an excess of fish oil or over too long a period. An oil emulsion containing lecithin actually acts as a vibrion proliferation factor. Continuous distribution of rotifers over 24 hours with 8 mg of oil emulsion per litre of enriching solution is sufficient. Another solution can be to limit the duration of enrichment to six hours, using 50 mg.l^{-1} of oil emulsion.

13.5.4 Specific treatments

Antibiotics are currently used in mollusc and crustacean hatcheries. Although they are sometimes used for experimental purposes, their use should be suppressed in fish larvae rearing in order to minimise the risk of the appearance of resistant pathogenic strains. On fisheries production sites, the use of antibiotics must be strictly reserved for the treatment of infectious diseases affecting growing fish.

An alternative treatment has been proposed – the input of probiotics. These preparations contain viable cells, usually lactic acid bacteria or Bacillus spores (Chapter 18). These bacteria are not capable of developing in the rearing environment and they must be added continuously into the production environment or via the enrichment solution (at about 10^4 cells.ml of water^{-1}day^{-1}). Their use is still experimental and positive results have been observed that vary depending on the products: for example, an improvement in rotifer production or in growth rate of turbot larvae. Limitations to bacterial proliferation have also been observed, in particular that of vibrions. In cases of experimental infection, turbot larvae survival can also be improved.

Lactic acid bacteria are present naturally in the intestinal flora of fish. Up to 10^4 lactic cells per gram of digestive tract can be counted in a species of bream and up to 10^4 lactic cells are naturally present per gram of turbot larva. Some srains which have been isolated and reintroduced each day as probiotics in the enriching solution at levels of 10^7 cells.ml^{-1} can result in concentrations of 10^7–10^8 cells.g^{-1} of larva. The lactic acid flora can thus exert a real 'barrier effect', limiting in particular the mortality of experimentally infected larvae. Other strains of lactic acid bacteria isolated in fish flora secrete substances that inhibit vibrion growth (antibiotics). Much work is still required before it will be possible to transfer from experimental study to aquaculture practice, but these stocks appear to be a great asset for the control of bacterial equilibrium in rearing units.

13.6 RESULTS

Methods of improving the food quality of rotifers and artemias now allow efficient feeding of fish larvae and production of juveniles of satisfactory quality. Large differences exist between species. For a robust fish such as sea bass, survival at the end of the first month often exceeds 90%. For a more delicate species such as turbot, a survival rate of 20–30% is considered satisfactory. For the majority of species already studied, provision of juveniles to growing on installations is carried out in excess. Production costs are still high and could be reduced. In addition, the methods which have been developed until now allow the production of new species to be envisaged, with minimal research into adaptations.

However, two significant problems remain to be resolved:

- feeding, and thus production, of juveniles relies heavily on the provision to hatcheries of artemia cysts produced naturally. A lack of cysts has already happened in certain economic conditions and will become pivotal with the increase in global aquaculture if other foods for larvae are not developed;
- juveniles produced by intensive production often have deformities that appear to be caused by the feeding or rearing conditions during the larval phase, but the determining factors have not yet been found; as the requirement for quality becomes greater, the numbers of these malformations must be reduced.

13.7 CONCLUSION

Despite the small number of planktonic species used to reconstitute the first levels of the food chain, it is now possible to produce prey of all the required sizes. The nutritional value of these prey varies little and is often insufficient to start with, but can be greatly modified, especially in terms of its lipid content. Using these techniques ensures sufficient reliability for the functioning of the hatcheries in practical terms. However, sustained growth of aquaculture production in the medium term does not appear to be possible, at least for species whose larvae are very small, using these techniques alone. The appearance of totally different methods implicating the increasing use of inert foods appears to be unavoidable.

13.8 BIBLIOGRAPHY

Gatesoupe, F. J. (1999) The use of probiotics in aquaculture. *Aquaculture*, **180**, 147–165.

Lavens, P. and Sorgeloos, P. (2000) The history, present status and prospects of the availability of *Artemia* cysts for aquaculture. *Aquaculture*, **181**, 397–403.

Reitan, K. I., Rainuzzo, J. R., Oie, G. and Olsen, Y. (1997) A review of the nutritional effects of algae in marine fish larvae. *Aquaculture*, **155**, 207–221.

Robin, J. H. (1995) The importance of n-6 fatty acids in the culture of marine fish larvae. *ICES Mar. Sci. Symp.*, **201**, 106–111.

Suantika, G., Dhert, P., Nurhudah, M. and Sorgeloos, P. (2000) High-density production of the rotifer *Brachionus plicatilis* in a recirculation system: consideration of water quality, zootechnical and nutritional aspects. *Aquacultural Engineering*, **21**, 201–214.

Watanabe, T. and Kiron, V. (1994) Prospects in larval nutrition. *Aquaculture*, **34**, 223–251.

14

Artificial food for fish larvae

In intensive salmon culture, the entire production cycle, from fry to spawning fish, is based on the distribution of dry feeds in the form of crumbs or pellets. These feeds provide all the nutrients required for growth and reproduction. Because they are very easily stored and distributed, they have supplanted the live food and fresh/moist feeds used previously. The salmonid's situation is not, however, a general one. For many species, and in particular for marine species, only juveniles and adults can be reared on dry food. The larvae of these species, which have body masses of the order of a milligram and not a hundred milligrams as in trout fry, must currently be fed with live prey such as rotifers and artemia nauplii (Chapter 13). Because of problems already described (Chapter 12), they can only be weaned onto an artificial diet after several weeks. For the rest of the rearing process, the situation is the same as in salmonid farming.

Research has been carried out over the past 20 years to reduce the period over which live prey must be used, using a better understanding of larval behaviour and physiology and improvement in microparticle formulation. The current potential and limitations of the use of artificial diets for freshwater and seawater larvae are presented in this chapter.

14.1 CONDITIONS FOR USING ARTIFICIAL FOOD

14.1.1 Methods of weaning

The age at which weaning is carried out varies considerably depending on species. It also appears to depend more on larval size and the status of rearing methods than on any particular physiological stage. For carp, size at first feeding has changed in the course of a few years from 50 mg to 2 mg, i.e. at the time when the mouth opens. Although it still gives lower initial growth rates, artificial food now gives just as good survival rates (greater than 90%) as live food. This is also the case for other

Table 14.1. Early weaning of some marine fish larvae.

Species	Total length (mm)	Body mass (mg)	Age since hatching (days)	Artificial food (% of diet)	Result
Sea bass	4	0.2	5	100	good
	9	3	20	100	good
Gilthead sea bream	3	0.1	3	50	good
	7	1	25	50	good
	–	–	20	80	good
Sole	4	0.4	2	100	mediocre
	7	3	10	100	good
	–	3	10	100	good
Cod	4	0.6	4	100	mediocre
Red sea bream	4	–	10	90	good
Red drum	3	0.2	5	partial	good
Tropical sea bass	2	–	2	100	mediocre
	–	–	–	partial	good

freshwater species where the body mass of larvae when the mouth first opens is between 5 and 10 mg (white fish, sheatfish) or greater (common grayling). For these species, artemia nauplii provide higher specific growth rates than artificial food over a one or two week period, but thereafter they give similar results.

Very significant progress has also been made over the past 10 years with marine fish such as the sea bass, turbot and sole, the larvae of which weigh 0.2–0.5 mg at the start of their feeding life. It has been possible to lower the age at weaning from 60 days to 30–40 days under production conditions. Very clear gains in survival and especially growth have also been noted, due particularly to the use of specialised feeds. In sea bass, for example, average body masses of 2 g can be obtained a month earlier (after three months instead of four); this has a considerable impact on the economics of the hatchery phase. In the same species, it has been shown under laboratory conditions that it is possible to utilise artificial food even earlier, from 20 days old with larvae of 2 mg, without decrease of survival but with a slight, acceptable delay in growth, not exceeding one week in a 3-month cycle. Currently, it is usually possible to wean larvae from a mass of 2–3 mg, at least under laboratory conditions; this mass is attained at different ages depending on the species and rearing conditions (Table 14.1).

Weaning can be carried out in several ways:

- by having an intermediate feeding phase using frozen or lyophilised zooplankton which is half-way between live artemia nauplii and artificial food. Frozen zooplankton are preserved in blocks of ice, which are placed directly in the tanks, releasing prey as they melt. This plankton is often made up of copepods collected in lagoon stations or from ponds; adult artemias can also be used;
- using simultaneous distribution of live prey and dried feed over a period of one

to several weeks. Mixed feeding can be started very early, before the larvae reach a body mass of a few milligrams. Live prey, including *Brachionus* and artemia, can represent 10–50% of the daily ration;

- using the same method as above, but shortening the overlap between live prey and dried food to two days or even one. This process ensures better size homogeneity within the population.

14.1.2 Distribution methods

Two stumbling blocks to be avoided are overfeeding the larvae and pollution of the rearing environment.

As with live prey, food particles must be presented in large numbers around the larvae. Small particles firstly float and then fall to the bottom of the tanks, following a path which depends on the pattern of water currents. Usually, the larvae are not interested in floating particles or those lying on the bottom but only in those that pass by in their immediate vicinity. In order to maintain a sufficient concentration of particles in suspension, the number of meals must be multiplied, and the period over which feeding is carried out extended over a large part of the daily cycle. The larvae only use a small amount of their body reserves and are very sensitive to starvation and feeding. Their growth potential, which is much higher than that of the juveniles, can only be expressed under conditions where consumption as a percentage of body mass is also much higher than that of the juveniles. However, unlike large fish, larvae cannot ingest their daily ration in only two or three meals. The resting time for food in their digestive tract is very short, usually about an hour and not about 10 hours, as in large fish.

In practical terms and taking into account the inevitable losses, continuous feeding to excess is thus advised. It is almost essential to use feeders, either rotating or conveyor belt. Supplementary feeding by hand is also valuable, especially in the first few days after changing the food, since the behaviour of larvae in relation to the food can be observed.

14.1.3 Adaptation of rearing devices

The risk of polluting the environment is much greater with dry or moist foods than with live prey. This risk is even higher when the particles break up and dissolve more rapidly in the water and when the temperature is higher. The dangers are the proliferation of micro-organisms, blocking of filters and the degradation of water quality. The larvae are more sensitive than the juveniles to the presence of ammonia and nitrites and to an oxygen deficit.

Most classic fish culture tanks used for rearing juveniles are shallow with flat bottoms and a central outflow, and are not suitable for larval rearing. The structure of weaning tanks for larvae requires particular developments adapted to the distribution of artificial food, while taking into account the behaviour of the larvae, which are very sensitive to hydrodynamics: a tank shape which facilitates cleaning and emptying without damaging the larvae and oversized biological filters where water is recycled. For marine fish larvae, first feeding is most often carried out in

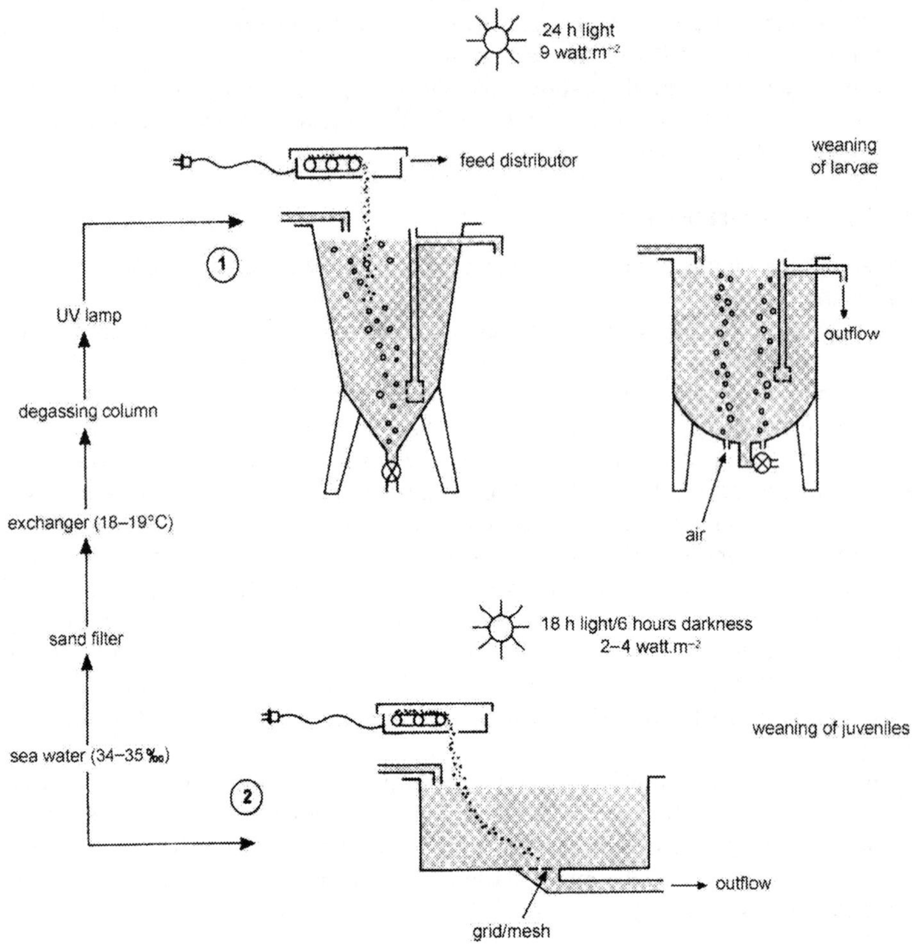

Figure 14.1. Set-ups for weaning larvae 1 and weaning juveniles 2 of marine fish.

deep tanks with a hemispherical or conical bottom, which allows better concentration and collection of waste products (Figure 14.1). Different solutions adapted to the requirements of each species in relation to its stage of development have been proposed: periodic transfer of the larvae from a dirty tank into a clean one, rotating rearing enclosures, removable outflow filters and zones for waste collection. An installation made up of double rearing tanks, which has been used for rearing carp larvae, allows the automatic transfer of larvae from one to the other (Figure 14.2).

Whatever rearing set-up is used, the water dynamics within the tanks must be adapted for the use of artificial foods. Water turnover must be rapid in order to eliminate dissolved substances. It is desirable to lengthen the path of the particles falling from the surface to the bottom of the tank by using a lateral current.

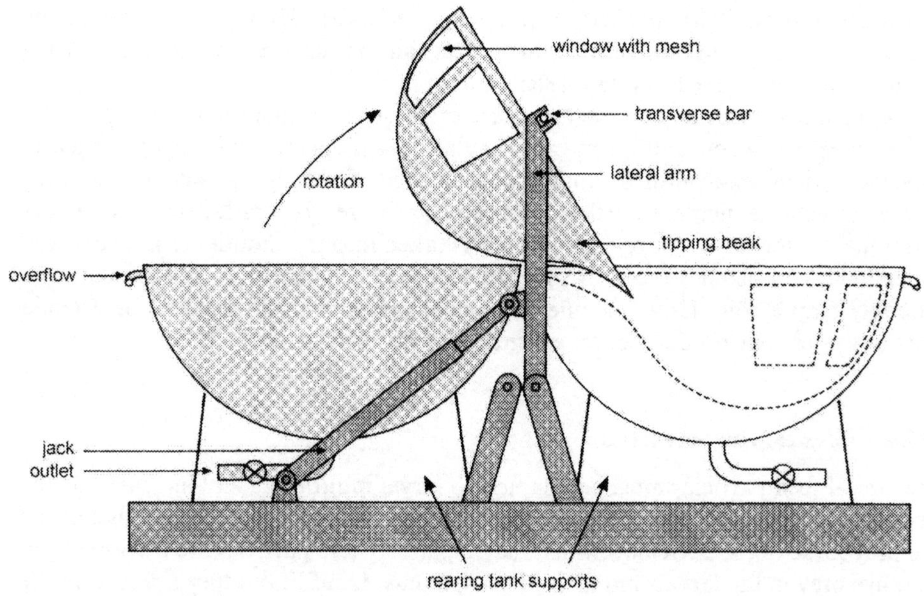

Figure 14.2. Rearing set-up for carp larvae with automatic transfer mechanism (after Charlon, 1990).

However, water speed must remain very moderate in order to avoid disturbing the larvae.

14.2 CHARACTERISTICS OF ARTIFICIAL FOOD FOR LARVAE

14.2.1 Objectives and constraints

The objective is to get the larvae to accept the particles in as short a time as possible; the younger the animal, the shorter the time must be. Rearing conditions which favour detection, capture and intake of particles must be found. Visual and chemical stimuli are important since they affect the rate of acceptance of artificial feed. Food particles must be of the right size and of such a stability in water that they can be broken down in the intestinal lumen but not disaggregate or dissolve in the water.

Larvae appear to locate food particles mainly visually. In most species such as the sea bass, sole and turbot before metamorphosis, perch and yellow pickerel, the larva shows interest in a particle by adopting a particular posture which consists of folding its body into an S-shape before darting towards the prey. The mouth is opened very rapidly which causes aspiration of the volume of water immediately in front of the animal and attraction of the prey into the buccal cavity. The larva can then either swallow the prey or spit it out. Light conditions (intensity, wavelength, location of light sources as well as the physical characteristics of the insides of the tanks) are important for particle catching by the larvae. They are sensitive to the contrast between feed and environment. Certain foods are coloured red using canthaxanthin in order to imitate artemia nauplii.

Attractants, especially amino acids, are often added to the food. Their presence in the water at low concentrations can stimulate larval appetite, which is particularly important when semi-purified foods are being used. Thus glycine, alanine, arginine as well as betaine increase particle consumption by sea bream larvae. The factors that influence the intake or rejection of foods taken into the mouth are not very well understood. Consistency (hard or soft), surface texture (smooth or rough) and taste probably play a role. However, the occasional presence of air bubbles or artemia cysts indicates that food selection is sometimes not very strict.

14.2.2 Physical characteristics

The size of feed particles must be adapted to larval mouth size and increased as the larva grows. In general terms, the size of dried foods must be smaller than that of the live prey (brachionus, artemia) which they replace. These particles are deformed less than live prey in the larva's mouth and oesophagus. In addition, they are susceptible to swelling up in the water and their size at the moment of intake must be taken into account. The particles must be very well calibrated around the optimal size. Too-large particles and aggregates that are formed with sticky feeds are not consumed and are a source of environmental pollution. Particles which are too small and the dust often associated with powdered foods are not ingested either and worsen the pollution. This is also the case for fragments of particles that break down in the water. The features of these three feed categories are shown in Table 14.2.

The particles must also be as homogeneous as possible in composition, in order that the quantity of food taken in at each meal is balanced from a nutritional point of view. This is probably one of the most difficult points in the manufacturing of larval feed, especially when they are very small. Particles of ingredients must be

Table 14.2. Composition of some weaning feeds.

				Start of utilisation (number of days since hatching)	
	Size (μm)	Protein content (% of dry matter)	Lipid content (% of dry matter)	Feed alone	Mixed feeding
Rotifer substitutes	50–100	60–65	20–25	–	10–15
Artemia substitutes	100–200	60–65	20–25	20–30	10–20
Crumbs	200–400	52–55	12–15	30–40	–

much smaller than the feed particles. The segregation of ingredients rich in proteins or lipids into different particles results in differences in density and thus behaviour in the water of these particles. Larvae that catch only certain ones risk having an unbalanced diet.

14.2.3 Manufacturing methods

Particles larger than 200 μm distributed to larvae which are greater than 30 to 100 mg in body mass are similar to the crumbs used in salmonid farming. They are obtained by crushing and sieving larger pellets produced by a pelleting device or an extruder. These granules are prepared using normal ingredients previously sieved using a 150 μm sieve and possibly enriched with attractants.

Particles of between 50 and 200 μm, currently called microparticles, which are destined for the smallest larvae, require particular manufacturing processes. Because of their nutritional quality and size, normal crumbs are not suitable since, in small dimensions, they are neither sufficiently homogeneous nor stable enough in water. Three fabrication processes are used, at least experimentally (Figure 14.3).

- Microbinding: soluble or finely-sieved main ingredients are mixed in an aqueous environment with a carbohydrate (carragheenan, alginate) or protein (gelatin, zein) binding agent which traps the soluble substances and glues together the insoluble ones. The product is dried or lyophilised, then crushed and sieved. This type of microparticle is usually easily accepted but is not very stable in water.
- Micro-coating: an external coating is added to the graded microparticles. This coating can be obtained by soaking in a protein solution (casein, zein) which is then precipitated, or by projection of a layer of fatty material over the particles using a current of hot air.

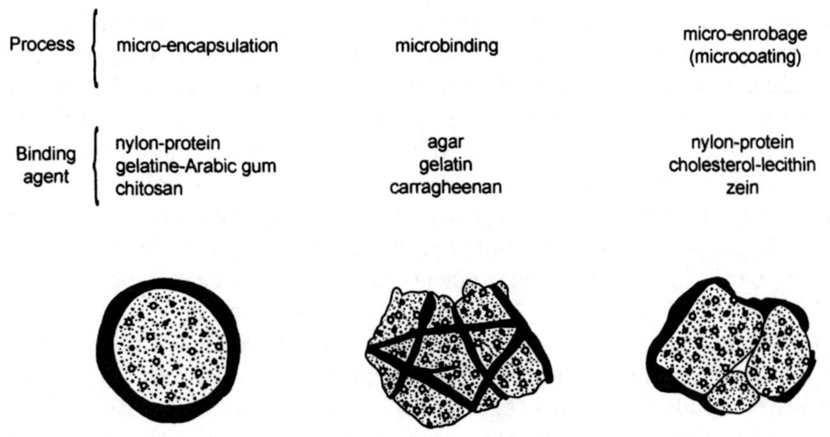

Figure 14.3. Diagram of food microparticles for larvae, manufactured using different processes.

- Micro-encapsulation: this process has been inspired by the methods of presenting drugs and occurs in many different forms. A liquid phase which brings together the nutritive elements in solution or suspension is emulsified in another non-miscible liquid phase. An insoluble film is formed at the interface by a polymerisation reaction (for example nylon) or by phase separation (gelatin–gum arabic–water). The particles obtained after drying usually have very smooth and regular surfaces. The thickness and continuity of the film around them affect their utilisation by the larvae. A thick coating makes them very stable in water but sometimes indigestible.

14.3 COMPOSITION OF FEED PARTICLES

14.3.1 Nutritional requirements

The nutritional requirements of larvae that have a body mass of less than one milligram are very poorly understood. Based on the composition of the yolk, live prey and the larvae themselves, it is assumed that the requirements for proteins, essential fatty acids (EFA) and vitamins of the youngest larvae are higher than those of the juveniles. However, these hypotheses are difficult to prove using live prey which do not easily lend themselves to the testing of deficiencies. Available experimental data mainly concern requirements for n-3 EFA (Chapters 7 and 13). It has been possible to study these using experiments with different enrichment levels of long chain polyunsaturated n-3 fatty acids (HUFAs) of live prey which are deficient in these fatty acids (*Brachionus* reared on baker's yeast and artemia with the freshwater type HUFA composition). Requirements determined for several species of marine fish and expressed as percentage of dry matter (DM) of food are higher (1.5% to more than 3%) than those of juveniles (0.5% to 1%).

For carp larvae of 1.5 to 2 mg, semi-purified diets only contain protein sources such as casein and sodium caseinate and casein hydrolysate may be used as soon as the first meal. These diets produce modest growth (body mass of 100 mg after 21 days' feeding instead of 200 mg with complex diets), but higher survival rates, greater than 90% after three weeks. Improvements are still required, in particular the amino acid (AA) profile, which is far from that of the larvae but which has already provided an approach to studying the requirements for EFAs and phospholipids (PL).

In contrast to marine fish larvae, carp larvae have low requirements for n-3 EFA (less than 0.1% DM of the food) and requirements for n-6 EFA are similar to those of the juveniles (1% of DM). 'Requirements' for PL in carp larvae appear to be met by a 2% content of dietary PL, phosphatidylinositol giving better survival results than purified phosphatidylcholine. In addition, it has been shown that the form in which nitrogen is supplied is important, a mixture of casein and casein hydrolysate giving better results for growth and survival than one or the other on its own. Independently of amino acid content, criteria such as solubility and degree of hydrolysation of dietary proteins must also be taken into account when formulating diets.

For larvae that weigh several milligrams and previously fed with live prey, it has been possible to study certain requirements using artificial diets. Requirements for essential amino acids (EAA) are correlated with the profile of EAA in the body (Chapter 6). The use of artificial diets after weaning has also allowed the requirement for PL to be proven. This requirement is currently the only known qualitative requirement which differs between larvae and juveniles. The capacity for PL synthesis is sufficient in the latter, while it appears to be insufficient in the larvae. It is accepted that the effect of PL on survival and growth does not correspond to a simple suppression of deficiency in either choline or inositol, nor in n-3 or n-6 EFA. The requirement appears to be met by 2–3% purified phospholipids, with phosphatidylcholine and phosphatidylinositol seeming to be more effective than phosphatidylethanolamine.

14.3.2 Meeting requirements and choice of main ingredients

For larvae that still cannot be fed on artificial feed from hatching, it is currently difficult to know whether the limiting factors are the quantities ingested, food digestibility, nutritional deficiency effects or an imbalance between nutrients. In early weaning foods, attempts are made to avoid possible limitations by using diverse raw materials. Certain animal products such as shrimp, squid or beef liver meal are used in larval feeds, but seldom for larger fish (Appendix C, Table C13). Fish ovaries which are rich in PL and n-3 EFA are also used. These products are often lyophilised to better preserve their original quality. As indicated above, very fine crushing is required before their incorporation into microparticles. Cheaper raw materials are also used, such as 'unicellular organism protein' (powders of yeast or bacteria cultivated on various substrates) and fish hydrolysates products (Table C.13).

Feed for larvae also contain pigments (canthaxanthin), attractants (AA) and antioxidants (in particular vitamin E and vitamin C incorporated at high levels, in order to protect the microparticles that have a high surface/volume ratio for oxygen from the air to act on during storage). Various other additives have been tested, such as mammalian pancreatic enzymes, but without any conclusive effects on rearing results.

14.4 CONCLUSION

Manufactured feeds will probably become increasingly important in hatcheries. The development of knowledge about the digestive physiology of larvae and improvement of microparticle manufacturing techniques should allow better adaptation of nutritional input to the digestive and metabolic peculiarities of the larvae. Better feeds could reduce the use of live prey, or even completely replace it, with progressively earlier weaning. This development will very probably involve the modification of hatchery organisation to take into account the constraints that are unique to artificial foods.

14.5 BIBLIOGRAPHY

Charlon, N. (1990) Technologie: automatisation de l'élevage larvaire en eau douce. *Aqua Revue*, **33**, 19–22.

Charlon, N., Durante, H., Escaffre, A. -M. and Bergot, P. (1986) Alimentation artificielle des larves de carpe (*Cyprinus carpio* L.). *Aquaculture*, **54**, 83–88.

Geurden, I., Radünz-Neto, J. and Bergot, P. (1995) Essentiality of dietary phospholipids for carp (*Cyprinus carpio* L.) larvae. *Aquaculture*, **131**, 303–314.

Jones, D. A., Kamarudin, M. S. and Le Vay, L. (1993) The potential for replacement of live feeds in larval culture. *J. World Aquac. Soc.*, **24**, 199–210.

Kanazawa, A. (1993) Essential phospholipids of fish and crustaceans. In: Kaushik, S. J. and Luquet, P. (eds.) *Fish Nutrition in Practice*. INRA, Paris, 519–530.

Person-Le Ruyet, J., Alexandre, J. C., Thébaud, L. and Mugnier, C. (1993) Marine larvae feeding: formulated diets or live prey? *J. World Aquac. Soc.*, **24**, 211–224.

Person-Le Ruyet, J., Fisher, C. and Thébaud, L. (1993) Sea bass (*Dicentrarchus labrax*) weaning and ongrowing onto Sevbar. In: Kaushik, S. J. and Luquet, P. (eds.) *Fish Nutrition in Practice*. INRA, Paris, 623–628.

Radünz-Neto, J., Corraze, G., Charlon, N. and Bergot, P. (1994) Lipid supplementation of casein-based purified diets for carp (*Cyprinus carpio* L.) larvae. *Aquaculture*, **128**, 153–161.

Tandler, A., Berg, B. A., Kissil, G. W. and Mackie, A. M. (1982) Effects of food attractants on appetite and growth rate of gilthead seabream, *Sparus aurata* L. *J. Fish Biol.*, **20**, 676–681.

Teshima, S. I., Kanazawa, A. and Sakamoto, M. (1982) Microparticulate diets for the larvae of aquatic animals. *Mini Review Data File Fisheries Research*, **2**, 67–86.

Watanabe, T. and Kiron, V. (1994) Prospects in larval fish dietetics. *Aquaculture*, **124**, 223–251.

Yufera, M., Fernandez-Diaz, C. and Pascual, E. (1995) Feeding rates of gilthead seabream (*Sparus aurata*) larvae on microcapsules. *Aquaculture*, **134**, 257–258.

15

Digestive physiology of shrimps

Numerous anatomical descriptions of the digestive apparatus of crustaceans have been published over the past centuries. They poorly represent the complexity of specialised organs involved in digestion of species in this group. In addition, they do not take into account – except in a very partial manner – the immensity and diversity of the subbranch of the Crustacea. The first studies of digestive physiology are rather ancient but form the basis of our current understanding. However, they did not become universally applicable until more recent times. While studies of nutritional requirements of crustaceans, using species of commercial interest and especially penaeid shrimps, have multiplied during recent decades, the same effort has not been deployed on fundamental research; comparative physiology remained the main line of interest for biologists and zoologists, and in the phyla of interest to aquaculture to which we are limited, only a few studies have been devoted to enzymology and digestive physiology. In this chapter, the term 'shrimp' is used to cover all decapod crustaceans commonly referred to as 'shrimp' or 'prawn'.

This chapter is devoted to the decapods and concentrates on the penaeid family. The so-called inferior crustaceans, including the Branchiopods, are not considered, despite the considerable importance of *Artemia salina* in aquaculture (Chapter 13).

In many studies, aspects of anatomy and physiology are ignored. The unique characteristics of the crustaceans in contrast to higher animals must be emphasised in order to allow a better comprehension of the principal mechanisms that form the basis of their nutrition; technicians of shrimp, giant freshwater shrimp or crayfish have to take them into account. These characteristics, of which the last five will be specially mentioned, are:

- existence of moults which obligates cycles of laying down of reserves; reutilisation of nutrients;
- existence of periods of physiological fasting during these moults;
- renewal of internal walls of part of the digestive tract;
- unique anatomy of mastication and absorption organs;

- specific mechanisms for digestion and absorption;
- existence of certain unique enzymes, alongside enzymes that are very similar to those found in vertebrates;
- involvement of the digestive gland in various functions and in particular in nutrient storage.

15.1 ANATOMY AND HISTOLOGY OF THE DIGESTIVE TRACT

This tract (Figure 15.1) is classically divided into five parts: mouth, oesophagus, stomach, hepatopancreas and midgut, and hindgut. While this subdivision is entirely justified, the naming of the different parts by reference to the higher vertebrates is much less so and sometimes leads to a certain amount of confusion. Thus, the stomach in crustaceans only remotely resembles that of vertebrates and the digestive gland or midgut gland or hepatopancreas, which is neither a liver nor a pancreas, is not even the equivalent of that found in certain fish in which the liver is penetrated by pancreatic acini.

15.1.1 Mouth

As in other arthropods, the mouth is surrounded by several pairs of appendages which are specialised for chemoreception and prey capture and manipulation: maxillae, maxillulae, mandibles and maxillipeds. These appendages allow the animal to bring food close to mouth, to commence breaking up the food, especially in the decapods, using the maxillipeds to sort particular sizes of food which are the right size for the buccal cavity, and finally to swallow them. The mouth itself has a relatively hard *labrum* which mainly plays the role of a value. Note that in very

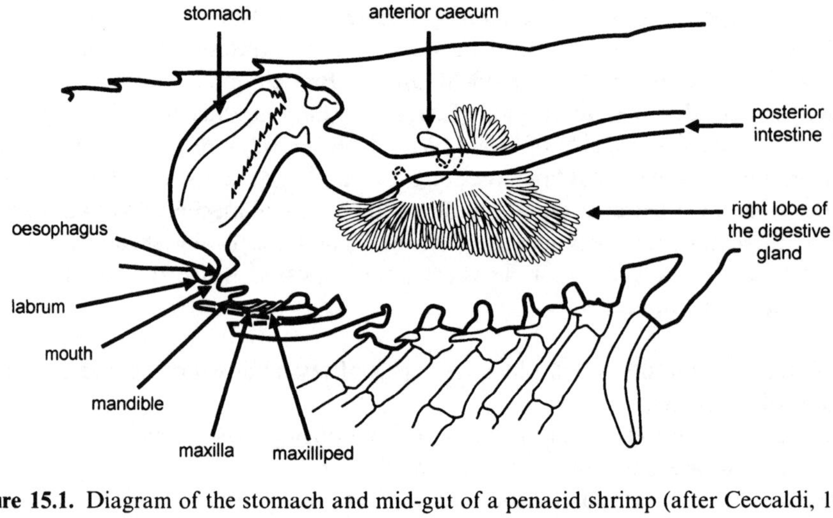

Figure 15.1. Diagram of the stomach and mid-gut of a penaeid shrimp (after Ceccaldi, 1994).

young larvae, the antennae, antennules and mandibles are used primarily for swimming.

15.1.2 Oesophagus

As in other animals, the lumen of the digestive tract is part of the external environment and its walls constitute a continuation of the epidermis. In crustaceans, this situation is particularly obvious since the walls of the anterior as well as the posterior parts of the digestive tract are covered by a thin layer of chitin, a major constituent of the exoskeleton. This cuticle is renewed during each moult. The oesophagus thus appears to be a conduit with walls made up of supple chitino-protein complexes. It is relatively short and straight in species important to aquaculture; its cross section is generally X-shaped. Waves of contraction run through it that are quite similar to, though simpler than, those found in the vertebrate oesophagus.

15.1.3 Stomach

In crustaceans the stomach, like the oesophagus, forms part of the anterior gut in broad terms (Figure 15.2). Its exterior structure is relatively simple and relatively rigid. During larval development, the presence of hard structures (bristle, spines and brushes) on the internal walls of the stomach have been described, which masticate the food in the same way as teeth do in vertebrate mouths. The development of these structures results in the 'gastric mill'. In the decapods, the stomach is made up of two parts separated by a marked constriction which has a sort of valve; its complexity varies between groups. Extending the usage of the terminology employed for the higher vertebrates, these two parts have been called the cardiac chamber and the pyloric chamber. Like the oesophagus, the cross-section of the cardial chamber is X-shaped as a result of the presence of folds in the form of grooves and thickenings of the walls. The antero-ventral part of this chamber contains a ridge with a row of hard, pointed projections called ossicules, or teeth where they are large (Figure 15.2). These calcified and articulated structures are moved by the action of specific muscles situated on the external walls of the stomach bag. Their number and complexity are characteristic of the species – there are 14 in the penaeids.

Neurophysiological studies have demonstrated a complex innervation (six groups of different nerves) allowing the movement of the different parts of the gastric mill, contractions of the cardio-pyloric valve as well as peristalsis of the main part of the pyloric stomach. Chemoreceptors and mechanoreceptor organs also play their part in these processes. All these parts, along with the specialised muscles, constitute the gastric mill, a crushing organ which also has a powerful unpaired dorsal tooth. The teeth and ossicules are shed at each moult, at the same time as the exoskeleton, the walls of the oesophagus and the posterior intestine. The complexity of the gastric mill varies between families and appears to be inversely linked to that of the mandibular organs: the more effective they are at breaking down the food, in addition to simple prey capture and manipulation, the simpler the gastric

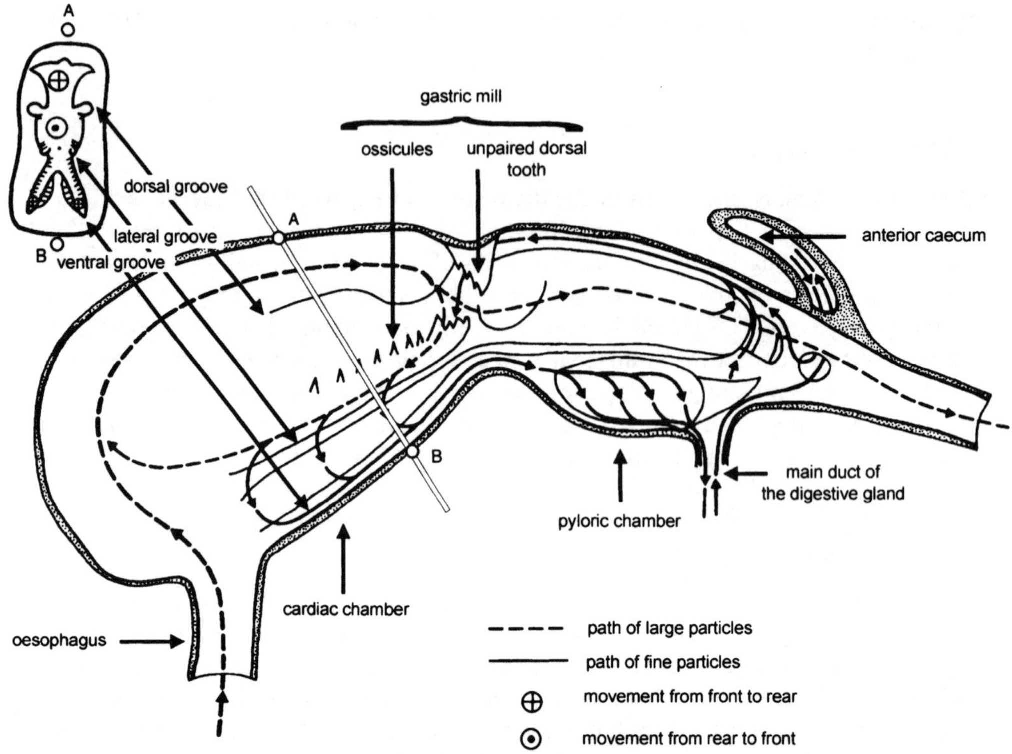

Figure 15.2. Diagram of the probable circulation paths of food particles and fluids within the stomach (after Dall and Moriarty, 1983). The section in the top left corner corresponds to a section through the organ from A to B.

mill, and vice versa. It is well developed in the brachyurids (crabs), moderately developed in the penaeids and very poorly developed in the carids, which include the common prawn and the giant freshwater prawn (*Macrobrachium*).

Calcified structures extend into the pyloric sac that forms the start of the midgut or mescuteron. Hard structures also exist here in the form of comb teeth as well as bristles which are sometimes very complex, and tubercules. The sac itself also contains tight folds. Folds, spines and bristles only allow the finest particles of the food bolus to pass; in other words, there is crushing by the teeth, compression by the stomach movements and filtration through the combs. The pyloric stomach compresses, sorts and filters. The finest particles thus cross a ventral, posterior filter sometimes called the filter-press from whence they are directed towards the hepatopancreas, an absorptive organ. Because of this, some zoologists consider crustaceans to be filter-feeding organisms, under the same heading as other marine invertebrates such as bivalve molluscs. If mastication and filtration were carried out by the mandibles, a large part of the food would be lost by dissolving into the surrounding environment.

The path of food particles to the interior of the stomach is complex and cannot be reduced to simple sequences: crushing, compression and filtration. Despite the difficulties involved in observation, detailed studies have been able to describe part of the particles' path. The most likely paths are shown in Figure 15.2. Food coming from the oesophagus moves firstly from front to back in the upper groove until it reaches the gastric mill. After grinding and a first mixing with digestive enzymes, the chyme returns to the forward part, but during this movement the fine particles separate off towards the rear in the ventral gutter until they reach the filter-press. The finest particles, only a few micrometres in size, can pass this and reach the opening of the digestive gland or hepatopancreas in the pyloric chamber, where they are digested and then absorbed through the internal lumen of the tubules. Instead of falling into the ventral channel, particles which have not been sufficiently crushed are subjected to another cycle: they reach the anterior part of the cardiac chamber, then pass again towards the rear in order to go through the gastric mill again. This reprocessing of food can continue until it is sufficiently crushed. Particles that are too hard or indigestible, however, finally enter the pyloric chamber, but they cannot pass across the filters and pass into the midgut; they constitute the faeces.

The other unique features of the stomach are an external musculature, a quasi-autonomic innervation and a particular kind of coating. In most species, the pH of the stomach contents is neutral or slightly alkaline, which is not surprising when the pH of seawater is about 8.3. The organ does not possess any glands or cells that secrete acids or enzymes. Slightly acidic pHs have, however, been indicated in the lobster. Despite this, crustacean stomachs do not correspond to the usual definitions for vertebrates to which physiologists are accustomed.

15.1.4 Digestive gland

The hepatopancreas, is a massive organ made up of two symmetrical lobes. It is situated in the dorsal part of the body, immediately below the heart. It represents 2–6% of the body mass and in the adult is formed of hundreds of blind-ending tubules opening into two symmetrical chambers which open into the pyloric chamber of the stomach. Each tubule has a double network of longitudinal and circular muscles that allow peristaltic and contraction movements, ensuring the movement of the liquid phases within the tubules.

In the larva, the hepatopancreas becomes differentiated in the form of diverticuli that appear from each side of the primitive stomach, which is then a simple bag. During larval development, these tubes elongate and multiply greatly. Connective tissue surrounds all these structures and a network of muscle fibres forms to shape the definitive gland. The fundamental structure of the tubules never differs very much from the primitive organisation and the gland remains as a collection of blind-ending tubules that converge towards the mesenteron, connected by a collecting bulb (Figure 15.3).

Within the tubules, as in the acini of secretory glands or in the intestinal villi of vertebrates, a distal zone can be distinguished in which there is cellular differentiation from embryonic cells, then one or more functional zones, secretory and

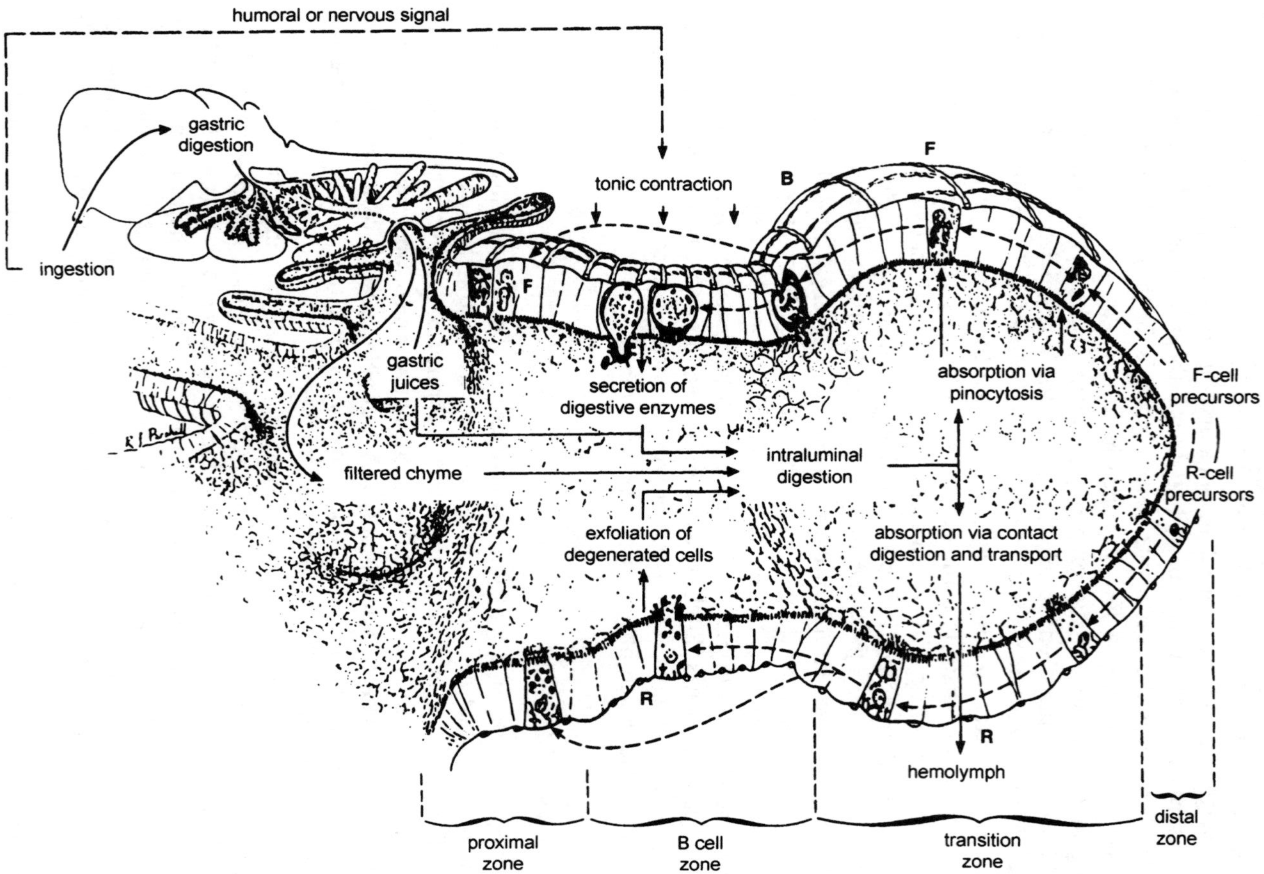

Figure 15.3. View 'in section and perspective' of tubules of the digestive gland (after Dall and Moriarty, 1983).

absorbent and finally, at the proximal part of the tubules, a zone in which the cells degenerate and are rejected into the lumen of the digestive tract. Nomenclature, classification and function of the cells making up the tubules have yet to be agreed unanimously. However, a certain consensus has been reached around the interpretations described below.

E cells (embryonic) in the apical proliferation zone give rise to three main cell types: F cells (fibrillar), B cells which are secretory (from the German *Blasenzellen*, cells in bulbs) and R cells (from the German *Restzellen*, reserve cells). Other cell types have been described in the tubules, in particular M cells (midget cells). According to the most recently accepted theory, there might be only two types of cell family produced from the proliferation zone: F and B cells on the one hand, and M and R cells on the other. F cells might be intermediate between E and B cells and M cells intermediate between E and R cells. F cells appear to secrete enzymes. At a subsequent stage, that of young B cells, they are the site of endocytosis. When they are mature, these same cells are characterised by a very large central, ovoid vacuole which occupies almost the whole cellular space; thereafter they break down and their contents are released into the tubular lumen. B cells are thus secretory. R cells contain significant amounts of lipids, as well as glycogen, calcium, zinc and copper and thus certainly play a role in storage, for nutrients either originating from food or from the destruction of the animal's own tissues during the biological cycle prior to moulting. These cells, like their precursor M cells, also appear to be capable of significant absorption. The energy metabolism of crustaceans relies much more on lipid utilisation than on a glycogenic cycle.

Movements of particles within the interior of each tubule are ensured by the successive contraction and relaxation of the organ and of each tubule. The mechanisms involved are still not fully understood, since the fluids contained in the tubules are partly discharged into the stomach to allow the start of enzymatic digestion while, apparently almost simultaneously, the chyme being prepared for digestion is aspirated into the tubules for enzymatic digestion and can then straight away be followed by the absorption of the products of biochemical digestion. Recent observations on larvae show rapid passage of food particles in the digestive process from some tubules that contract strongly towards tubules on the opposite side, due to the action of the valves situated beside the filter-press. It is not known if similar mechanisms exist in the juveniles.

15.1.5 Midgut

This rectinlinear part of the digestive tract extends from the pylorus to the rectum. Although its surface area is increased by the presence of quite large caeca or diverticuli and the apex is covered in well-developed microvilli, it is not comparable to the vertebrate intestine with its convolutions and villosities. Nerve cells, haemocytes and probable endocrine cells can be distinguished in the epithelium. The great uniqueness of the epithelium in this area is its ability firstly to secrete mucus which coats the solid waste produced by the stomach and then, by means of an annular structure, a pellicule of chitin forms a cylinder which envelopes the faecal mass; this is the peritrophic membrane of the excreta. This permeable membrane does not impede the absorption of residual nutrients from the faeces. Its role has not yet been elucidated, but may be analogous to that of the glycocalyx in vertebrates.

15.1.6 Hindgut

The distal part of the digestive tract is short and straight and probably plays the role of a rectum. However, its epithelial cells contain many mitochondria, as is found in the distal part of the fish intestine and suggests a similar role in ionic regulation. Its musculature suggests that the decapods are, like terrestrial isopods, capable of drinking through the anus.

15.2 FUNCTION OF THE DIGESTIVE APPARATUS

15.2.1 Chemoreception and feeding behaviour

The anatomy of the buccal appendices and the mouth itself results in very characteristic feeding behaviour. Decapod crustaceans are 'nibblers' that can only swallow their prey whole under exceptional circumstances, for example when they are provided with feed in the form of very small diameter cylinders. In other cases, the time required to gnaw the prey is longest when the prey is hard and the particles large. Whether it is mainly achieved by the peri-buccal appendages or finished by the stomach, this mechanical phase of food utilisation is required for good digestive efficiency.

The manner in which crustaceans make for their prey and decide to swallow them is partly determined by the external chemoreceptors located on the appendages. The chemical nature of the substances that play the role of attractant, phagostimulant or the inverse, of repulsion, is not yet known except only very partially (it has been shown that amino acids, several organic bases, nucleotides and nucleosides as well as hexoses all play a role). Feeding strategy certainly varies a lot between carnivorous, detritivorous and omnivorous crustaceans, although this usual form of classification is very blurred.

15.2.2 Digestive transit

The different phases of digestive transit have been mentioned alongside the anatomical descriptions of the different parts of the digestive apparatus. The four most unique stages are the repeated mastication of particles until they are sufficiently crushed, filtration of the chyme, the to-and-fro movement in the tubules of the digestive gland and finally the coating of the faeces by the peritrophic membrane.

Very few quantitative studies of transit times have been carried out. It is still not possible to state rules as has been done for fish. However, it is clear that entire transit times are remarkably fast: in 20 g tropical penaeids, at 28°C, faeces emission starts about 90 minutes after food is taken and ceases about four hours after ingestion.

Transit regulation is also poorly understood; the existence of peptides with hormonal activity, as found in the digestive tract of vertebrates, has been suspected but not demonstrated.

15.2.3 Digestion and absorption

In crustaceans, it is quite difficult to separate enzymatic digestion, i.e. hydrolysis of macronutrients, from absorption (transport of nutrients across the walls of the digestive apparatus) because only part of the digestion occurs in the lumen. It must also be remembered that the entire digestion is enzymatic, as there is no acid secretion to allow the chemical digestion of proteins, peptides or 'labile' carbohydrates.

15.2.4 Digestive enzymes

Despite old and recent research, an exhaustive list of crustacean digestive enzymes has not been made using modern techniques, even for one species. This section therefore refers to generalities.

It has long been considered that the mode of enzyme secretion is very different to that of vertebrates, with none being secreted in the form of zymogen (Chapter 4). However, the existence of zymogens has recently been clearly demonstrated, although their mode of activation is still not known. There are probably equivalents of the membrane-bound enzymes of the vertebrate intestine (which hydrolyse fragments of macromolecules), in particular in the R cells of crustaceans, but they are hardly known even when several activities of this type are demonstrated. The role of lysosomal enzymes, which act during intracellular digestion, is certainly greater than in the higher animals, but it is difficult to distinguish them from the enzymes produced by the lysis of B cells.

The main known enzymes are listed in Table 15.1. It is very similar to that established for agastric fish.

Among the proteases, which are always well represented, pepsin is never present. In contrast, trypsin is synthesised in large quantities and represents up to one-third of the soluble proteins in the digestive gland. The molar masses appear to vary depending on species (from 25 to 36 kDalton) and the functional properties of the molecules are not exactly the same as those of vertebrate trypsins. Crustacean trypsin is more active on non-denatured proteins than on the same, denatured molecules. Some authors even refer to pseudo-trypsins or trypsin-related enzymes, although Indian shrimp trypsin is 50% analogous to that of cattle. It is the main protease found in crustaceans. The activity of chymotrypsin has long been underestimated, because it has been measured using an unsuitable substrate. It is now known that it is very high in the penaeids. Some 'collagenases' are in fact chymotrypsins. Another, very different enzyme, astacin, has been found in numerous crustaceans, while it is still unknown in vertebrates. It is a metallo-protease and not a serine-protease and its spectrum is very wide.

The activity of these endoproteases is complemented by that of a group of exopeptidases which are very similar to those of the vertebrates: carboxypeptidases

Table 15.1. Main digestive enzymes of decapod crustaceans.

Category	Enzyme	Comments
Endoproteases	trypsin	very high activity
	chymotrypsin	high activity, wide spectrum
	astacin*	very wide spectrum
Exoproteases or Exopeptidases	carboxypeptidases A and B	
	aminopeptidases, dipeptidase	
Glucidases	amylase	
	chitinase	
	laminarinase	poorly understood
	cellulase	presence uncertain but probable
	other glucuronidases	
Glucosidases	maltase	probable
	chitobiase	probable
	laminaribiase	probable
	cellubiase	
	saccharase	
	mannosidase	
	other glucosidases	rather unspecific
Lipases	lipase equivalent to pancreatic lipase (?)	poorly understood
	esterases	numerous
Various	ribonuclease	
	deoxyribonuclease	
	phosphatase	

* For some years this enzyme was called 'protease with a weak molecular mass' because it was wrongly believed that the latter was 11 kD. The name 'astacin' is probably not definitive.

A and B, aminopeptidases and dipeptidases. In the lobster, the main proteolytic activity is due to the cathepsins, lysosomal enzymes which are released by the lysis of cells of the digestive gland.

There are four glucosidases: amylase ($\alpha 1 \rightarrow 4$ glucanase), chitinase ($\beta 1 \rightarrow 4$ N-acetyl-glucanase), laminarinase ($\beta 1 \rightarrow 3$ glucanase) and cellulase ($\beta 1 \rightarrow 4$ glucanase). The first two appear always to be present. Chitinase activity seems to be reduced, however. It only achieves partial digestion of chitin by shrimps that ingest the moulted shells of individuals living in the same ecological niche, or the teguments of other dead or living crustaceans. The laminarinases that hydrolyse the laminarin of algae and some other marine organisms are also poorly understood (several enzymes are perhaps confused under this name), but they appear frequently, much more than the cellulases whose action has only been truly demonstrated in a wood-boring crustacean and several other primitive species. Several experimental results, however, point towards significant cellulase activity in the giant freshwater shrimp which is known to be relatively herbivorous.

Hydrolase activity is completed by that of enzymes which hydrolyse the resultant dimers: maltose is hydrolysed by a maltase (α-glucosidase), chitobiose by a chitobiase (β-N-acetyl-glucosaminidase) and cellubiose by a cellubiase (β-glucosidase). Laminaribiase has not been studied, as far as we know, but other enzymes which hydrolyse low molecular weight carbohydrates certainly exist and some have been demonstrated occasionally – saccharase, α-mannosidase, α-fucosidase, etc. Glucaronidases which take part in the digestion of carbohydrate residues from glycoproteins have also been found.

In comparison to the enzymes described above, crustacean lipases have been studied very little. The lipase-colipase couple that already exists in one branch of the invertebrates, the molluscs, does not appear to have been clearly found although lipase activities similar to that of pancreatic lipase in vertebrates have been found and even quantified. There are no bile salts in crustaceans, but they do have compounds with comparable tensioactive properties. These are usually formed from a fatty acid with 22 carbon atoms and a dipeptide containing mainly taurine, as in the combined bile salts of vertebrates. Many esterases have been found, but their location, specificity (probably weak as in the other phyla) and especially their precise role are poorly understood.

Among the various enzymes, the phosphatases, deoxyribonucleases and ribonucleases usually have high activity both in the herbivores with symbiotic flora and in the detritivores, i.e. in animals whose diet includes many bacteria and other microorganisms.

Although many enzymes have been described qualitatively, quantitative data are much rarer. It is generally assumed that there will be a recognisable relationship between the feeding behaviour of the animal in the wild and the enzyme activity responsible for the hydrolysis of the most frequently-found components of the diet. Thus 'carnivorous' species, such as the Japanese shrimp, would have a higher protease activity than omnivorous species, which would have higher amylase activity. This apparently logical relationship has not, however, been convincingly established. It appears that, in reality, activities vary enormously for a given species, in relation to the qualitative (and quantitative) composition of the diet, activity being higher if the diet is rich in the macronutrient involved, and vice versa. In other words, the activity of digestive enzymes adapts to the composition of the compound diet ingested. It should not be concluded, however, that the activity of digestive enzymes is correlated with nutrient requirements.

For this group of enzymes, the optimal pH range is between 5.5 and 9, with lysosomal enzymes close to 5.5. As is found in fish (Chapter 4), the optimal temperature range is probably close to the internal temperature of homeotherms and not the mean temperature of these ectotherms.

15.2.5 Digestive flora

The only major studies on the digestive flora of crustaceans refer to primitive taxa with herbivorous tendencies. In the decapods, the genera *Vibrio*, *Moraxella* and *Flavobacterium* in particular have been found. However, bacterial populations appear to be neither stable nor characteristic of the host species. No correlation between abundance or nature of the flora and the natural food of the species has been found. All this suggests that the nutritional role of the digestive flora is even more reduced in the crustaceans than in reared fish (Chapter 4), which is not surprising given the brevity of transit times.

15.2.6 Routes of absorption

The different mechanisms involved in the absorption of nutrients by the digestive apparatus are certainly the same as those found in other animals (Chapter 4), but the crustaceans show at least one difference from the vertebrates: the secretion of enzymes and absorption are carried out within the same organ, the digestive gland, the cells of which do not appear to be very specialised.

Absorption of macromolecules by endocytosis (Chapter 4), a standard phenomenon in primitive animals, also exists in the crustaceans; the first stage of this process has been demonstrated using electron microscopy in young B cells, i.e. old secretory F cells. The following stage (intracellular digestion) is visible, but the transport of nutrients toward the baso-lateral cell wall has not been described and the part of the food which is rejected into the lumen of the tubule during disaggregation of the B cells is ignored. The quantitative significance of this route of absorption remains unknown, but some authors estimate that it exceeds that of absorption using transport mechanisms.

The secretion of digestive enzymes by F and B cells leads to significant extracellular digestion. Therefore, from all the evidence there must be nutrient absorption by passive diffusion, active transport or facilitated diffusion. This absorption can be carried out *a priori* by the epithelial cells of the tubules with the exclusion of the embryonic cells which have no microvilli. It appears that it is the M cells as well as young R cells that contribute most to this function, as they have well-developed microvilli. During R cell development, the absorbed nutrients serve to constitute reserves. Nutrients carried in the haemolymph also form part of these reserves, but the nutrients of food origin certainly contribute the majority as has been shown by the development of formations situated between the nucleus and the brush border.

Transport of water and mineral ions appears to occur in the posterior intestine, as in fish; as in the latter, this function cannot be considered only from the nutrition angle but should also be considered with regard to osmoregulation.

15.3 DIGESTIVE BALANCE AND REGULATION

15.3.1 Digestibility of main nutrients

A few studies have been carried out on crustaceans, distributing animal prey or plant material to them. More recently, measurements of digestibility have been carried out using compound diets or the ingredients which make them up, firstly with palaemonid shrimps, then mainly with penaeid shrimps. With the exception of some old studies, measurements of digestibility have been carried out using the indirect method, i.e. using an inert marker and partial collection of faeces.

Although the emission of faeces enveloped in a nitrogen-rich peritrophic membrane could in theory increase the endogenous loss of nitrogen, the digestibility values measured are usually high. The digestibility coefficient (ADC) of proteins are often greater than 90%, sometimes greater than 95%. Those of lipids are of the same order of magnitude, at least in two species, the tiger shrimp (*Penaeus monodon*) and the blue shrimp (*P. stylirostris*). These values are comparable to those obtained in fish. Digestibilities measured for starch under rigorous conditions have relatively high values, of the same order of magnitude as mammals (70–94%), i.e. much higher than those known in fish, at least for raw starch.

Marked differences in digestibility have definitely been observed for the three main types of macronutrient, although only a limited number of observations have been made. However, among the classic sources of proteins, plant meals or cakes, animal meals and purified products, almost the same hierarchy of digestibility exists as in the higher vertebrates or fish. Similarly, for sources of starch, botanical origin and technological treatments have very similar effects on digestibility as are known for mammals and birds.

There remains, however, an area which is much less understood in crustaceans than in fish (Chapter 4) or, *a fortiori*, in the higher vertebrates; that of the regulation of digestibility. The constancy of digestibility of a nutrient, whatever the food supply of the same nutrient, has been observed but been little studied. It is known that digestibility can increase with the animal's age, in post-larvae at least. The effect of abiotic factors has been studied very little. It cannot be confirmed, therefore, that nutrient digestibility has the same constancy in crustaceans as is known to exist in vertebrates.

15.4 CONCLUSION

The study of the anatomy and digestive physiology of crustaceans has revealed profound differences from vertebrates as well as surprising and unique features. The digestive gland, with its unique structure and functions, has no parallel in the vertebrates, even in fish which have a 'hepatopancreas'. In contrast, there is a striking analogy between this phylum and fish larvae or even fish with no stomach. All these studies remain very partial and coordination between basic research and specialised research would certainly allow great progress in this sector.

Considered in its totality, the digestive system of crustaceans appears at first sight to have a comparable efficiency to that known for the vertebrates, for proteins, lipids and especially carbohydrates. This first impression must however be treated with caution: very few measurements have been carried out under rigorous conditions, and those concerning nutrients for which crustaceans have a specific requirement (in the broad sense) such as chitin, phospholipids or cholesterol are even fewer; possible interspecific differences remain almost unknown.

15.5 BIBLIOGRAPHY

Ceccaldi, H. J. (1994) Appareil digestif: anatomie et physiologie. In: Grassé, P.-P. (ed.) *Traité de Zoologie. Anatomie, Systématique, Biologie. VII. Crustacés 1. Morphologie, Physiologie, Reproduction, Systématique.* Masson et Cie, Paris, 487–527.

Ceccaldi, H. J. (1998) A synopsis of the morphology and physiology of the digestive system of some crustacean species studied in France. *Rev. Fish. Science.*, **6**, 13–39.

Dall, W. and Moriarty, D. J. W. (1983) Functional aspects of nutrition and digestion. In: Mantel, L. H. *et al.* (eds.) *Internal Anatomy and Physiological Regulation.* Academic Press, New York, 215–261.

Van Wormhoudt, A. and Bellon-Humbert, C. (1991) Bases biologiques de la culture des crustacés. In: Barnabé, G. (ed.) *Bases Biologiques et Écologiques de l'Aquaculture.* Lavoisier, Paris, 212–270.

16

Nutrition and feeding of penaeid shrimp larvae

The sequence of development of penaeid shrimp larvae was described by Hudinaga (1942), who had just achieved the laboratory rearing of Japanese shrimp, *Penaeus japonicus*. However, it is only since 1975 that studies of nutritional requirements and larval digestion have multiplied, studies destined to support the significant development of hatcheries throughout the world. It has rapidly become apparent that shrimps, which are pelagic organisms during their ontogeny, have slightly different feeding behaviour and nutritional requirements than post-larvae and juveniles which are generally benthic.

16.1 LARVAL DEVELOPMENT AND ONTOGENY OF THE DIGESTIVE TRACT

16.1.1 Endotrophic phase

Between spawning and hatching, embryonic development only lasts about 15 hours and the nauplius, an organism about 300 µm in size, appears at hatching. In contrast to other decapods, penaeid shrimps carry out all their larval development outwith the egg, which makes this zoological group an ideal model for studying the occurrence of the different digestive functions during ontogeny. The six nauplius stages (N1 to N6) occur in 48 hours; the animal still has no digestive tract, growth and ontogeny only being ensured by vitelline reserves. Lipids represent 20% of the dry matter of N1 and only 8% of that of N6; they constitute the principal energetic substrate of this endotrophic phase (Figure 16.1). The biochemical composition of the egg (Table 16.1), linked to the feeding of the broodstock, determines the capacity of the organism to develop correctly. A high concentration of n-3 series long chain unsaturated fatty acids (HUFAs), important constituents of biological membranes, is required for sustaining rapid cellular development during both embryonic and

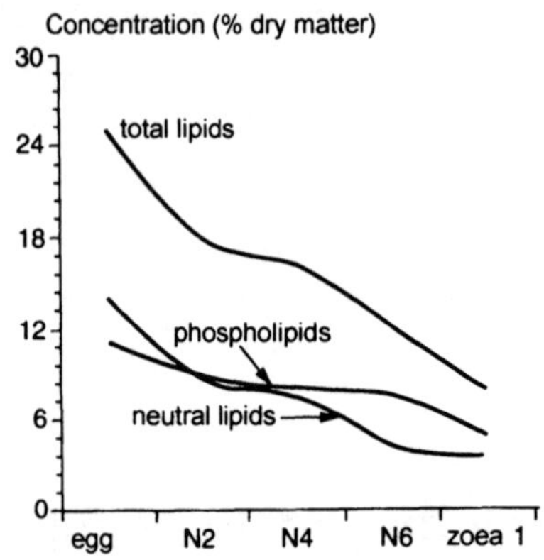

Figure 16.1. Variation in total lipid, phospholipid and neutral lipid content during the endotrophic phase in *Penaeus indicus* (after Cahu *et al.*, 1988).

Table 16.1. Concentrations of some components of the penaeid shrimp egg expressed in relation to dry matter (Cahu *et al.*, 1995).

Components	Concentration
Proteins	50–55%
Total lipids	22–26%
of which phospholipids	8–12%
neutral lipids	10–14%
n-3 HUFA	3.5–4%
α-tocopherol	0.3–0.6 mg/g
Ascorbic acid	0.4–0.8 mg/g

post-embryonic stages. This concentration has been evaluated at 4% of the dry matter in the egg. Vitamins C (ascorbic acid) and E (α-tocopherol) must also be present in the egg at a sufficient concentration to protect intracellular and membrane lipids from peroxidation.

16.1.2 Exotrophic phase

The organism develops through a series of moults which may or may not be accompanied by metamorphosis. A linear and functional digestive tract opens at the start of the zoea phase which consists of three stages (Z1 to Z3). The organism is therefore a filter feeder and mainly a herbivore. In the wild, the zoea feeds on phytoplankton, made up of very varied species and mainly diatoms. This is followed by the mysis stage (M1 to M3), preceding metamorphosis into post-larva (PL). The larvae orient

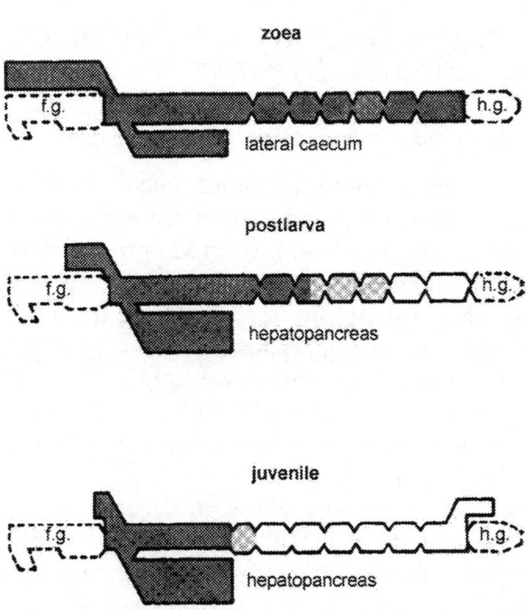

Figure 16.2. Distribution of alkaline phosphatase activity in the digestive tract of *Penaeus setiferus* during larval development. This enzyme is characteristic of sites of membranous digestion and absorption. Dark shading: high activity; light shading: low activity; white: no activity. f.g.: fore gut; h.g.: hind gut. The mid-gut is in the centre. Note that during ontogeny the lateral caecum transforms into the hepatopancreas. Adapted from Lovett and Felder, 1990.

towards a carnivorous diet, capturing various prey which, in the wild, are mainly copepods. Depending on the temperature, at between 25 and 32°C it takes 10–12 days from hatching to the first post-larval stage at the time animals became benthic.

This morphological and behavioural change is accompanied by huge anatomical and functional rearrangement of the digestive tract. This phenomenon has been described for *Penaeus setiferus* by Lovett and Felder (1990). In these larvae, the functions of synthesis and secretion of 'pancreatic' enzymes are carried out by the anterior and lateral digestive caeca situated in the proximal part of the digestive tract; absorption occurs along the whole length of the digestive tract. At the moment of metamorphosis, between stages M3 and PL4 (4-day-old PL), there is degeneration of the anterior caeca. Thereafter, the lobes of the hepatopancreas or digestive gland ramify from the lateral caeca and considerably increase in volume to give a definitive gland. Only after 3–4 weeks of post-larval development does the digestive tract become differentiated and acquire its adult form. An adult mode of digestion commences, which is quite different from that described for vertebrates (Chapter 15 and Figure 16.2). A significant increase in transit time in the digestive tract is observed after metamorphosis.

16.2 OCCURRENCE AND LEVEL OF ACTIVITY OF DIGESTIVE ENZYMES DURING DEVELOPMENT

16.2.1 General profile of enzyme activity during development

Digestive enzymes are present before the mouth opens: trypsin and amylase have been found from the nauplius stage and even in the egg. All the enzymes studied, trypsin, amylase, carboxypeptidase and esterase, present quite similar activity profiles throughout development: a rapid increase in specific activities occurs during the zoea stage, then a sharp drop occurs during the mysis stage, reaching a minimum several days after metamorphosis (Figure 16.3). Thereafter, activity increases progressively during the post-larval stage. The lowering of enzyme activity observed at metamorphosis coincides with the degeneration of the anterior digestive caeca. The final increase in activity corresponds to the formation of the many tubules of the hepatopancreas.

This activity profile of digestive enzymes in penaeid larvae is similar to that described during post-natal development in more evolved phylogenetic groups. Fish and even mammals show a marked increase in digestive enzyme activity during their first days. Thereafter, specific activities fall after a certain number of

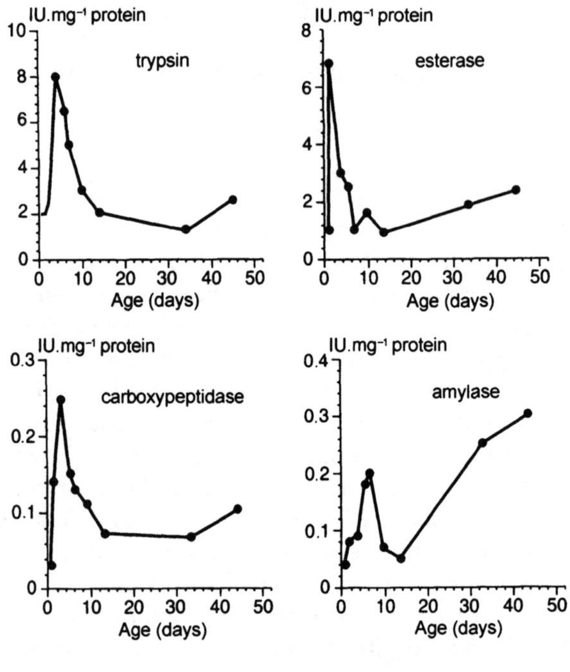

Figure 16.3. Variation in specific activities of some digestive enzymes during larval development in *Penaeus setiferus* (adapted from Lovett & Felder, 1990). IU: international unit of measure of enzyme activity.

days, a period which coincides with the appearance of adult-type digestion (Chapter 12).

16.2.2 Adaptation of enzymatic activity to the food

Experiments on feeding *Penaeus vannamei* larvae with artificial diets in the form of microparticles have shown a positive correlation between the amount of trypsin synthesised and the protein content (casein) of the feed (Figure 16.4a). In addition, the nature and molecular form of the nitrogen input affects the synthesis levels of different proteases: squid meal specifically stimulates chymotrypsin activity and concentrated soluble fish protein (FCP), a source of already hydrolysed proteins, causes a decrease in trypsin and chymotrypsin levels (Figure 16.4b). It also appears that amylase synthesis is stimulated more by native maize starch than by soluble starch. From the earliest stages (zoea), the larvae are capable of modulating their enzyme synthesis in relation to the concentration and nature of the nutrients. Similarly, larvae reared following a classical feeding sequence – including micro-algae during the zoea stage, then artemia nauplii in increasing quantities during the mysis stage – firstly secrete large amounts of amylase and then large amounts of proteases. The amylase/protease ratio, which is often used to describe a food as herbivore type or carnivore type, decreases tenfold between the zoea and mysis stages. An increase in chitinase activity linked to feeding on zooplankton is also observed from the mysis stage.

Feeding using microparticles produces lower enzyme levels in the larvae than those obtained in animals fed sequentially with algae and artemia. However, the addition of the algae *Chaetoceros gracilis* into the rearing environment at low concentrations (10,000 cells/ml) results in a high increase in enzyme activities in larvae fed with microparticles. This complementary algal supply would stimulate enzyme production and, according to some authors, would explain the improvement in growth obtained when feeding larvae with a mixed diet of microparticles and algae. This phenomenon evokes the beneficial action attributed to 'green waters'

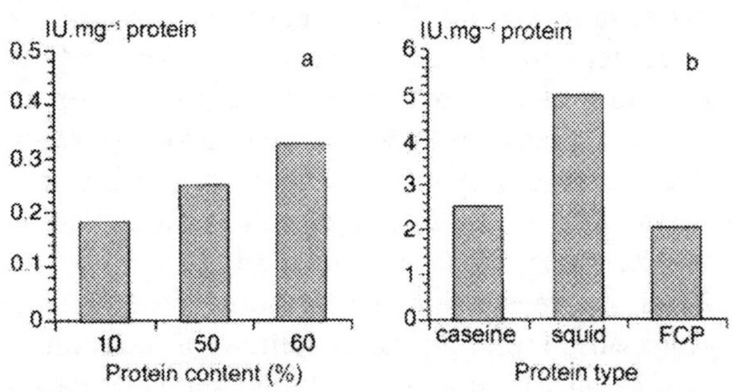

Figure 16.4. Response of specific activity of (a) trypsin to protein content and (b) chymotrypsin to the types of proteins in the diet of zoea stages of *Penaeus vannami* (after Le Moullac *et al.*, 1994). IU: international unit of measure of enzyme activity.

in fish larvae rearing units. However, no simple relationship between the amount of enzyme measured in the larvae and their growth has yet been found. Numerous hypotheses have been proposed to explain the beneficial effect of 'green water'.

Food composition only affects the activity of enzymes whose expression develops during ontogeny. For example, the specific activity of amylase and the amylase/protease ratio, which falls during the mysis stages, increases again considerably during the post-larval stages, even when the animals are fed only with artemia. Starch input at this stage only accentuates this phenomenon.

16.3 NUTRITIONAL REQUIREMENTS

In order to identify the nutritional requirements of larvae, many experiments have been carried out using purified or compound diets which now permit high survival rates. However, knowledge remains somewhat imprecise.

16.3.1 Protein nutrition

Very variable protein requirements have been reported. According to Kanazawa, they lie between 23% and 57% depending on the species being studied and the protein source being used. More precisely, it is generally accepted that proteins should represent 35–50% of the diet, as for juveniles.

Sources of protein classically used in the feeding of older shrimps are suitable for larvae: fishmeals, squid, crabs and mussels, soya cake, lactic yeast, casein, etc. Fish meals constitute the best source of proteins for shrimp larvae and can be used practically as a prime source for microparticles. In contrast, the concentrations of certain essential amino acids in crab protein concentrate, squid meal or casein are limiting and thus they should be associated with other proteins to permit appropriate larval growth and survival.

The optimal protein concentration of a diet is also linked to its carbohydrate content, which allows protein sparing. For a diet containing only 5% carbohydrate, the optimal protein content for *Penaeus japonicus* larvae is about 55%, while a carbohydrate concentration of 25% would allow the protein content to be lowered to 45% (Figure 16.5). An increase from 5% to 25% carbohydrate in a diet containing little protein (35%) improves larval survival but has no effect in a diet containing 55% protein. The larvae thus utilise carbohydrates as an energetic substrate much better than the juveniles, which allows protein sparing.

On the other hand, it appears that the larvae have quite a good capacity for utilising pure (crystallised) amino acids. Addition of essential AAs, especially arginine, to a diet based on casein increases larval survival. Lastly, peptides such as alanyl-glycyl-glycine or alanyl-valine appear to have a promoting effect on growth. However, it is not known whether the peptides act via an attractant effect, improve the nutritional value of the diet, or are actual growth factors.

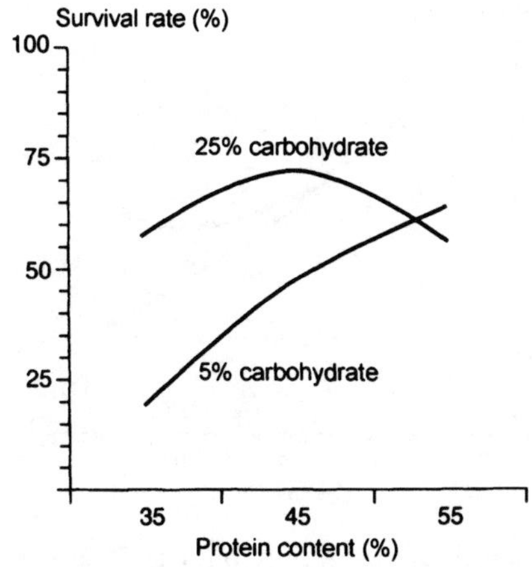

Figure 16.5. Effect of protein concentration in the diet on larval survival in *Penaeus japonicus*. Sparing effect on proteins of carbohydrates (after Teshima and Kanazawa, 1984).

16.3.2 Lipid nutrition

Among the lipids, the compounds found to be essential are cholesterol, the phospholipids (PL) and the essential fatty acids (EFA). As in the juveniles, the larvae do not synthesise sterols *de novo*; however, they may be able to effect a bioconversion of ergosterol and β-sitosterol into cholesterol, a compound which can thus be replaced by other sterols. They only have a reduced capacity for elongating–desaturating FAs and the EFAs are, as in juvenile crustaceans, polyunsaturated n-3 series FAs (HUFAs), especially 20 : 5 n-3 (EPA) and 22 : 6 n-3 (DHA) (Chapter 7). Phospholipids are essential because there is insufficient capacity for synthesising these molecules from FA or triacylglycerols (TAG).

The optimal concentration of cholesterol is close to 1%, total absence of sterols and an excess of cholesterol (5%) resulting in high mortality (Figure 16.6). Optimal concentrations of n-3 HUFAs and PL are interdependent. In the Japanese shrimp, the best combination is obtained using a diet containing 0.5–1% n-3 HUFA (EPA + DHA) and 3–6% PL (Figure 16.7). Dietary PL is involved in the emulsification of ingested neutral lipids, triacylglycerols and cholesterol, in the digestive tract. The most effective ones in the larvae are phosphatidylcholine and phosphatidylinositol which are present in commercial soya lecithin. Some of the lipid is supplied by protein sources and another part by oils (including cod liver oil which is rich in HUFAs) and soya lecithin. A concentration of 8% food lipids is usually sufficient to ensure good larval growth. Increasing this concentration to 16.5% has no effect on growth or survival.

16.3.3 Vitamin nutrition

Water soluble and liposoluble vitamins are essential for crustacean larvae as they are for vertebrates and their absence leads to rapid death of the larvae. However,

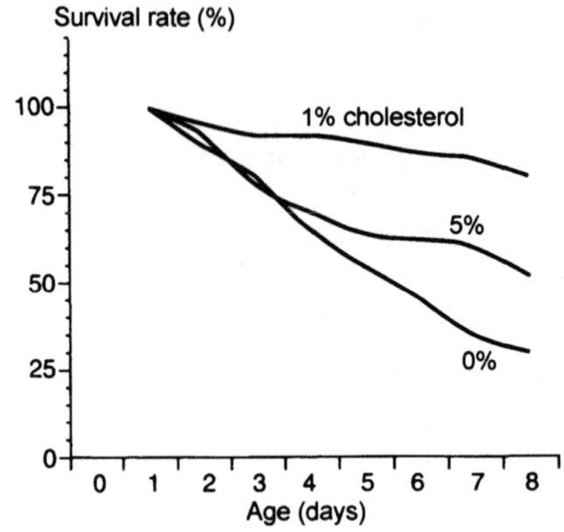

Figure 16.6. Effect of cholesterol concentration on larval survival in *Penaeus japonicus*. Experiment carried out using purified diets containing 5%, 1% or 0% cholesterol (after Teshima *et al.*, 1983).

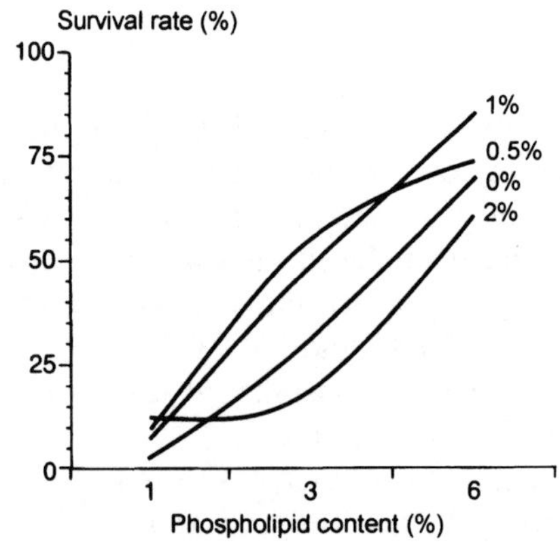

Figure 16.7. Effect of long chain polyunsaturated fatty acids and phospholipids on larval survival in *Penaeus japonicus*. Experiment used purified diets containing 2%, 1%, 0.5% or 0% long chain polyunsaturated fatty acids (HUFA) (after Kanazawa *et al.*, 1985). Reprinted by permission of Elsevier Science.

requirements are very difficult to quantify and vitamins must be added in excess to the food in order to prevent losses through leaching. A concentration of 1% of the food has been recommended for vitamin C, but this enormous overestimation was due to spectacular levels of leaching. The amounts used can be considerably reduced by utilising stable forms (ascorbyl phosphate). It has recently been shown that 100 mg of ascorbic acid/kg of food is sufficient to ensure survival and growth of larvae when this vitamin is supplied as L-ascorbyl-2-phosphate. However, doses of about 2000 mg.kg^{-1} give post-larvae better resistance to environmental stress and to bacterial infections.

16.4 PRACTICAL FEEDING

The nutritional requirements of penaeid shrimps are completely met by the classical feeding sequence made up of unicellular algae during the zoea stage, to which *Artemia salina* nauplii (Na) are added during the mysis stage. The most widely-used algae are *Chaetoceros* sp., *Tetraselmis* sp., *Skeletonema costatum* and *Isochrisis galbana* at a density of 50,000 to 100,000 cells/ml. Artemias are distributed at a rate of 10 Na/larva/day for M1 larvae and the quantity is increased progressively up to 100 Na/larva/day for PL4. This method currently allows a survival of 80% in rearing units under controlled conditions. The extensive Japanese method based on the development and maintenance of phyto- and then zooplanktonic blooms, in a very large volume (20–50 m^3) results in very variable production.

Many different microparticles or microcapsules are now available throughout the world and are used in hatcheries. In most cases, they are used as a partial substitute for natural food, at a level of 50–70%. However, in some hatcheries they are used as the only food. According to Jones *et al.* (1993), these microparticles should correspond to three other criteria, apart from nutritional value in its strictest sense.

- Acceptability: the microparticles should have a density similar to that of live prey and be of the correct size to be ingested. Their diameter is around 50–100 μm for the first stages and 125–250 μm for mysis stages.
- Stability: the microparticles should remain stable and sustain minimal leaching prior to ingestion. Different encapsulation processes have been developed and binding agents such as carragheenin, zein, gelatin or alginates are utilised. This treatment should ensure the maintenance of the different components of the microparticle and thus limit the pollution of the rearing water.
- Conservation: it should be possible to store the microparticles for several months without any alteration.

Recommended daily rations are provided with the products (Figure 16.8) and the daily ration should be divided into five or six distributions per day.

16.5 CONCLUSION

Knowledge concerning the digestive physiology and nutritional requirements of penaeid shrimp larvae has progressed sufficiently in the past 20 years to allow the development of artificial feeds which can be used from the first larval stages. There is currently almost no equivalent for the young larvae of marine fish, despite a much larger body of knowledge about the nutrition of the latter.

However, the survival, growth and quality of hatchery larvae fed exclusively on manufactured diets could be improved. Farmers still prefer to use wild post-larvae, which are more resistant. Feeding probably partly explains this difference. Foods of the same biochemical composition are used for the different larval stages, while the

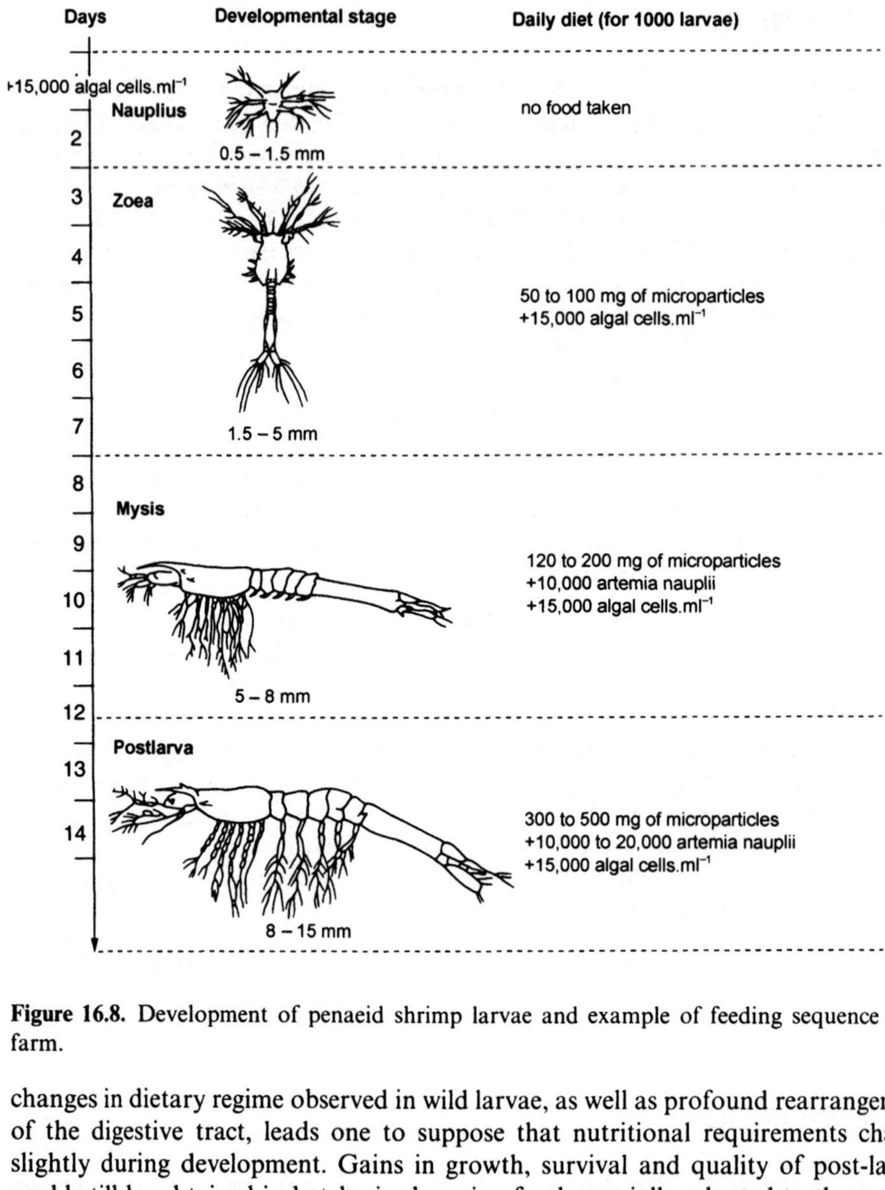

Figure 16.8. Development of penaeid shrimp larvae and example of feeding sequence on a farm.

changes in dietary regime observed in wild larvae, as well as profound rearrangement of the digestive tract, leads one to suppose that nutritional requirements change slightly during development. Gains in growth, survival and quality of post-larvae could still be obtained in hatcheries by using foods specially adapted to the specific requirements of each stage.

16.6 BIBLIOGRAPHY

Cahu, C., Cuzon, G. and Quazuguel, P. (1995) Effect of highly unsaturated fatty acid, alpha-tocopheol and ascorbic acid in broodstock diet on egg composition and development of *Penaeus indicus*. *Comp Biochem. Physiol.*, **112A**, 417–424.

Cahu, C., Sévère, A. and Quazuguel, P. (1988) *Variation de la composition en lipides de* Penaeus indicus *au cours de son développement larvaire*. ICES Annual Congress.

Jones, D. A., Kamarudin, M. S. and Le Way, L. (1985) The potential for replacement of live feeds in larval culture. *J. World Aquaculture Society*, **24**, 199–210.

Kanazawa, A., Teshima, S.-I. and Sakamoto, M. (1985) Effects of dietary lipids, fatty acids, and phospholids on growth and survival of prawn (*Penaeus japonicus*) larvae. *Aquaculture*, **50**, 39–49.

Le Moullac, G. and Van Wormhoudt, A. (1994) Adaptation of digestive enzyme to dietary protein, carbohydrate, and fibre levels and influence of protein and carbohydrate quality in *Penaeus vannamei* larvae (Crustacea, Decapoda). *Aquat. Living Resour.*, **7**, 203–210.

Lovett, D. L. and Felder, D. L. (1990) Ontogenic change in digestive enzyme activity of larval and postlarval white shrimp *Penaeus setiferus* (Crustacea, Decapoda, Penaeidae). *Biol. Bull.*, **178**, 144–159.

Teshima, S.-I. and Kanazawa, A. (1984) Effects of protein, lipid, and carbohydrate levels in purified diets on growth and survival rates of the prawn larvae. *Bull. Jpn. Soc. Sci. Fish*, **50**, 1709–1715.

Teshima, S.-I., Kanazawa, A. and Koshio, S. (1993) Recent developments in nutrition and microparticulate diets of larval prawns. *The Israeli Journal of Aquaculture-Bamidgeh*, **45**, 175–184.

Teshima, S.-I., Kanazawa, A., Sasada, H. and Kawasaki, H. (1982) Requirements of the larval prawn *Penaeus japonicus* for cholesterol and soybean phospholids. *Mem. Fac. Fish. Kagoshima Univ.*, **31**, 193–199.

17

Nutrition and feeding of shrimps in intensive and extensive culture

The study of the nutritional requirements of shrimps started with the original experiment of Kanazawa *et al.* (1970). In order to work on the Japanese shrimp, they had the idea of using a manufactured diet inspired by those given to a fish and two arthropods, namely the chinook salmon, the silk worm and artemia. This diet immediately allowed quite good growth and opened up the route for systematic studies that could soon be extended to other species. The rearing of Japanese shrimp developed in its country of origin from 1970 but rapidly stagnated, the domestication and rearing extending to tropical species that now supply almost the entire world market.

In theory, the requirements of the main species used in aquaculture should also be reviewed. In fact, while world production is largely dominated by the tiger shrimp, also called the giant shrimp (*Penaeus monodon*) and, to a lesser extent, by white shrimp (*P. vannamei*), 20% of production is divided between about 10 secondary species, the main ones being the Chinese shrimp (*P. chinensis* = *P. orientalis*), the blue shrimp (*P. stylirostris*) the Indian shrimp (*P. indicus*) and the Japanese shrimp. In practice, the specific nutritional requirements of these different shrimps are only known to a very limited extent (D'Abramo *et al.*, 1997).

The degree of intensification of culture should also be taken into account. From south-east Asia to Latin America, the whole range of 'semi-intensive' culture is found, differentiated by the number of shrimps reared per m^2 of rearing pond. The density varies between 10–15 to 20–25 animals.m^{-2}. Feed supply is in the form of pellets, sometimes in their extruded form (Chapter 21). The FGR is generally between 1.5 and 2.5 and yields vary from 1.5–3 t.$ha^{-1}year^{-1}$. Intensive techniques merit only a mention; one of the few still used today, on a small scale, is in Tahiti and allows production equivalent to 20–25 t.$ha^{-1}year^{-1}$ in two harvests. In semi-intensive aquaculture, it is difficult to know the proportion of the food provided directly by man. Addition of feed complementing the natural food plays a direct or indirect role (by stimulation of natural production), which can be minor or major depending on the circumstances. The respective parts played by the two sources develop with the age of the animals. The principles of nutrition and feeding described in this chapter, although meant for practical rearing purposes, are based almost exclusively on laboratory experiments where the animals are fed only formulated feed.

17.1 NUTRITION OF JUVENILES UNDER EXPERIMENTAL CONDITIONS

17.1.1 Protein requirements

17.1.1.1 Optimal protein level for the main species

Of all the nutrients required by the shrimp, proteins have been by far the most studied. Requirements, or at least optimal protein level, have been determined or estimated for about 20 different species. The results for nine species of economic interest are given in Table 17.1.

From these data, there seem to be distinct differences from one species to another. According to one interpretation given by several authors, species tending to be carnivorous would have higher requirements than those tending to be omnivorous or, *a fortiori*, herbivorous. At first glance this assertion appears to be logical, but it must be treated with caution. It may involve a superficial interpretation of differences due to various factors. Shrimp requirements have been measured using diverse protein sources: pure or AA supplemented casein, fishmeal, squid meal, or even a mixture of proteins from animals and plants. Many experiments were carried out without knowledge of protein digestibility. Lastly, the energy level varies from one trial to another in an unpredictable way.

Table 17.1. Crude protein content ensuring optimal growth under controlled experimental conditions for different species of the genus *Penaeus*.

Species	Content (%)	Protein source	Author
P. aztecus	30	squid and fish meal	Shewbart & Mies, 1973
P. chinensis	40	fish meal	Liu *et al.*, 1990
P. japonicus	42	casein	Koshio *et al.*, 1993
P. merguiensis	50	mixture	Aquacop, 1978
P. monodon	40	casein	Aquacop, 1977
P. schmitti	20–35	fish meal	Galindo, 1983*
P. setiferus	30	mixture	Andrews & Sick, 1972
P. stylirostris	30–35	mixture	Colvin & Brand, 1977
P. vannamei	30	mixture	Cousin *et al.*, 1993

* personal communication

17.1.1.2 Required revision

Because of the lack of simultaneous determination in a comparative experiment on the requirements of different species (a protocol which is difficult to put into practice), the optimal protein level must thus be re-examined, species by species, under standardised conditions. A recent experiment carried out on the Japanese shrimp indicated that nitrogenous excretion is minimal and growth optimal with a dietary protein content close to 40%, while previous work carried out in the same laboratory produced a value of 60%. This result brings *P. japonicus* closer to other penaeid shrimps in terms of protein requirements. It increases the doubts surrounding a number of 'historical' studies. The concept of an overestimation of protein requirements is widespread among food manufacturers who, without precise and accurate data, appear to be geared towards contents close to 40%, or even less. This value corresponds approximately to the requirements of species such as *P. monodon*, *P. stylirostris*, *P. vannamei* and *P. chinensis* under semi-intensive rearing conditions.

17.1.1.3 Adaptation to other factors

In contrast to the role played by species, stage of growth or environmental factors have not attracted great interest from scientists. Most trials have been carried out on juveniles of around 1g, sometimes on post-larvae and possible changes in requirements between body masses of 1 g and 25 g have been ignored. At the start of the 1970s, Japanese scientists recommended the same food for *P. japonicus* from the juvenile stage, or even from the end of the larval stage, until the adult stage. This involved a feed manufactured from very high-quality raw materials (squid, krill and fish meals) according to standards that provided comfortable error margins. This method of working is no longer used and, for *P. monodon*, Taiwanese industry has developed a range of feeds which are listed in Table 17.2. The effect of shrimp size taken into account here is probably a transposition of knowledge acquired from other animals. Other manufacturers take into account the fact that, in semi-intensive rearing, the young shrimp feeds mainly on natural production and have proposed the use of food with lower protein content at the start of rearing than the feeds used for older shrimps.

Table 17.2. Change in protein content of feed in relation to development (recommended levels).

Developmental stage or individual biomass	Protein content (% DM)
Post-larva, up to 1 g	42
2 to 5 g	41
5 to 10 g	40
10 to 20 g	37
20 to 25 g	35

It should be noted that for shrimp, as for fish, it is assumed that temperature and salinity have no effect on the optimal protein level.

17.1.2 Requirements for essential amino acids

With regard to amino acids (AA), studies carried out using the C^{14} labelled acetate or glucose (Chapter 6) has allowed essential amino acids (EAA) to be identified in 10 species of crustacean, shrimps and crabs. The list of these EAAs is the same as for fish (Table 6.1).

The problem of EAA balance an has only been touched upon. As reference proteins some authors have chosen shrimp muscle, others proteins from the Manila clam (*Ruditapes philipinarum*), a mollusc whose flesh constitutes an excellent food for shrimp. Formulations based on the EAA profile of shrimp muscle have been proposed, in view of the lack of sufficient knowledge about quantitative requirements for each EAA. Progress in this area is being made, but slowly because of the technical difficulties encountered, especially the leaching of pure AAs from experimental feeds that are consumed slowly (Chapter 16) and remain in contact with the water over a long period. Recently, supplementation trials using encapsulated arginine have allowed a quantitative approach to the determination of EAA requirements for *P. monodon.* It is to be hoped that this method will soon be extended to other EAAs of economic importance. Thereafter, formulation could be based not on whole proteins but on the group of requirements involving total proteins and EAAs.

Akiyama *et al.* (1989) have provided data on availability of EAA in the ingredients as well as recommended levels for each EAA. However, two theoretical considerations should be mentioned: protein requirement in crustaceans is possibly less directly correlated with protein accretion than in vertebrates, due to the synthesis of chitin, a polyholoside containing 43% raw protein; in addition, the significance of phosphagen (synthesised from arginine) in the energetic metabolism very probably decreases the well-known correlation between the EAA of body proteins and requirements for these same EAA. In other words, the determination of real requirements for EAA and 'requirements' for non-essential AA is likely to be more challenging in the case of shrimps than for vertebrates.

17.1.3 Energy and carbohydrates

17.1.3.1 Energy

The utilisation of dietary energy in shrimps is still very poorly estimated. At an experimental level, the term digestible energy (DE) is generally used, obtained by calculation from coefficients that are not always appropriate. Measurements of the digestibility of proteins and especially lipids and carbohydrates are still too few in number. The rational use of the two latter macronutrients should allow some economy of proteins in the diet, which the shrimp, like fish, apparently uses in significant amounts as an energy source. It thus appears to be urgent to evaluate the daily energy requirements for each species, developmental stage and temperature, to mention only the main factors (Cuzon & Guillaume, 1997). Acquiring these data should allow the establishment of recommendations on the protein/energy ratio which can be expressed for example in grams of protein per megajoule of digestible energy (DP/DE, Chapter 5).

17.1.3.2 Available carbohydrates

These nutrients have become increasingly important since the formulation of diets less rich in crude protein, leaving more room for carbohydrates. The tolerance to the supply of some carbohydrates, simple sugars, disaccharides or polysaccharides is known, at least in some species. Shrimps, like fish, demonstrate a certain degree of hyperglycaemia after carbohydrate intake and the negative effects of this phenomenon on growth are sometimes spectacular (although links between cause and effect have not been clearly established). Few studies have been devoted to the digestive or metabolic utilisation of complex carbohydrates.

Starches are now incorporated at levels of 20–45% in feed for *P. monodon*. The starches used are from rice, wheat, maize, potato and sago (extract from the pith of a palm from the Far East), or cassava. Recent laboratory studies tend to demonstrate that, in contrast to fish, some shrimps tolerate a high level of starch in their diet very well, especially if these starches are raw. In *P. vannamei*, digestibility is of the order of 70–90% for raw starches and 93–96% for pregelatinised starches. Incorporation levels of 20–35% of raw starch in the diet appears to be compatible with good performances, at least for this species. It thus appears to be logical to search for protein sparing by starch in shrimp as is found in fish with lipids. The metabolic utilisation of glucose, however, remains to be quantified because of the difficulty encountered by the shrimp in regulating its glycaemia.

17.1.3.2 Unavailable carbohydrates

Crustacean carapaces are composed mainly of chitin, a polymer of N-acetyl-glucosamine. Some studies have noted a beneficial effect of adding glucosamine to the diet; although it can be synthesised by the shrimp, it appears to be semi-essential. This role could explain the value of shrimp meal, as chitin is digestible to a degree. Despite the presence of a chitinase, chitin digestibility appears to be low, less than 10% in *P. stylirostris*. The reingestion by neighbouring shrimp of moulted material from animals that have just moulted is exacerbated by the distribution of poorly-balanced diets, while it is reduced by the use of good quality feeds. This phenomenon has also been attributed to a requirement for glucosamine, but the semi-essential nature of this amino-sugar has not been confirmed. The beneficial effect of shrimp meal could have other reasons.

Cellulose appears to be partly digestible in tropical carids such as the giant freshwater prawn which possess cellulase. However, this is not the case for penaeid shrimps in which the fibres only act as a filler. Manufacturers of shrimp feeds sometimes attempt to keep the supply of raw fibre at a very low level (less than

1% in Philippine feeds used for *P. monodon*). It is difficult to say if this constraint confers a real advantage to the diet.

17.1.4 Lipids

17.1.4.1 Lipid as energy sources

Several studies appear to show that shrimps do not tolerate lipids very well, with growth slowing down at levels above 10% and certainly at 15% of the diet. As a result, these nutrients have been given little consideration as sources of energy. The negative effect of high levels of lipids on the physical quality of pellets formed by pressing is another reason for the lack of interest in these macronutrients. However, a recent study appears to show that in *P. chinensis* there is no reduction in growth with a diet containing 15% lipid. It would be desirable to confirm this and to extend it to other species. In parallel, the digestibility of these lipids in these new conditions should be determined, since it is known only that the ADC of lipids is of the order of 85% for levels of incorporation of 4–10% of the feed. In addition, the effects of an increased supply of lipids over an entire rearing period remain to be verified, while studying the possible repercussions on the composition of muscle and hepatopancreas.

17.1.4.2 Requirements for essential fatty acids

In the shrimps, the same families of fatty acids (FA) are essential as for the vertebrates, that is either n-3 or n-6 series (Chapter 7). In addition, the species studied until now (mainly *P. japonicus*) only have a very low capacity for utilising C18 polyunsaturated fatty acids to synthesise HUFAs: EPA, DHA and arachidonic acid. This capacity for bioconversion appears to be even lower than that of marine fishes. There is a higher requirement for n-3 series (and thus EPA and DHA), at least in *P. japonicus*, although there is also a less well understood requirement for arachidonic acid. These fatty acids are abundant in oils and meals of marine origin. A supply input of cod liver oil provides these FAs, in addition to those provided by sources of animal protein of marine origin, and allows the requirements (estimated at 0.8% of the diet), of larvae and juveniles to be easily met.

17.1.4.3 Cholesterol and phospholipids

One of the most unique aspects of shrimp nutrition is the requirement for cholesterol which appears to act as a vitamin for arthropods. In crustacean muscles, total sterols represent up to 41.5% of the lipids and, in the shrimp, cholesterol represents 90% of the total sterols. In marine shrimps and in *Macrobrachium*, it is a component of circulatory or membrane lipoproteins and serves as a substrate for the synthesis of various compounds, including moulting hormone. While crustaceans are incapable of synthesising the sterol nucleus, they can convert certain phytosterols (in particular ergosterol and β-sitosterol) into cholesterol, the nutritional efficiency of phytosterols

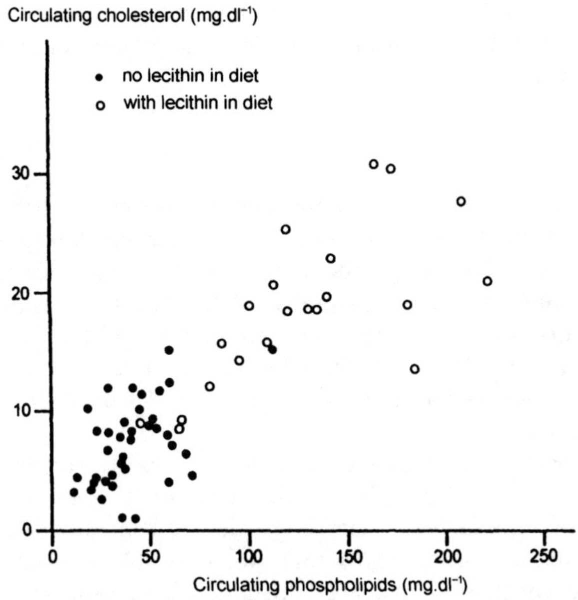

Figure 17.1. Relationship between circulating cholesterol and phospholipid in the lobster (after d'Abramo *et al.*, 1982). By kind permission of Springer Verlag.

being somewhat less than that of cholesterol. Nevertheless, the beneficial effects of supplementation with cholesterol have been demonstrated using more or less purified experimental diets. They are reduced using manufactured feeds including various ingredients, in particular shrimp meal, algae powder and even oil seed meals (copra, soya) that provide sterols which can be used by crustaceans.

Another unique aspect of crustacean lipid nutrition concerns the phospholipids (PL) which do not act only as sources of phosphorus, HUFAs, inositol or choline. These are firstly the emulsifiers and agents required for the formation of micelles in the digestive tract; they are also the components of the lipoproteins required for the transport of cholesterol between absorbent cells and the target organs. These phospholipids, whose synthesis by crustaceans is certainly slow, are sometimes presented as growth promoters. They are not all equally effective, phosphatidylcholine usually being the most active. The 'requirement' for phospholipids is all the higher when dietary cholesterol input is low; the relationship between phospholipid-cholesterol in the haemolymph of animals fed with or without lecithin is particularly clear in the lobster (Figure 17.1). The role of phospholipids as growth promoters is controversial in the lobster, but it has been clearly demonstrated in the shrimp where these compounds are necessary for the post-larva and juvenile and also for the spawners. In the latter, when incorporated into manufactured diets, they improve the number of eggs and hatching rate.

In general terms, phospholipids are supplied in the form of lecithins at concentrations of 0.5–3% of the diet. Phospholipids are sometimes incorporated in the food purely for technological reasons. Thus, 0.5% soya lecithin containing at least 65% phospholipid facilitates lipid emulsion and allows the homogenisation of a mixture of various ingredients.

17.1.5 Vitamins

In rearing units, several diseases or abnormalities have been attributed to vitamin deficiencies: black death disease, 'moulting death' syndrome, blue disease, reduction of growth, etc. Current knowledge about vitamin nutrition in shrimps is still very scarce, and is mainly limited to qualitative data. The list of organic molecules required by shrimps, regrouped under the term vitamins in the widest sense of the word, is the same as for fish, with the addition of cholesterol. Requirements for purine bases, which are found in artemia (a lower order crustacean), probably do not exist in the penaeids.

Quantitative requirements for vitamins are very poorly understood because of leaching which occurs when feed pellets, which may or may not be stable, remain for a certain length of time in the water before they are ingested. Measurements of leaching carried out on pellets show that water soluble vitamins dissolve most rapidly, in particular those which are added in a free form in the premix and especially vitamin C. This phenomenon can be aggravated by the destruction in sea water of unstable vitamins in a basic environment, which is again the case for ascorbic acid. For these reasons, researchers firstly used considerable overdoses. The situation changed with the appearance of stable forms of vitamin C, such as ascorbyl polyphosphate. This form is easily hydrolysed and then absorbed in the shrimp intestine and allows the commencement of precise studies on quantitative requirements. Thus, in *P. vannamei* and *P. monodon*, doses of 100–500 ppm are sufficient for maximal survival and growth, while certain researchers had been using levels of ascorbic acid reaching 2–3% of the diet!

Requirements estimated for other vitamins during growth–survival trials are shown in Table 17.3.

17.1.6 Minerals

Mineral nutrition has been little studied in shrimps, although crustaceans lose inorganic elements, in particular calcium, in significant amounts at each moult. Like all fish, crustaceans are capable of absorbing calcium, phosphorus and probably many other elements dissolved in seawater. The presence of an exoskeleton or carapace does not seem to notably decrease this capacity, as was supposed by some authors. The requirement for a dietary source of calcium has not been demonstrated in marine crustaceans. Instead, efforts have been made to estimate the minimum available phosphorus required by the shrimp, as the phosphorus–calcium ratio is of little significance. Monophosphate forms are better utilised as a result of their solubility and pH of dissociation. The shrimp, which does not possess a true stomach, only efficiently uses phosphates that dissociate in a slightly basic environment, which is not the case for bicalcium phosphate. It is for this reason that

Table 17.3. Quantitative requirements for vitamins of some shrimps and proposed standards (mg.kg^{-1}) (after Conklin, 1997).

Vitamin	*P. japonicus*	*P. monodon*	*P. vannamei*	*Macrobrachium*	Recommended levels
Thiamine	60	15	–	–	60
Riboflavin	–	22	–	–	25
Niacin	–	7	–	–	40
Pyridoxine	60	–	–	–	50
Pantothenic acid	–	–	–	–	75
Biotin	–	–	–	–	1
Folic acid	–	–	–	–	10
Vitamin B_{12}	–	0.2	–	–	0.2
Choline	600	–	–	–	600
Myo-inositol	400	–	–	–	400
Vitamin C (stable form)	99	209	120	104	200
Vitamin A	2.4	–	–	–	5000 UI
Vitamin E	–	–	100	–	100
Vitamin D	0.2	0.1	–	–	0.1
Vitamin K (liposoluble form)	–	–	–	–	5
Vitamin K (hydrosoluble form)	–	35	–	–	–

shrimp diets currently include sodium or potassium monophosphate. It has also been noted that the penaeids have some capacity for utilising phytic phosphorus.

With regard to trace elements, certain authors recommend the addition of a mixture at 0.1% of the feed: others, relying on the supply from the seawater, content themselves with the addition of copper (minimum of 60 ppm in the food), because crustaceans have a particular requirement for this metal which is found in haemocyanin, the equivalent of haemoglobin in the crustaceans.

17.1.7 Miscellaneous

The input of nutrients in the true sense of the word is not the only consideration; it appears certain that food attractants play a practical role, although their nature has only been studied for a very few species. In other areas, input of growth factors has been recommended, in particular squid meal, although the active molecule has not been isolated.

17.2 FEEDING OF SHRIMPS IN A COMMERCIAL REARING SYSTEM

17.2.1 Importance of processing

Because of the nature of its buccal appendices and mouth, the shrimp is only rarely able to swallow whole feed particles distributed to it. It dissects and swallows them progressively, which inevitably leads to the leaching which has already been mentioned. Meals are most often taken at night, at dawn or dusk and so, without an automatic distributor, it is difficult to provide them with feed in tune with their feeding rhythms. The feed remains in the water longer, leaching is accentuated and its consequences are aggravated. Even if a feed has an ideal composition, its nutritional value is uncertain if it is actively leached after immersion. In extreme cases, the pellets might disaggregate seconds after their immersion and transform into a soup which cannot be consumed by the shrimps, even though it reinforces the natural production and contributes indirectly to feeding them.

All this explains the role of pellet stability in shrimp feeding and the importance of applied studies that have allowed these techniques to be mastered.

17.2.2 Main feed types

The different techniques used for feed manufacture for aquatic animals are described in Chapter 21, to which we shall refer for more details. It seems necessary at this point, however, to define the two categories of shrimp feeds: pelleted and extruded feed. Pellets can be formed 'dry' (industrial process) or moist.

17.2.2.1 Industrial dry pelleted food

Pelleted feeds used to feed shrimp can be made by pressing a mixture of ingredients through a thin, compressing die (2.7–3.5 mm thick), with the injection of steam. The pellets are thus 1.7–2.5 mm in diameter and 5 mm long. They are hard and retain their shape in water as a result of binding agents which are incorporated into them: clays, polymers, lignosulphonates, pregelatinised starches, wheat gluten, etc. Urea-formaldehyde-type compounds are very efficient but their use is prohibited by European and American legislation. Binding agents are sometimes used in combination to reinforce their action and they are often required to work at a temperature of at least 90°C in the press. Some factories have 'maturation chambers', mixers in which the ingredients are blended in contact with steam over a period of several minutes. Formulae containing wheat gluten benefit greatly from this treatment which confers much higher stability on the pellets than those made in processes without maturation chambers. However, it should be noted that the precise methods used are often kept secret by the manufacturers.

17.2.2.2 Moist pelleted foods

These type of pellets are formed by pressing a paste through a thin grid, using machinery widely used in the food or pharmaceutical industry. 'Spaghetti' is obtained to start with, which is then dried and can be crushed, screened and calibrated to the desired granulometry. These pellets are highly valued by the aquaculturists, who use them mainly during the nursery stage (after metamorphosis). Their water stability is relatively high if the formula contains gluten, as this forms a network, trapping other feed particles. In addition, this process allows 'spaghetti' of 0.8 mm or even 0.6 mm in diameter to be formed directly, which can simply be broken up after drying. Thus, the ideal diameter is obtained, which allows the shrimp to ingest the particle directly without having to tear it to pieces. At present, this type of technology has not been transferred onto an industrial scale.

17.2.2.3 Extruded particles

The cooking-extrusion process is now used on an industrial scale, although less than the pelleting process. It allows the use of starches that partially gelatinise inside the extruder, thus effectively binding the particles. A similar effect can be obtained by texturising soy proteins. The application of this process to shrimp feeding necessitates some adaptations in order to obtain an agglomerate of less than 3 mm in diameter because of the phenomenon of expansion when exiting the die. Technological improvements have been made to the die (holes 1 mm in diameter) and to the extruder pipe and the process is currently used more and more, despite its cost.

The cooking-extrusion process possibly represents the solution of the future. It allows particles to be obtained whose texture remains supple and which become saturated while remaining very stable. If the size of the extruded feed is appropriate (around 1 mm in diameter), the shrimp can ingest it whole, as it does with an agglomerate obtained using the moist pelleted feed process.

17.2.3 Rations

These are adjusted in relation to individual mean biomass of the shrimps and taking into account, in theory at least, the species and temperature. An example is given for animals of body masses between 0.2 g and 5 g in Table 17.4.

17.2.4 Feed distribution method

The method used for feed distribution is an essential factor in the success of a hatch or a rearing unit. Shrimps have a slow feeding behaviour and do not move towards the food as soon as it is distributed. Theoretically, several meals per day are required.

Table 17.4. Example of feeding table. This feeding regime allows the farmer to obtain feed conversion rate of about 1.8, in 2000 m^2 ponds.

Mean individual mass of shrimps in ponds (g)	Feeding level as a percentage of biomass
0.2–1.0	20–17
1.0–2.0	17–14
2.0–3.0	14–11
3.0–5.0	11–9

In intensive rearing the practice is to have four manual distributions per day, simply by walking round the pond. Control feeding trays can be placed on the bottom of the pond, which allows the farmer to note what is left and to adjust the quantity distributed from one meal to the next. Another solution consists of using automatic, programmable distributors which can provide more than 20 meals per day, at a rate of one machine per $100\,m^2$ pond, for example. This method of distribution can partly compensate the poor stability of the pellets in water, thus ensuring better food utilisation. However, in semi-intensive rearing, with ponds of several hectares, food distribution is generally carried out at one time, towards the end of the afternoon, according to quantities determined from a predictive growth-survival model.

Freshwater shrimps pose very particular problems. Instead of moving around the ponds in groups like penaeids, they have very strict territorial behaviours and the food must be distributed over the territory of each animal, i.e. uniformly over the entire area.

17.2.5 Feed efficiency

Feeding efficiency is not easy to measure, even under experimental conditions, because of the imprecision involved in measuring intake and the reingestion of moulted material, etc. Nevertheless, the concept of the FCR prevails among farmers who regularly relate the quantity distributed to the biomass produced over a period of 3–6 months. This criterion gives them an indication of the overall production efficiency.

In shrimps, the FCR depends on both the type of rearing unit and the distribution method. Values obtained from experimental rearing units (without natural food), as in intensive rearing units, are usually between 1.8 and 2.2, i.e. much greater than those found in fish and probably due to the production from moulted material, which is not taken into account. In semi-intensive rearing systems, the FCR can be higher (2.2 to 2.5), but they can also fall to around 1.6 or 1.4 or even less in the best performing systems. It is well known that in these cases natural production contributes greatly to the nutrient supply of the animals. In some ways, it can be said that in semi-intensive rearing, the FCR will be high if pond management favours the shrimps' competitors, as it will be low if it favours the development of organisms which are eaten by the shrimps. Under experimental conditions, values of only 1.3 have been obtained, which leads to hope of more progress in the future.

The distribution method is of primordial importance in the improvement of feed utilisation, all the more so when the ponds have a low surface area, rendering them easier to manage. The improvement which is noted empirically after multiplication of the number of meals can be explained by a lowered level of waste (managed using control feeding trays) and an improved availability of food, which reduces competition between the animals. The nutritional value and stability of the pellet also contribute to the improvement in feed efficiency, along with the management of water quality in the rearing ponds which aims to provide the shrimps with optimal growth conditions.

17.3 CONCLUSION

Knowledge about shrimp nutrition has progressed more slowly than for fish, except perhaps in the field of lipids. This results in part from experimentation difficulties caused by the feeding behaviour of the shrimp, leaching of feeds which results from this, as well as the phenomenon of moulting. In addition, optimal supply under experimental conditions can differ from those which are suitable for recommendation in rearing ponds.

Despite this, a certain amount of original data, and even points which appear paradoxical to mammalian nutritionists, have been obtained. Into this category fall some digestibility of phytic acid or chitin, a better efficiency of raw starches than pregelatinised ones and a requirement for cholesterol and phospholipids, etc. These unique characteristics are few, however. In many areas (proteins, essential fatty acids, vitamins ...), shrimps have requirements which are *sensu lato* similar to those of fish, although not similar to those of all vertebrates. In fact, the uncertainties often outweigh the unique features. It is surprising that practically nothing is known about the nutritional repercussions of a fundamental phenomenon such as moulting.

While shrimp rearing has progressed from an archaic artisanal phase to an industrial scale of considerable importance in a few decades, the formulation and manufacture of the required feeds on a large scale have been achieved without too much difficulty, despite the crustacean's peculiarities. This may be because the rearing involved is semi-intensive and so the supplemented feed does not have to be as complete and balanced as in intensive rearing systems. However, the results have given overall satisfaction. The rapid development of economic and ecological concerns now requires movement in new directions: the serious problems of pollution necessitate the immediate use of low polluting diets; the decrease in profitability requires a push towards more precise studies of nutritional requirements and the characteristics of the main reared species.

17.4 BIBLIOGRAPHY

Andrews, J. W. and Sick, L. V. (1972) Studies on the nutritional requirements of penaeid shrimps. In: *Proc. 3rd Annual Meeting World Mariculture Soc.*, 402–414.

Aquacop (1977) Reproduction in captivity and growth of *Penaeus monodon Fabricius* in Polynesia. In: *Proc. 8th Annual Meeting World Mariculture Soc.*, 927–945.

Aquacop (1978) Study of nutritional requirements and growth of *Penaeus merguiensis* in tanks by means of purified and artificial diets. In: *Proc. 9th Annual Meeting World Mariculture Soc.*, 225–233.

Colvin, L. B. and Brand, C. W. (1977) The protein requirement of penaeid shrimp at various life cycle stages in controlled environment systems. In: *Proc. 8th Annual Meeting World Mariculture Soc.*, 821–840.

Conklin, D. E. (1977) Vitamins. In: D'Abramo, L. R., Conklin, D. E. and Akiyama, D. M. (eds.) *Crustacean Nutrition: Advances in World Aquaculture, Vol 6.* World Aquaculture Society, Baton Rouge, LA, 123–149.

Cousin, M., Cuzon, G., Blanchet, E., Ruelle, F. and AQUACOP (1993) Protein requirements following an optimum dietary energy to protein ratio for *Penaeus vannami* juveniles. In: Kaushik, S. J. and Luquet, P. (eds.) *Fish Nutrition in Practice*, INRA, Paris, 599–606.

Cuzon, G. (1998) Nutritional review of *Penaeus stylirostris*. *Reviews in Fisheries Science Special Issue*, **61**, 129–141.

Cuzon, G. and Guillaume, J. (1997) Protein and energy. In: D'Abramo, L. R., Conklin, D. E. and Akiyama, D. M. (eds.) *Crustacean Nutrition: Advances in World Aquaculture, Vol 6*. World Aquaculture Society, Baton Rouge, LA, 51–70.

D'Abramo, L. R., Bordner, C. E. and Conklin, D. E. (1982) Relationship between dietary phosphatidylcholine and serum cholesterol in the lobster *Homarus* sp. *Mar. Biol.*, **67**, 231–235.

D'Abramo, L. R., Conklin, D. E. and Akiyama, D. M. (1997) (eds) *Crustacean Nutrition: Advances in World Aquaculture, Vol 6*. World Aquaculture Society, Baton Rouge, LA.

Guillaume, J. (1996) Protein and aminoacids. In: D'Abramo, L. R., Conklin, D. E. and Akiyama, D. M. (eds) *Crustacean Nutrition: Advances in World Aquaculture*, Vol 6. World Aquaculture Society, Baton Rouge, LA, 26–50.

Kanazawa, A. (1990) Protein requirements of shrimp. In: *Advances in Tropical Aquaculture*. IFREMER, Plouzané, 261–270.

Kanazawa, A., Shimana, M., Kawasaki, M. and Kashiwada, K. I. (1970) Nutritional requirements of prawn. I. Feeding on artificial diet. *Bull. Jap. Soc. Sci. Fish*, **36**, 949–954.

Koshio, S., Teshima, S.-I., Kanazawa, A. and Watase, T. (1993) The effect of dietary protein content on growth, digestion efficiency and nitrogen excretion of juvenile kuruma prawns, *Penaeus japonicus*. *Aquaculture*, **113**, 101–114.

Lee, O. C. (1971) Studies on the protein utilization related to growth of *Penaeus monodon* Fabricius. *Aquaculture*, **1**, 119–211.

Liu, F., Liang, D., Sun, F., Li, H. and Lan, X. (1990) Effects of copper on the prawn *Penaeus orientalis*. *Oceanol. Limnol. Sinica*, **21**, 404–410.

Shewbart, K. J. and Mies, W. I. (1973) Studies on nutritional requirements of brown shrimp: the effect of linolenic acid on growth of *Penaeus aztecus*. In: *Proc. 4th Annual Meeting World Mariculture Soc.*, 277–287.

Teshima, S.-I., Kanazawa, A., Sasada, H. and Kawasaki, M. (1982) Requirements of the larval prawn, *Penaeus japonicus* for cholesterol and soybean phospholipids. *Mem. Fac. Kagoshima University*, **31**, 193–199.

Part IV

Feeding of fish: Applications

18

Raw materials and additives used in fish foods

The various raw materials or ingredients used for the manufacture of feeds destined for fish and crustaceans can be classified in several ways depending on factors such as their origin, composition, certain nutritional or physico-chemical properties, or economic criteria. In databases or feed composition tables, the classification used is based mainly on the origin of the products. It is this classification which is used in this chapter as well as for the tables in Appendix B. Only 'semi-purified' materials such as oils, or purified ones such as starches, salts or amino acids have been grouped according to their main chemical characteristic.

A raw material is usually a food in itself, i.e. a source of nutrients. These are the nutrients which are given, along with those of the cellulose compounds (Chapter 8) in the tables of food composition (Appendix B). However, certain ingredients are used to confer particular properties rather than to render the formula more balanced: to improve palatibility, colour attractiveness or stability in water, for example. In other cases, the primary constituents have negative properties. In practice they contain a certain amount of compounds that may be harmful to varying degrees, whose nature and properties are defined in Chapter 19 but whose presence is mentioned in this chapter.

The main characteristics of raw materials which are sources of nutrients, attractants, carotenoids, fibre and antinutritional factors are described in relation to their importance in aquaculture feeds and the permitted additives are considered as ingredients in the broad sense.

18.1 MAIN RAW MATERIALS OF ANIMAL ORIGIN

18.1.1 Advantages and disadvantages

These materials only represent a small part of the feed for terrestrial farm animals, but their content nearly always exceeds 50% in the diet of aquatic animals. While products of animal origin are used less and less in the feeding of terrestrial omnivores, their use has increased for fish. Compared to purely plant raw materials, they generally allow carnivorous species to perform better, even when care is taken to formulate diets equivalent to the usual nutritional recommendations. The reasons for this are diverse and not always obvious: digestibility, the profile of essential amino acids (EAA), palatability, richness in A group vitamins, content of growth factors or essential fatty acids (EFA) are usually high in products of animal origin and especially those from marine animals. Also worth mentioning is the absence of cellulose-type compounds (fibre) and the rarity of antinutritional factors *sensu stricto*.

Last but not least, ingredients from terrestrial animal origin such as meat meal or meat and bone meals may contain a recently discovered pathogen: the prion, responsible for bovine spongiform encephalopathy (BSE). That is the reason why these raw materials are not allowed to be used in animal feeds in the European Community although fish do not appear to be suspectible to prions.

18.1.2 Fishmeal and other products of marine origin

18.1.2.1 General

These meals constitute sources of protein that are well adapted to fish, since they are rich or very rich in essential amino acids whose profile corresponds remarkably to the requirements of vertebrates and in particular to those of fish. The oils contained in these products are an excellent energy source and are high in long chain polyunsaturated fatty acids (HUFAs), the EFAs which are most important to fish. These raw materials are also very good sources of essential minerals (calcium, phosphorus, magnesium and trace elements) and vitamins (vitamins B_{12}, A, D_3, choline, inositol and also, to a lesser extent, all other vitamins with the exception of ascorbic acid). They also often contain carotenoids.

These characteristics explain the overwhelming part played by fishmeals in relation to other raw materials in foods for aquaculture. However, these products are only available in limited amounts and are expensive. In addition, in the world market aquaculture is usually in competition with other types of farming (poultry, pigs and even pets).

18.1.2.2 Manufacture of fishmeal and quality control

The main principle (Figure 18.1) consists of separating water and oil from the dry matter. The raw material is firstly chopped and then transported through a cooker which is usually a heated screw system (about 20 min at 90°C). At this stage, a first separation occurs between the solid phase (coagulated proteins and bony parts) and the liquid phase (water and oils).

Pressing after transfer completes this operation and the cake obtained contains no more than 50% water. The liquid phase is decanted and centrifuged to separate

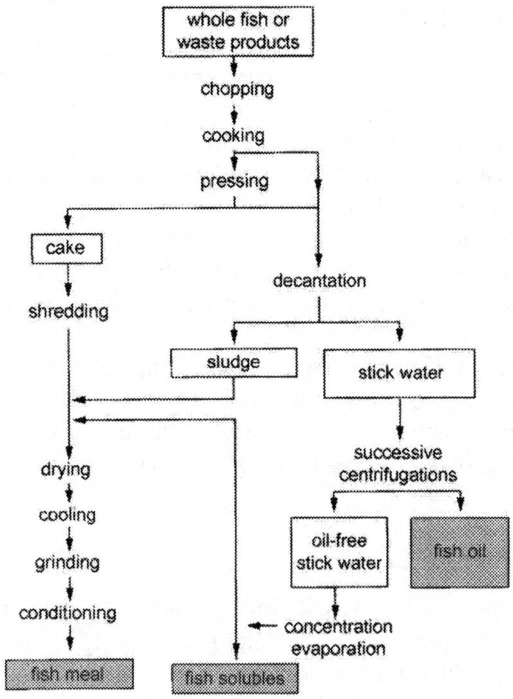

Figure 18.1. Diagram of fishmeal manufacturing process.

the oils. Press liquor is concentrated, evaporated and the fish solubles generally reincorporated into the cake.

In addition to the freshness of the raw materials, drying constitutes an essential factor which affects the quality of the final product. It can be direct or indirect (without direct contact between the product and flames or the heat source). The maximal temperature of the primary constituent is of major importance for the quality of the meal. A 'good' meal should not have been heated above 90°C. Low temperature (LT) meals have been on the market for several years now; they include products whose maximal cooking or drying temperature does not exceed 70°C. For all of these, the residual moisture level has a maximum of 10%, ensuring that the product is sufficiently stable for marketing.

The main component of quality which is influenced by treatment temperature is protein digestibility. It can be measured either in fish or crustaceans for which the fish meal is being made, or in mink, carnivorous animals that are much easier to experiment with than aquatic animals. *In vitro* digestibility tests do not provide as reliable results as *in vivo* tests; some can, however, be used for a preliminary evaluation of meal quality – this is true for the Torry test and multi-enzyme tests. These tests do not take into account the possible harmful metabolic effects of compounds that can appear during the degradation of the fish before or during treatment.

Biogenic amines have a positive effect on the deterioration of the fish before the meal is made. Performances of fish receiving meals are negatively correlated with their biogenic amine content. However, the measurement of the four biogenic amines is tedious and is rarely carried out. Gizzeroline is a poorly defined compound, related to histamine but much more toxic; it causes erosion of the gizzard followed by 'black vomit'. It can be measured using a simple *in vivo* test using turkey poults or chicks. Measurement of the total volatile basic-nitrogen, TVBN, is supposed to provide similar information but, used alone, it does not provide very reliable data. Among other proposed *in vitro* criteria is the measurement of sulphydryl groups and disulphide bonds. None of these provides information which can be highly correlated with *in vivo* performances.

Different manufacturing techniques exist similar to this general scheme. They are currently developing towards a decrease in the energetic costs of processing as well as towards optimisation of the quality of the final product. Improvement in digestibility, which allows a decrease in waste, constitutes a major objective.

18.1.2.3 Different categories

Fishmeals All these meals, based on whole fish, are similar in chemical composition (66–71% protein, 9–12% lipid and 12–15% ash, cf. Appendix B). However, despite this fact, they may differ greatly in their digestibility or more in the availability of their nutrient. The main producing countries have defined a quality policy based on the criteria which are unique to these meals. Several qualities can be distinguished.

- Norwegian quality: these meals are based on hake, sand eel, capelin, herring, smelt, blue ling or cod. Listed in order of increasing quality, the meals are standard, Norseamink® and Norse LT94®.
- Danish quality: these are mainly based on sand eel, but also contain sprat, pout, mackerel and pilchard. The different quality meals are standard, special A and B, 'aquaculture' and LT.
- Chilean quality: based on sardine and pilchard, these meals differ from the ones above mainly in the mode of quality control which is applied *a posteriori*. This is the gizzard erosion test (on young chicks) that reveals the presence of histamine and gizzeroline. In relation to these results, the meals are classed as standard, Corpesca® type, special and LT.
- Iceland quality: mainly based on capelin, these comprise standard and LT meals.

There is also significant production of American (menhaden), Canadian (herring), Peruvian (anchovy), South African (pilchard) and Japanese (sardine) meals.

- Offal meals (France, Morocco, Senegal): these are produced using the offals from filleting and canning and sometimes from bycatch. They are richer in ash (22–24%) and contain only 55–64% protein (Appendix B). In addition, their composition tends to be more unpredictable and their quality lower than those made from whole fish.

- On board factory (United States, Japan, Russia, Poland): these are manufactured on board factory ships using the waste from white fish filleting (hake ...) or using fish of no value for human consumption. No fish solubles are added to them as the equipment required for their concentration cannot be carried on board. These are called 'white fishmeals' and are used in foods destined for eels and also are sought after for shrimp feeding. Their chemical composition is close to that of offal meals. They have the advantage of being manufactured from very fresh products, but the manufacturing processes are more difficult to manage than in land-based factories. The absence of fish solubles does not have any known effect.

Soluble fish 'protein' concentrates This is a product obtained by the hydrolysis of coarsely-ground fish proteins using plant enzymes such as papayine; the process is carried out in reactors either continuously or discontinuously. Bony parts are separated off from the hydrolysate, which is then delipidated by centrifugation and spray dried. This type of product, of which the most well known is SFPC (soluble fish protein concentrate), is particularly low in minerals and rich in proteins (Appendix B). They do not contain proteins in the true sense of the term, but mainly peptides and amino acids which render them almost completely soluble. This property can be inconvenient in foods for aquatic animals, but the nitrogen digestibility is always very high. Depending on the quality desired, lipids are added after drying, in variable quantities. They are not necessarily entirely of marine origin.

Fish autolysates These are similar to those described above, but hydrolysis is carried out by enzymes present in the fish tissues. Management of hydrolysis is more delicate and is often carried further than in soluble concentrates. The price of this type of material generally excludes it from formulation of normal fish feeds.

Fish silage These are liquid or semi-liquid products obtained from fish offal or trash fish conserved at an acidic pH and do not always constitute silage in the proper sense. There are several processes based on the addition of an acid solution (formic, hydrochloric or sulphuric, for example) in order to obtain a pH of 4 to 4.5. This pH inhibits the bacterial proliferation which leads to putrefaction, while allowing a certain amount of hydrolysis as a result of the tissue enzymes. Hydrolysis carries on during storage until the enzymes are inactivated by heating. Complete liquefaction can take hours or weeks, depending on temperature and enzymatic activity of the raw material. The product cannot be used as it is, but must be mixed with meal and used in the form of dried or moist pellets. Since fish are sensitive to diet acidity, the silage must be neutralised before use, depending on its level of incorporation. Thus, 1.5–4% slaked lime or soda is added to obtain a pH close to 5.5. Tryptophan is unstable in an acidic environment and so the final product is relatively deficient in this essential amino acid. This must be taken into account during formulation. Other disadvantages have been reported: excess of 3 carbon chains, the metabolic utilisation of which is limited, and effects of added ions on plasma equilibrium.

It should be emphasised that the products described in the last three sections often come under the same heading of fish hydrolysates.

18.1.2.4 Other products of marine origin

Shrimpmeal These are in fact meals made from shrimp 'heads' (cephalothorax). They are rich in calcium (although quite low in phosphorus), but are powerful attractants and provide an abundant supply of carotenoids as well as EFAs. Krill meals (rarely found on the market) are made from the whole animal but their fluorine content limits their level of incorporation.

Squid meals Rich in proteins and containing relatively low levels of minerals, these are sought after for their attractant qualities and for the presence of growth factors for crustaceans.

18.1.3 Animal meals of 'terrestrial' origin

18.1.3.1 Introduction

All products listed below are now prohibited for animals (including fish) by the regulation of the European Community. In the future, since processes involving heating at higher temperatures appear to destroy prions, they could be allowed for feeding some species such as aquatic animals.

18.1.3.2 Meat and bone meal (MBM)

These are manufactured using offal products from abattoirs and butchers. The quality of meat and bone meals is very variable and is often limited owing to an excess of minerals, the ash content reaching 35% in meals made from bony meat. In the best cases, they contain 45–60% protein of quite good biological value, although they contain limited amounts of sulphur AAs. The cooking temperature must always be high for health reasons, and so does not allow the production of proteins which are as digestible as those obtained in fish meals. Between 8% and 10% of the product is lipid, including mainly saturated or mono-unsaturated fatty acids and a minute amount of HUFAs, except arachidonic acid.

18.1.3.3 Connective tissue meal

These are meals made from collagen, the residue from the manufacture of tallow or lard. Protein content is high (70–80%), with a good essential amino acid profile. These meals differ greatly from the true meat meals with which they are often mixed, since they have much lower lipid (1–3%) and mineral (10–20%) contents.

18.1.3.4 Hydrolysed feather meal

This is produced by poultry abattoirs and contains mainly keratin. This protein is totally indigestible in its natural state and is hydrolysed using a powerful hydrothermic treatment. The protein content of these meals is high (80–85%), but their biological value is very low because of their extreme lack of methionine, lysine and histidine. In contrast, cystine content is high. These meals can only be used to complement other protein sources. Their lipid content requires that they be mixed with offal other than feathers.

18.1.3.5 Poultry offal meal

These meals are made from waste including blood and viscera, constituting a type of meat meal which is very variable in composition but widely used in certain countries. Their essential amino acid content is significant when mixed with plant products.

18.1.3.6 Blood meal

This is obtained by dehydration (drying in a drum or spray drying) of blood from the abattoir. Protein content is high (85%), but biological value is quite low: blood meal contains low levels of methionine, isoleucine and arginine but great excesses of leucine. More interestingly, it is also particularly rich in available lysine if it is treated by atomisation, a process that guarantees overall good quality. It is incorporated increasingly frequently in fish foods because of its binding properties. The starting product can also be coagulated. Thus, after separation of the plasma used elsewhere and drying by steaming, a powder of red blood cells is obtained that is richer in proteins and has even greater binding powers.

18.1.3.7 Other terrestrial animal ingredients

In the laboratory, other sources of protein of animal origin are sometimes used; these are usually purified, such as casein which is of excellent biological value, or gelatine which is a very powerful binding agent. On an industrial scale, dry skim milk is still used (denatured, usually with fishmeal) or lactoserum.

18.2 MAIN RAW PLANT MATERIALS

18.2.1 Advantages and disadvantages

There are numerous raw materials of plant origin. They are less expensive than those of animal origin, so manufacturers tend to substitute them for the former. They often have a certain binding power, which may be associated with the presence of digestible or indigestible substances. They also constitute sources of group B vitamins. However, they are much less well adapted to the requirements of fish – their n-3 HUFA content is nil and they are less palitable. Their main energy source, starch, is not always well tolerated by fish; they often contain a significant amount of complex membrane carbohydrates (e.g. pectins, hemicellulose, pentosans, cellulose, lignin), which are often indigestible. Lastly they can also contain various antinutritional factors (Chapter 19).

18.2.2 Oilseed meals

Cakes are coproducts of oil extraction, which are less rich in protein than animal ingredients (30–50% protein). Their mineral content is lower but they include a significant quantity of cell wall compounds as well as, frequently, antinutritional factors.

18.2.2.1 Soybeanmeal or soya (glycine max)

This is by far the most widely used because of its availability on the market, its consistency, price and especially its good nutritional value for most farmed fish. It is rich in protein (49% if it is dehulled, 44% if it is not). Its essential amino acid profile is good despite a deficiency in methionine. This cake contains many antinutritional substances, in particular antitrypsic factors. The latter are largely destroyed by rigorous thermal treatment. The manufacturing requirements are greater for soybean destined for fish than for that destined for terrestrial animals. Lower in cellulose and complex carbohydrate than other oilseed meals, when correctly cooked it constitutes a very important raw material for aquaculture. There are also more purified soya products than meals on the market; these are richer in protein and have lower fibre and undesirable carbohydrate contents, but their price is too high as a food for fish.

18.2.2.2 Rapeseed or cotton cake

Rape (*Brassica napus* and *Brassica campestris*) is increasingly abundant on the market. Its proteins are quite well balanced and, although it still contains a significant amount of fibre, new varieties are much less toxic than the old ones (Chapter 19). However, protein digestibility, energy value and even limits of use in relation to fish remain poorly understood, even for the most widely-farmed species.

18.2.2.3 Cotton meal

Cotton meal (*Gossypium* sp.) is not sought after in Europe as it is not widely available on the market and also because of its fibre content, lysine deficiency, presence of gossypol and also sterculic acid residues. However, because it is rich in proteins, it is widely used in developing countries as an ingredient in aquaculture feed. New varieties with increased cotton yields and no gossypol raise the chances of greater use of this cake.

18.2.2.4 Other oilseed meals

Sunflower meal (*Helianthus annuus*) is low in antinutritional factors (polyphenols) and relatively rich in methionine. However, it is deficient in lysine and has a high fibre content, which limits its use in fish feed formulation.

Groundnut/peanut meal (*Arachis hypogea*) is rich in proteins (48–50%) that have a high arginine content but are deficient in lysine and methionine. The presence of aflatoxin has often been reported and has discredited this cake which nevertheless is very digestible and contains almost no antinutritional factors.

Oil palm meal (*Elaeis guineensis*) and *copra* (*Cocos nucifera*) meals only contain 19% and 22% protein respectively, while their fibre contents exceed 50%. They are of little interest to aquaculture except in the local context.

18.2.3 Cereals and by-products

These are products whose protein content generally varies between 10% and 20% and which mainly constitute energy sources. They have low mineral contents, except for phosphorus in its phytic form. They are good sources of group B and E vitamins. Wheat (*Triticum aestivum*) or corn (*Zea mays*) flours are rich in starch (62–72%) but this is rather indigestible in its crude form, especially if it is incorporated in large amounts for carnivorous fish. In contrast to cakes, cereals are quite low in proteins and also have low EAA contents, particularly lysine. Heat treatments improve starch digestibility which can thus become a significant source of energy. These same starches, as constituents of raw cereals or preferably in cereals which have been heat treated, are also sought after for their physical properties. In the different feed processes making (pelleting and especially cooking-extrusion), they have a beneficial action in the same way as purified starches.

Without any obvious reasons, other cereals are little used in aquaculture feeds. They certainly could be used more, despite some possible disadvantages: oats could pose a problem because of its husk, some sorghums are rich in tannins and might be toxic – this has yet to be studied in fish.

Wheat bran (the cereal husk) and middlings (intermediate between flour and bran) contain much fibre, but also group B vitamins and vitamin E. If they are finely ground, the middlings constitute good binding agents and are also used as bulk agents. Rice bran (*Oriza sativa*) is widely used in the intertropical belt despite its high content of membranous compounds of abrasive nature.

18.2.4 Pulses

These are the seeds of leguminous plants whose culture has been promoted in recent years in European Union countries: broad beans (*Vicia faba*), lupins (*Lupinus* sp.) and peas (*Pisum sativum*). These products contain 25–45% protein, the composition of which is similar to soya. While lupins are relatively rich in oil but deficient in starch, beans and peas are almost as poor in lipids as cakes but contain more starch than protein. The pea, which is currently the most abundant on the European market, is available shelled, either raw or cooked. Its antinutritional factors (lectins) do not pose a problem for fish. Although their use requires further study, the pea derivatives cited above are certainly of great technological significance with regard to their very high binding capacities.

18.2.5 Alfalfa and leaf meals

Lucerne meals are well known in Europe where they are sought after as a balanced source of protein, certain vitamins and carotenoids. They are little used in aquaculture since, in fish, their proteins are poorly digested, their carotenoids ineffective and their fibre content is often excessive. In addition, care must be taken with the effects of saponines, especially in crustaceans. In the intertropical zone there are many products which are more or less equivalent such as ‘ipil-ipil’ flour (*Leucana glauca*), the toxicity of which (mimosine content) must be carefully controlled.

18.2.6 Raw materials from unicellular organisms

These raw materials originate from unicellular or filamentous algae (*Spirulina*, *Spirogyra*, *Cladophora*, *Chlorella*), lower fungi and in particular yeasts (*Candida*, *Saccharomyces*, *Torula*, *Kluyveromyces*) or bacteria (*Methanobacter*, *Pseudomonas*, *Micrococcus*, etc.). This type of product is of great experimental interest and appears, *a priori*, to have a promising future. In practice, the suitability of these organisms for culture above the ground on various substrates or waste products and the theoretical possibility of rapidly modifying their nutritional quality using genetic techniques and the culture conditions constitute an advantage. It should be noted, however, that alongside short-lived industrial scale attempts, the yeasts (brewer’s, lactic and torula) are still the only sources of unicellular protein present in significant quantities on the market. They constitute good balanced sources of protein. They are also very rich in E and B group vitamins. A particular yeast (*Phaffia rhodozemia*) is produced as a source of carotenoid pigments. The presence of growth factors, which has been suspected for a long time, has never been confirmed. However, the stimulant role of parietal glucans with regard to cellular immunity appears to be tangible. With regard to brewer’s yeast, its bitterness is repellent to fish and so only debiltered yeasts can be used.

18.3 PURIFIED PRODUCTS

18.3.1 Fats

18.3.1.1 Fish oils

Fish oils are well digested even at high levels. They are very rich in HUFAs and in particular EPA and DHA, as well as liposoluble vitamins (A and D). Some are good sources of astaxanthin (red oil from capelin).

Their quality depends on the origin of the raw material (species, time and catching areas, etc.) and also on the freshness of the product used and the extraction process itself. These are fragile products which must be protected absolutely from oxidation by antioxidants (BHT, BHA, ethoxyquine).

18.3.1.2 Animal fats

Tallow and lard are hardly used in aquaculture. Their high melting point (due to the abundance of saturated fatty acids) renders them solid in the fish's digestive tract and insensitive to lipase action and so relatively indigestible. In addition, they contain practically no n-3 HUFAs. At present their use is not allowed in the European Community.

18.3.1.3 Plant oils

These constitute energy sources comparable to fish oils and the presence of antinutritional factors, synthesised by the plants, is exceptional. However, they do not contain vitamin A or D, astaxanthin or, especially, HUFAs. The only essential fatty acids that they contain are either linoleic acid, usually a non-limiting factor in diets, or linolenic acid which can easily be converted into HUFAs by freshwater fish (Chapter 7). These oils are generally too expensive to be used in animal feeds. However, in certain economic contexts, and depending on the extent to which the fishes' requirements for essential fatty acids are covered by other ingredients, they can be used as an energy complement. It must be ensured that the plant oil input does not modify the quality of the final product, either at the organoleptic level, nor in dietetic terms, in order not to harm the brand image of the farmed fish.

18.3.2 Starches

Starches are used both for their binding properties and their energy value. They differ in their plant origin, which determines the proportions of amylose and amylopectin as well as the size of the starch grains, and in the technological treatment to which they are subjected, which breaks down the structure to a certain extent and gelatinises the starch (Chapter 8). Cereal starches are more digestible than those of pulses, which are in turn more digestible than tuber starches, and in all cases cooking largely facilitates enzymatic degradation. The term 'pregelatinisation' is often used and lacks precision; the user should ask for information about the physico-chemical properties of a product in order to choose and optimise the technological process, pressing or cooking-extrusion.

18.3.3 Purified amino acids

Among the several amino acids produced synthetically or biologically which are currently available for animal feeds, only two appear to be effectively used for feeding fish: methionine and lysine. The tonnage used in this sector still remains low in relation to elsewhere because of the widespread use of marine proteins.

Incorporation of these compounds into the diets of aquatic animals has posed more problems than for terrestrial animals because of the solubility of the molecules. However, this problem is only crucial for crustaceans, where the granules may remain in the water for a long time leading to leaching, which has led to the belief that there is a lack of utilisation of purified amino acids by the animal.

HCl lysine is produced biologically and poses no metabolic problem since it only includes the L isomer. The two D and L forms of methionine present in the commercially-synthesised product (DL) have approximately equal activity, while MHA, a synthetic analog of methionine, is not efficient in fish (Chapter 6).

18.4 ADDITIVES

Listed below are purified compounds other than amino acids and carotenoids (considered in Chapters 6 and 11 respectively) whose incorporation into foods is allowed in low amounts for various reasons.

18.4.1 Attractants

18.4.1.1 Role

In the strict sense, attractants are substances that orient animals at a distance towards a 'prey'. Three other types of compound are classed as attractants in the widest sense: those that immobilise the animal close to the prey (arrestants), those that incite tasting (incitants) and those that lead to ingestion (stimulants). This distinction remains rather theoretical in fish, in which the gustatory and olfactory senses only differ in the location of their receptors.

Attractants play a major role in species in which feeding behaviour relies more on olfaction than on vision. Despite the current trend for replacing an increasing part of fish meals with other protein sources, the former are always preponderant in the formulation of most feeds destined for feeding farmed fish. These meals, as well as those of shrimp or squid, are particularly attractive for fish and there is currently no problem with appetibility with these foods. This will not be true when the primary marine raw materials are incorporated at low levels into the formulae. As a first step it should be possible to guard against using raw materials which have a noticeably repulsive effect on the fish (non-debiltered brewer's yeast for example). A second step would be to add attractant substances, purified or unpurified, in order to mask an unfavourable effect or more simply to increase the appetibility of the food mix.

Addition of attractant substances is already used in the laboratory to promote the ingestion of simplified or purified experimental diets. In production systems, these substances have a role to play in improving feed intake in a difficult environment, after stress or when medicinal feeds are being used.

18.4.1.2 Types

There is a substantial literature on the subject of fish attractants. Chemical compounds possessing attractant powers can be divided into three main groups: L series amino acids (series D ones sometimes have no effect and are sometimes repellent); betaine (or glycine-betaine) and other molecules which have a pentavalent nitrogen atom; and lastly, the nucleosides and nucleotides (inosine and inosine-monophosphate in particular).

These attractants appear to be very specific – some substances which act as attractants for a given species can either have no effect or be repellent for another. Synergistic effects are very difficult to predict (for a given species, the mixing of two weakly attractive substances can turn out to be extremely effective). Because of this it is difficult to define 'recipes' that will always work. As examples:

- L series amino acids (alone or in mixtures) have been described as active for the Channel catfish, sea bass, plaice, dab and European and Japanese eel;
- glycine-betaine, either associated or not with the preceding compounds, has an attractant effect on sole, plaice, dab, sea bream and eel;
- inosine is especially active in turbot, brill and yellowtail.

The use of rehydratable foods allows the extemporaneous addition of potentially attractant substances. This could be carried out on the farm to resolve problems of lack of appetite (transitory or chronic) in stock.

In practical terms, the two usual sources of natural attractants are first, fishmeals (rich in nucleotides) and second, invertebrate meals (shrimp, squid), which are rich in nitrogenous bases.

18.4.2 Specific additives *sensu stricto*

The list of additives that can be incorporated into animal feeds comes under strict European legislation for reared species or groups of species, fish often appearing in the 'other species' category. Maximal authorised doses and conditions of use are specified, as well as whether the additives may be incorporated alone or in association with another substance. This list does not include drugs, which can only be incorporated by veterinary prescription, but does include nutrients such as vitamins and oligo-elements.

No antibiotics, growth factors or acidity regulators are authorised for inclusion in fish feeds. Only the following four categories are considered in this chapter:

- antioxidants;
- emulsifiers, stabilisers, thickeners and gelling agents (group E);
- preservatives (group G);
- binding agents, antilumping and coagulants (group L).

The carotenoids form a fifth group (group F) – see Chapter 11.

18.4.2.1 Antioxidants

Antioxidants are substances that are easily oxidised and, as a result of this, protect other compounds which are sensitive to oxidation. They are used especially to break or at least slow down the chain of reactions involved in peroxidation. This chain starts with the formation of free radicals at the level of the double bonds in fatty acids and ends with the formation of stable aldehyde and ketone type compounds which can be toxic (Chapter 7). These reactions do not only occur *in vivo* but also in feeds and they especially affect oils rich in HUFAs (Chapter 19). Peroxidation is catalysed by metallic ions such as iron and copper; it damages not only fatty acids but also many other labile nutrients such as vitamin A and carotenoids.

Peroxidation is slowed down by several types of molecules that affect the chain of reactions using various mechanisms: sequestering oxygen (ascorbic acid) metal chelaters (citric acid, EDTA, amino acids), and proton donors (tocopherols, ethoxyquine, BHA, BHT, galates). About 15 antioxidants are authorised for all species and all feeds, the most widely used being, in addition to ascorbic acid and tocopherol which are nutrients (Chapter 8), BHA (butylhydroxyanisol), BHT (butylhydroxytoluene = butylmethylphenol) and ethoxyquine (dihydroethoxytrimethyl quinoline). It is recommended that antioxidants which are preventive (oxygen sequesters, chelaters) and curative (proton donors) should be associated, for example ascorbic acid and α-tocopherol, BHA or BHT. Antioxidants are frequently added to some raw materials used in fish nutrition (fish, shrimp or squid meals, or fish oils) and they are almost always added during feed manufacture.

18.4.2.2 Binding agents

These are numerous and widely used, taking into account the requirement for obtaining a final product which is stable in water, a physical property which is only remotely related to stability in air. They almost always involve more or less complex polysaccharides:

- algal extracts: alginates (which form stable gels in the presence of calcium ions), brown algae, agar-agar and carragheenan, extracted from red algae;
- terrestrial plant extracts: pectins, arabic gum, guar gum and carob pod extracts;
- animal extracts: chitosans;
- bacterial extracts: xanthan gums;
- others: carboxymethylcellulose and other transformed celluloses, chemical industry products, lignosulphites produced by the paper industry, bentonite and other clays.

It should be noted that several ingredients such as molasses, starches (either raw or transformed), wheat gluten, casein, collagen and gelatine, which are sources of carbohydrates or proteins, are also used for their binding properties.

18.4.2.3 Preservatives

These are used to preserve moist foods. Potassium sorbate is a widely-used antifungal compound. Organic or mineral acids are also added during the manufacture of fish silage.

18.4.2.4 Probiotics

This involves 'biologically effective microbial supply', made up of several bacteria (*streptococcus sp.*, lactic bacteria) or revivable spores, present as they are or on various media (lactoserum, germinating grains, etc.) which act favourably on growth, via mechanisms which are still poorly understood but most probably by modification of the intestinal flora. In fish, clear effects have been observed in various species during the first stages of rearing using live prey. For later stages, the effectiveness of such substances remains to be demonstrated.

It should be noted that probiotics are not true additives in the legal sense, nor are they drugs or medication.

18.4.2.5 Enzymes

When a digestive enzyme required for the breakdown of a given substrate is missing from an animal, it could appear attractive to add it to the feed along with the poorly digestible compound. Numerous attempts have been made to do this but have often resulted in failure. The concept of adding phytase to foods in order to improve the utilisation of phytic phosphorus and eventually to reduce the amount of phosphorus waste is currently being contemplated; it is carried out successfully in some terrestrial rearing units with pigs. The significance of these inputs is limited to the utilisation of plant raw material which are rich in phytic acid (Chapter 19).

18.5 CONCLUSION

The number of raw materials currently used in the feeding of aquatic animals is low. This results firstly from the unique requirements of fish (including high optimal level of proteins, requirement for particular essential fatty acids, relatively low tolerance to starch, role of attractants) which excludes certain ingredients from the formulae. Animal meals, despite their high price, are very significant for both fish and crustaceans, while it is more difficult to use plant products. However, in small amounts or in certain contexts, many ingredients can be used, especially for relatively omnivorous species such as carp, tilapia or Channel catfish. The list of products that can be used in aquaculture nutrition will thus logically become longer in the future.

In several areas, our knowledge about the value of various raw materials is still limited: protein or energy digestibility is often ignored (to cite only two points), as well as the true effects of antinutritional factors. These gaps can only be filled by applied research, which may be long term, even if limited only to the main aquaculture species.

18.6 BIBLIOGRAPHY

Anderson, J. S., Lall, S. P., Anderson, D. M. and McNiven, M. A. (1993) Evaluation of protein quality in fish meals by chemical and biological assays. *Aquaculture*, **15**, 305–325.

Hara, T. J. (1982) *Chemoreception in Fishes.* Developments in Aquaculture and Fisheries Science, **8**, Elsevier.

Larbier, M. and Leclercq, B. (1992) Les matières premières utilisées en aviculture. In: *Nutrition et Alimentation des Volailles.* INRA, Paris, 255–301.

NRC (National Research Council) (1993) *Nutrient Requirements of Fish.* National Academy Press, Washington DC.

Pike, I. H. and Hardy, R. W. (1997) Standards for assessing quality of feed ingredients. In: D'Abramo, L. R., Conklin, D. E. and Akiyama, D. M. (eds) *Crustacean Nutrition: Advances in World Aquaculture, Vol 6.* World Aquaculture Society, Baton Rouge, LA. 475–492.

19

Antinutritional factors

Undesirable compounds in raw materials are very varied in nature. They can include pollutants that are contaminants of the raw materials, such as pesticides or heavy metals, toxic compounds that appear in the basic product (for example in fish which are damaged to varying degrees before being transformed into fishmeal) or during transport or storage of the manufactured product. When considering plant materials, antinutritional factors *sensu stricto* are involved; these are compounds of certain plants that hinder the animal's digestive or metabolic utilisation of nutrients. If this concept is widely accepted, it is possible to include certain membranous plant compounds which, when consumed to excess, always decrease digestibility in monogastric animals. These substances exert an effect in opposition to attractants and are often classed as antinutritional factors.

In this chapter, the emphasis will be on antinutritional factors *sensu stricto* since, on the one hand, their content is not mentioned in standard tables and, on the other, their action is more difficult to take into account than that of nutrients during formulation.

19.1 GENERAL

19.1.1 Classification

Compounds with antinutritional properties are widespread in plant products and rare in animal or micro-organism products. They include:

- antinutrients in the true sense of the term, which limit the utilisation of specific nutrients such as antitrypsic factors;
- compounds that limit the efficiency of the digestive processes which may or may not be specific, such as phytic acid;

- compounds that act at the metabolic level, corresponding more to toxic substances than to antinutritional factors in the true sense. Examples are the antithyroid compounds found in the crucifers;
- substances with multiple modes of action, including appetite suppression.

Another possible classification is based on the chemical composition of the 'factors'. In the common factors, only a small number of chemical families are found: organic acids and amino acids (AA), peptides or proteins, glucosides, saponins, polyphenols and alkaloids.

Fungal or algal toxins form another category which is chemically extremely heterogeneous. There is no classification system for antinutritional factors in animal products. However, a brief summary of the most important undesirable compounds found in basic animal ingredients will be given.

19.1.2 Incidence and methods of counteraction

Antinutritional factors are sometimes metabolites or compounds which play a role in plant physiology. This is true for phytic acid and, more specifically, phytin (calcium and magnesium bi-phytates) which are stored forms of phosphorus and metals in the plant. This is why phytic acid and phytates are found mainly in the storage organs, grains and tubers. For the same reason, genetic selection can *a priori* only have quite a weak effect on this characteristic while the degree of maturation or even the culture conditions create a certain amount of variability.

In other cases, no physiological role has been attributed to these compounds and the conclusion has been reached that they are a means of defence for the plant against predators. This is true for very many alkaloids and can lead to all sorts of problems for animals. In extreme cases where acute toxicity is involved, the plant is no longer classed as edible and is regarded as poisonous. However, the threshold between plants rich in 'unfavourable' factors and those which are toxic is often a matter of dose or level of incorporation. In every example it is usual to encounter varieties with varying degrees of toxicity within one species. It is therefore theoretically possible to create by genetic techniques types low in 'antinutritional' factors. Obtaining varieties which are free or practically free from antinutritional factors, when possible, has provided a radical solution to the problems encountered by nutritionists.

Depending on the chemical nature of the compound, it is sometimes possible to decrease the antinutritional effect using appropriate treatment. Although organic acids are particularly heat-stable, a beneficial effect can almost always be obtained from heat-treating compounds which are protein in nature. Similarly, the liposoluble nature of some compounds means that they are eliminated during oil removal (pressing) and thus they are almost totally absent from cakes.

19.2 MAIN ANTINUTRITIONAL FACTORS

19.2.1 Phytic acid

This compound is formed from an inositol esterified by six molecules of phosphoric acid and constitutes *a priori* a source of phosphorus and inositol, a pseudo-vitamin, for fish (Chapter 9). Vertebrates do not possess the enzymes capable of effectively hydrolysing the ester bonds of this compound. The nutrients within phytates are thus unavailable, at least to animals that do not possess an active digestive flora, which is generally the case in fish. In addition, during its transit through the digestive tract, the molecule of phytic acid can form complexes with various molecules and in particular form salts with divalent metals such as calcium, magnesium, manganese, zinc, cobalt, etc. Thus it is possible for the molecule, if it is not already in the form of a salt, to become an 'antinutrient'. The harmful effect of phytic acid continues since the molecule can also act on certain digestive enzymes, for example decreasing the digestibility of dietary proteins.

All these effects are well known in higher vertebrates but have not been studied in detail in fish. Nevertheless, they have been observed in salmonids where the depressive effect of phytates on protein utilisation is clear. In contrast, in penaeid shrimps, phytic phosphorus is relatively available; no phytase has been found, but hydrolysis using another phosphatase seems possible.

Because the phytic acid molecule is so stable, there is no technological method which can be used to even partially hydrolyse it. In order to take into account its presence and its low availability, the classic method has been to assume that phytic phosphorus is less available than phosphorus in animal products or mineral salts (Chapter 10). Interactions with trace elements are also covered in this chapter. The incorporation of purified phytases into feeds has recently been proposed; this would allow supplements of mineral phosphorus to be decreased and hence reduce the amount of faecal phosphorus released into the environment. Other organic salts such as oxalic acid which have similar properties exist, but their practical importance in aquaculture nutrition is minimal.

19.2.2 Antitrypsic factors

These compounds are proteinaceous in nature and are very widespread in seeds, but they have an unfavourable effect when present in high concentrations in feeds. Thus in cereals, where they regulate the mobilisation of proteins for the seedling, their antinutritional role can be considered negligible. In contrast, in certain leguminous plants and in particular in soya (*Glycine max*), they reach such a high concentration that raw soya cake has long been considered as unfit for animal consumption. In this species there are at least two distinct antitrypsic proteins, the Kunitz factor and the Bowman-Kirk factor. Their mode of action has been very well studied in mammals and birds as well as in *in vitro* experiments. There is firstly inhibition of trypsin activity as well as, to a lesser extent, that of chymotrypsin, and elastase by formation of stable complexes, which leads to a very clear decrease in protein digestibility. After a certain period of time, a compensatory mechanism is set up and trypsin secretion is activated by a hypersecretion of cholecystokinine (Chapter 4) followed, in most species, by hypertrophy of the pancreas. This compensation is far from sufficient and protein digestibility of the diet remains low. In addition, the presence of these factors can lead to a deficiency in sulphur amino acids, even in balanced diets, as a result of the high cysteine content of the trypsin.

Antitrypsic factors are thermolabile. A large number of technical studies have been devoted to treatments that would allow their destruction without any significant alteration to the nutrient value of proteins. Temperature, pressure, humidity and time required are now well known. Fish appear to be very sensitive to antitrypsic factors, which are also found in other leguminous plants (for example beans) but at much lower concentrations. In practice, only cakes particularly low in antitrypsic factors can be used in fish feeds.

19.2.3 Other proteinaceous antinutritional factors

Antiamylases exist; these have an inhibitory effect on starch digestion in a similar way to that of antitrypsins on proteins. However, their role is minor. Among other factors which are proteinaceous in nature are enzymes such as urease, another thermolabile antinutritional factor found in soya which does not appear to be of great importance (it is toxic as a result of the liberation of ammonia in the blood and this is certainly of limited importance in ammonotelic aquatic animals), although it serves as a good indicator that the cakes are cooked. Other enzymes have antinutritional activity, such as myrosinase (that will be mentioned with glucosinolates), and lipoxydase which acts as an anti-vitamin A.

Lectins constitute another group of proteinaceous compounds which are ubiquitous. They have been found in both higher plants and micro-organisms, or lower animals. These compounds are often called haemagglutinins since they attach to blood cells which they agglutinate. In fact, their characteristic property is a particular affinity for certain polysaccharides, including those found in cellular membranes. Lectins appear to have an antinutritional effect by attaching to enterocytes, whose absorbent functions they inhibit. Several leguminous plants, in particular the pea (*Pisum sativum*) are rich in lectins. Cooking is a means of inactivating these factors although thermoresistant lectins do exist (ricin). Their harmful effects on fish remain almost unknown.

19.2.4 Glucosinolates and antithyroid factors

These compounds are mainly found in the Brassica family (Cruciferae) and are present in all parts of the plant. There is a large number of compounds which are either molecules synthesised by the plant or products of degradation which appear during technological treatments. To date, more than 70 compounds have been identified. Put together, these can represent up to 8% of the dry matter of the seed, twice as much as found in cakes. The two botanical species, rape (*Brassica napus*), mainly cultivated in Europe, and 'wild turnips' (*Brassica campestris*), culti-

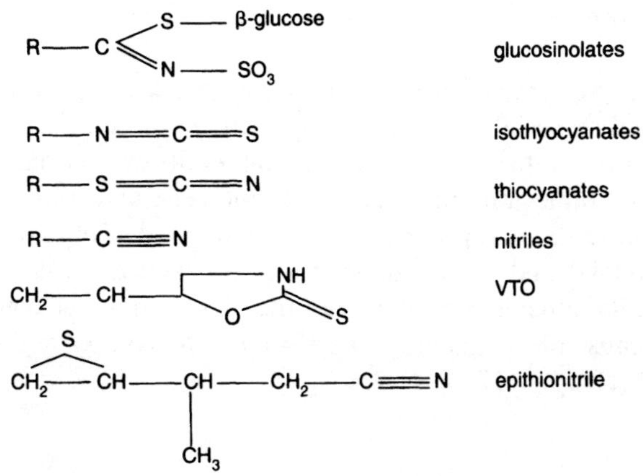

Figure 19.1. Chemical formulae of glucosinolates and some compounds derived from them (after Larbier and Leclercq, 1992).

vated in Canada, are often confused under the generic name of rape (colza), but do not have the same content; the turnip is richer than rape in glucosinolates.

All glucosides are formed from a molecule of glucose and a sulphur compound consisting of a lateral chain which is variable in nature. Glucosinolates are not themselves very toxic, although they can influence feeding behaviour by their acrid taste. Most 'antithyroid' effects result from molecules produced by the hydrolysis of glucosinolates under the effect of myrosinase, which is not an enzyme but an enzyme complex. The formulae of the compounds most commonly formed in this way are shown in Figure 19.1. The lack of appetite frequently manifested by animals when given foods containing rapeseed meal is mainly due to the isothiocyanates, while the antithyroid effect is due mainly to the 5-vinyl-2-thio-oxazolidone or VTO, also called goitrine. This molecule inhibits the attachment of iodine to thyroid hormones. This effect cannot be compensated for by input of iodine supplements in the diet and results in hyperplasia of the thyroid, or goitre.

There are several methods for making cruciferous cakes less toxic, in particular by the thermal inactivation of myrosinase. None is completely satisfactory and the problem was completely changed when geneticists, after obtaining 'zero erucic' (in which the oil contains almost no erucic acid), selected 'double zero' varieties, which were also very low in glucosinolates. The Canadian 'double zero' rapeseed, which is in fact a rape low in glucosinolates, was named 'canola'. The sensitivity of fish to glucosinolates and their derivatives seems to be quite variable from one species to another, even within the same group such as the Pacific salmon.

19.2.5 Saponins

These compounds, as with glucosinolates, are glucosidic in nature. They are found in a large number of plant species in the whole plant and especially in the leaves. One of the most well known, present in lucerne (*Medicago sativa*) consists of a molecule of glucose which has been esterified by medicagenic acid. There are tens of molecules of this type.

Saponins have two types of property: they can reduce the palatibility of a diet by their bitterness and, especially through their affinity for sterols, interfere with the intestinal absorption of these compounds, and limit the recycling of bile salts. Saponins are very stable, but there is significant genetic variability and lucernes containing low amounts of saponins are now cultivated. The toxicity of saponins to fish is controversial and poorly understood in practice. In crustaceans, experimental data are still almost non-existent, but the toxicity must be higher *a priori* than in vertebrates, because of the inability of arthropods to synthesise the sterol nucleus, which is essential (Chapter 17).

19.2.6 Gossypol

This compound is a polyphenol. It is abundant in the microscopic glands of cotton seeds (*Gossypum* sp.). Its effect appears to be complex: after ingestion, there is a decrease in protein digestibility. In addition, in trout and if the dose is high, growth is greatly reduced and multiple signs of toxicity appear in the kidney, liver and spleen. Humid heat treatment of the cake allows the formation of complexes in which 90% of the gossypol is inactivated. Very productive strains have recently been obtained, which have few glands and thus no gossypol.

There are many other glucosides which have harmful effects, in particular cyanogenic glucosides, the hydrolysis of which releases a violent poison, cyanhydric acid. They are present in cassava and some seeds of leguminous plants, etc.

19.2.7 Tannins

Tannins are polyphenols which belong to two families of which only one, the condensed tannins (Figure 19.2), includes the true antinutritional factors. These are the macromolecular polymers found in graminae such as sorghum and in certain legumes such as broad beans (*Vicia faba*) or in some crucifers such as rape. They are localised in the seed coat.

As with glucosinolates and saponins, the antinutritional effect can be doubled by the reduction of appetite; this is particularly clear in the case of 'bird resistant' sorghums. The true antinutritional effect results from the formation of non-specific complexes between tannins and certain proteins, especially digestive enzymes. It thus involves inhibition (in the broad sense) which decreases digestibility. Condensed tannins are molecules which are easily denatured by heating, but simply dehulling the seeds largely eliminates them. In addition, for beans, geneticists have obtained varieties low in tannins, in which the proteins are more digestible.

19.2.8 Alkaloids

Strictly speaking, alkaloids are nitrogenous molecules derived from amino acids (lysine, ornithine, tyrosine, phenylalanine and tryptophan). In a wider sense, they

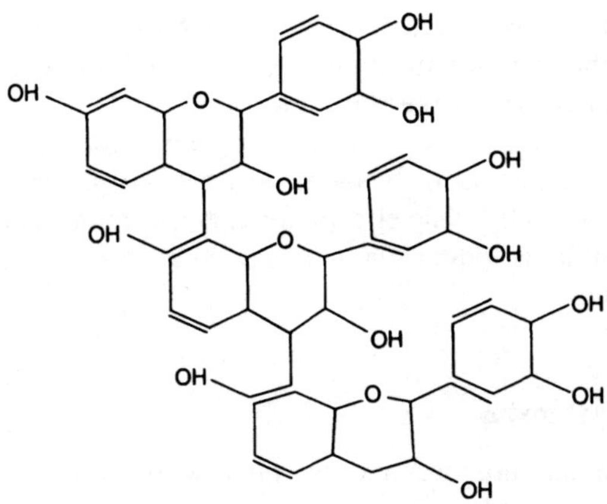

Figure 19.2. Chemical structure of condensed tannins (after Larbier and Leclercq, 1992).

form a very large family of molecules which always include at least an atom of nitrogen, usually as part of a heterocycle. Several thousand are known and it is estimated that a quarter of all higher plants contain them; they are rarely found in food plants and are unknown in lower plants. Some, such as caffeine and nicotine are well known for their effects on the nervous system, while others are poisonous in varying degrees. All lupins used in foods for people or animals contain alkaloids (lupinine, lupanine, etc.) which have such a pronounced bitterness that they must be extracted by soaking in water in order to make the seeds or flours consumable. Since the nineteenth century, varieties containing almost no alkaloids have been obtained from four species including the white lupin (*Lupinus albus*) which is mainly cultivated in Europe and the blue lupin (*Lupinus angustifolius*) which is mainly cultivated in Australia. These 'sweet' varieties have now replaced the bitter ones. They are very well accepted by fish and produce excellent performance. Seeds of sweet lupins were one of the first sources of plant protein used to feed pond-reared fish.

Numerous other alkaloids also exist. In the Fabaceae family (Leguminosae), there are vicine and convicine (broad bean). Harmful toxicity has been obtained experimentally in fish using several alkaloids, but not, to our knowledge, using alkaloids from normal raw materials.

19.2.9 Other antinutritional factors from higher plants

Further antinutritional factors can be added to the list of those found in fish feeds, although their occurrence may be low, effects poorly understood and economic importance unknown. There are, for example:

- oestrogenic factors in plants, which vary chemically and can affect female maturation;

- fatty acids with toxic effects such as erucic acid from crucifers or sterculic acid in cotton oil (an inhibitor of desaturation reactions of fatty acids) which are almost completely eliminated from cakes (and meals);
- non-protein amino acids, poisons blocking protein synthesis by competing with protein-forming amino acids for RNA messengers. These particular amino acids sometimes have additional toxic effects, for example mimosine which is present in a tropical leguminous fodder plant, *Leucaena glauca*;
- various antivitamins, etc.

19.2.10 Fungal or algal toxins

Fungal toxins are compounds secreted by microscopic fungi, commonly called moulds. Some of these molecules are very toxic; the carcinogenic effect of aflatoxins, for example, is 100 times higher than that of the most active carcinogenic manufactured compounds. The danger of mouldy plant products has been known for a long time and it is known that certain ingredients of tropical origin (peanuts, oilcake) are very often contaminated, sometimes only slightly, since it is difficult to prevent the development of mould in a warm, humid environment. The nature of mycotoxins has only recently been determined and the different molecules, present in very small quantities, are difficult to measure accurately. For these reasons, estimation of the toxicity of products spoiled by micro-organisms remains difficult, especially for fish. We shall thus limit ourselves to a very succinct account.

19.2.10.1 Aflatoxins

These are quite common compounds which are synthesised by *Aspergillus flavus*. The most commonly contaminated raw material is peanut oilcake, but many others may also be, including cotton or copra cake, cassava and cereals. Aflatoxin molecules fluoresce under ultraviolet light and this is used to detect them; molecules with blue fluorescence are called B aflatoxins and those with green fluorescence, G aflatoxins. The structure of aflatoxin molecules is now known (Figure 19.3). Their toxicity varies depending on animal species. Thus aflatoxin B_1 appears to be the most toxic. Coho salmon and American catfish are more resistant than trout, which contract liver cancer when they receive a ration containing a level as low as 0.4 ppb (mg.t^{-1} of food) of aflatoxin B_1 over a long period. From all studies to

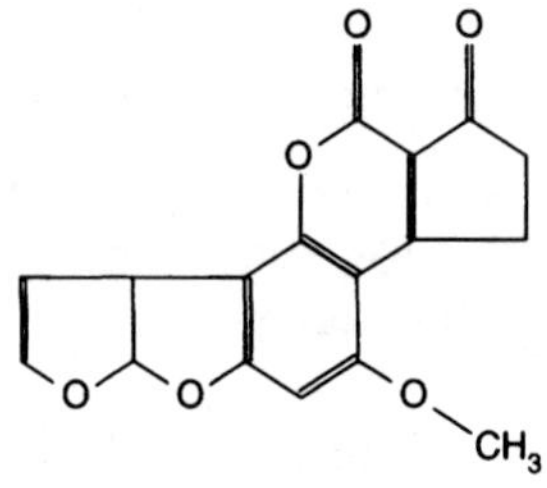

Figure 19.3. Chemical formula of aflatoxin (after Larbier and Leclercq, 1992).

date, trout appear to be the most sensitive to aflatoxins. In addition, the effect of aflatoxin is reinforced by sterculic acid, which renders cotton oilcake contaminated with *Aspergillus flavus* very dangerous. It should be noted that the concentration leading to chronic toxicity is of the same order of magnitude as the detection threshold.

19.2.10.2 Other fungal toxins

These are numerous and generally less well known than the aflatoxins. Trichotecenes secreted by moulds of the genus *Fusarium* and zearalenone secreted by *Fusarium roseum* are widespread on mouldy cereals, as well as ochratoxins secreted by *Aspergillus* and *Penicillium* species. Ochratoxin A has a toxicity close to that of the aflatoxins, for trout. For most other toxins, the effect has not been observed or has not yet been studied in fish. Note, however, that a trichotecene – vomitoxin – does not cause vomiting in trout, but it does totally suppress food intake.

19.3 ANTINUTRITIONAL FACTORS IN ANIMAL PRODUCTS

It is much more difficult to describe antinutritional factors in the true sense in the case of animal products than for plant products, although antitrypsic factors have been found, for example in fish meal. Similarly, avidine, a protein found in raw egg white, is a true antivitamin. Nevertheless, several types of harmful compounds exist in raw animal materials.

19.3.1 Thiaminase

The presence of an enzyme that hydrolyses thiamine has often been indicated in raw fish, more specifically in freshwater fish. It has caused disease in fish farms using moist foods based on ensilage or minced fish. This enzyme is inactivated by heating, and accidents can be avoided simply by reducing the duration of contact between enzyme and substrate, i.e. only using freshly-prepared feed and increasing the content of pure thiamine.

19.3.2 Histamine

Histamine is a biogenic amine derived from histidine which is synthesised by animals and plays a role in allergy phenomena in higher animals. It is also produced after death in fish by bacterial activity, in particular that of *Proteus morganii*. In addition, histamine content is one of the criteria widely used to evaluate fish meal quality, or at least the freshness of the product used for its manufacture. It cannot be used alone, however, since histamine content increases at the same rate as the fish spoils and then decreases when deterioration is very advanced. This content can reach particularly high levels in tuna flesh. Illness has been observed in humans after consumption of tuna rich in histamine. In contrast, this same product used as a fodder fish does not appear to cause histamine shock in the fish to which it is being fed.

Gizzeroline, which acquires its name from the erosion it causes in the gizzard of turkeys and chickens, is a derivative of histamine formed when fish meals are overheated. It attaches to the same receptor as histamine but is 200 times more active. There is not yet any definitive way of chemically measuring this poorly-defined amine which can only be quantified by using biological tests on chickens.

19.3.3 Peroxidised lipids

Unsaturated fatty acids and particularly long chain n-3 polyunsaturated fatty acids (HUFAs) are very sensitive to peroxidation which occurs naturally *in vivo* (Chapter 7). These reactions can also occur in raw materials, during heating due to pelleting or cooking-extrusion processing, or during storage of raw materials or feed. The unstable compounds that appear during the reaction, as with the stable, toxic compounds (aldehydes and ketones) which appear at the end of the chain (Chapter 7), thus constitute true antinutritional factors. They are usually measured globally and collectively listed under the term Thiobarbituric Acid Reactive Substances (TBARS). They act at three levels: they can cause loss of appetite in fish with regard to the feed; induce vitamin C and especially vitamin E deficiencies; and have truly toxic effects. The latter include digestive intolerances linked to inhibition of amylase, lipase and trypsin secretion as well as partial inhibition of the Krebs cycle enzymes and reduction in ATP synthesis in the mitochondria. However, most of the symptoms observed after adding peroxidised oils to the diet, including red blood cell haemolysis, muscular dystrophy, exudative diathesis or microcytic anaemia, are in fact characteristic of α-tocopherol deficiency. These symptoms result from the destruction of cell membranes following peroxidation of HUFAs and are inhibited by the addition of vitamin E. It must be remembered that the peroxidation of fatty acids in raw materials, feeds or even the food bolus is slowed down greatly by synthetic antioxidants and tocopherols, but not by tocopherol acetate which is inactive before hydrolysis (Chapter 9).

19.3.4 Algal toxins carried by fish

A large number of compounds which are toxic to man or fish are secreted by certain phytoplankton species during phenomena called 'red tides'. Their description is outwith the remit of this book. Nevertheless, one group of toxins meriting mention is the ciguatoxins and the maiotoxins. Ciguatoxins are secreted by a unicellular alga, *Gambierdiscus toxicus*, which develops on macrophytic algae, especially after the breakdown of coral reefs. The structure of ciguatoxin has only recently been discovered. It is a very complex molecule, very unusual in biological chemistry, which can block the sodium channels of cell membranes and is thus extremely toxic. The formula of maiotoxin, secreted by the same alga, is not yet known. These molecules accumulate along the food chain and concentrate preferentially in older littoral carnivorous fish.[1] They are thus apparently not very toxic to fish. However, it is very harmful for the farmer to use toxic fodder fish in aquaculture, or even fish meal manufactured from it, since ciguatoxin is thermostable.

Bacterial toxins can be present in spoiled animal products and lead to devastating illness. However, it is not usual for them to be included as antinutritional factors. They are among the compounds to which are attributed faults in feed security; this group also includes the algal toxins mentioned above.

19.4 CONCLUSION

Considerably less work has been carried out on the study of the effects of antinutritional factors of plants on fish than on higher vertebrates. In certain cases this is justified, since the raw materials are rarely used in aquaculture feeds for economic reasons or various technical reasons. In other cases (antitrypsic factors, aflatoxins), this study has revealed great sensitivity in fish, thus imposing rigorous control over the quality of primary ingredients. However, in the area of raw animal materials and especially those produced from the marine environment (the only source of n-3 HUFAs), knowledge is more advanced, although it still remains insufficient.

19.5 BIBLIOGRAPHY

Hendriks, J. D. and Bailey, G. S. (1989) Adventitious toxins. In: Halver, J. E. (ed.) *Fish Nutrition*. Academic Press, San Diego, 605–651.

Larbier, M. and LeClerq, B., (1992) Les matières premières utilisées en aviculture. In: *Nutrition et Alimentation des Volailles*. INRA, Paris, 255–301.

Liener, I. E. (1980) *Toxic Constituents of Plant Foodstuffs*. Academic Press, New York.

[1] In man, repeated consumption of contaminated fish leads firstly to benign symptoms, then to more serious ones and may lead to death.

20

Formulation of feeds for aquaculture

Since compound feeds for fish have been produced on an industrial scale, their manufacture has benefited from high-performance computing tools used for terrestrial species, especially in least cost linear formulation. However, in aquaculture, the efficiency – and hence the use of these tools – has been impeded by the lack of precision concerning nutritional or technical constraints. Recent concern about preservation of the environment and even the development of various techniques such as cooking-extrusion has widened the range of parameters that must be taken into account. The problem has become greater in this new context. Nevertheless, the formulation of feeds for aquatic animals is based on the same principles as those used for terrestrial animals, which are described in this chapter.

20.1 GENERAL PRINCIPLES

20.1.1 Basic objectives

When there is a wide range of raw materials available to feed a given animal, there is an infinite number of combinations that correspond to the formulae of utilisable feeds. Nutrition studies allow the determination of the animal's requirements and chemistry provides the content of nutritive elements in the raw materials; together, these allow optimised formulae to be calculated. However, the number of possible theoretical solutions remains infinite. Least cost formulation allows an additional, capital hurdle to be cleared, in the selection of the most economical solution.

20.1.2 Overall scheme

Finding the most economical feed which meets the animal's requirements is, however, only one of the aspects of the problem which must be resolved by the

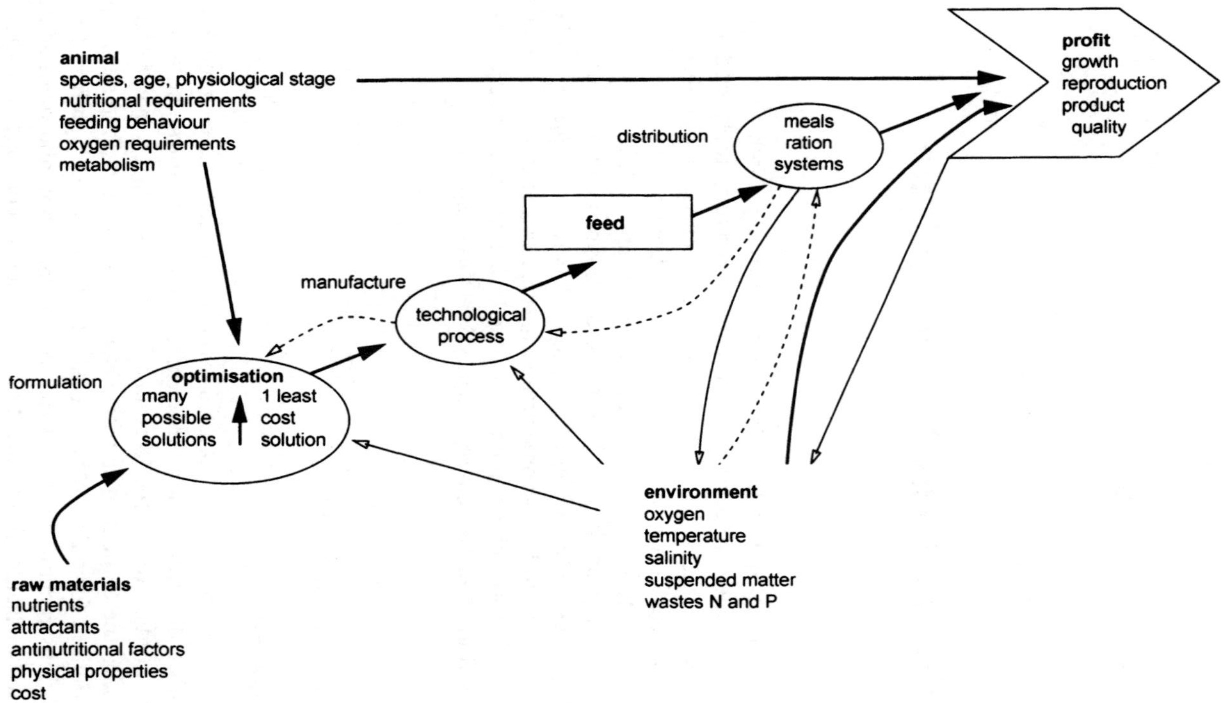

Figure 20.1. Problems involved in feed manufacture.

feed manufacturer, or more precisely, the formulator. The best food is not necessarily the one that best meets the animal's requirements at the lowest cost (purely from a nutritional aspect), nor the one which ensures the most profit for the feed manufacturer or the aquaculturists. The best feed is the one that ensures the highest profit for the group of people or enterprises involved, including manufacturers, aquaculturists and processors, as well as the community responsible for the preservation of the environment. Interrelationships exist between the specifications related to the animal, to the primary constituents and to the environment, as well as the implications of the various stages of manufacture and use of the feed; these are shown in Figure 20.1.

20.1.3 Features linked to the animal, feeds and the environment

Among the features of the animal are not only its nutrient requirements but also its feeding behaviour, aptitude for digesting certain feeds and other characteristics, particularly metabolic ones, which depend on the species and physiological status. Among the characteristics of feeds, not only nutrients – on which many studies have been focused – must be taken into account, but also antinutrients, palatibility (or unpalatibility) factors, as well as a certain number of physical properties. With regard to aquatic environmental characteristics, it is necessary to consider oxygen content, mineral and organic matter as well as suspended solids. In general, account must be taken of the repercussions of fish farming on these factors.

The environment has repercussions for physiology and dietary requirements and, in return, each of these factors has an effect on the environment. It is thus necessary to take into account the various feedbacks and interactions which result. Other parameters (in the broadest sense) such as the methods of manufacture or feed distribution should also be considered.

20.2 THE PRINCIPLE AND LIMITATIONS OF THE LINEAR LEAST COST FORMULATION

Linear least cost formulation is the main method currently used, analogue calculators having been abandoned. The general principle consists of solving a system of inequations, as shown in Box 20.1.

Assume that a series of primary materials M is available, for which the content a_{ij} of each significant nutrient is known, the index i relates to a nutrient (for example lysine) and the index j to the ingredient (for example fishmeal). If b_i is the requirement for the nutrient i (for example lysine requirement), in order for the sum of the nutrient i inputs to be greater than or equal to requirements, the formula is:

$$\sum a_{ij}x_j \geq b_i$$

Box 20.1 Mathematical presentation of the problem posed by least cost linear formulation

a_{ij} = raw material j content of a 'nutrient' (broad sense) i.
b_i = requirement (broad sense), i.e. recommendation standard for nutrient i.
B_i = maximum admissible for nutrient i.
x_j = % (or quantities) of raw material j.
n = number of primary ingredients available.
m = number of nutrients taken into account.
c_j = cost of primary ingredient j.

Inequalities relevant to nutritional constraints

$$a_{11}x_1 + a_{12}x_2 + \cdots + a_{1j}x_j + \cdots + a_{1n}x_n \geq b_1 \quad (1)$$

$$a_{21}x_1 + a_{22}x_2 + \cdots + a_{2j}x_j + \cdots + a_{2n}x_n \geq b_2 \quad (2)$$

— —

$$a_{m1}x_1 + a_{m2}x_2 + \cdots + a_{mj}x_j + \cdots + a_{mn}x_n \geq b_m \quad (m)$$

with

$$x_1 + x_2 + \cdots + x_n = 100 \quad (m+1)$$

For certain nutrients, the following can also be applied

$$a_{i1}x_1 + a_{i2}x_2 + \cdots + a_{ij}x_j + \cdots + a_{in}x_n \geq b_1 \quad (p)$$

$$a_{i1}x_1 + a_{i2}x_2 + \cdots + a_{ij}x_j + \cdots + a_{in}x_n \leq B_1 \quad (p+1)$$

Inequations or equations relevant to technical constraints (examples)

$$x_j > d$$

$$x_k \leq e$$

$$p < x_o < f$$

$$x_h = g$$

These inequations or equations are used for very varied goals.

The system with $m + 1 + N$ inequations or equations, where N represents the total number of $(p + 1)$ inequations, relevant to the technical specifications, includes an infinite number of solutions.

Least cost linear formulation only allows the retention of one solution, that which minimises the function

$$C = c_1x_1 + c_2x_2 + \cdots + c_jx_j + \cdots + c_nx_n$$

i.e. the formula which is the least costly. The qualitative term 'linear' assumes additivity of the contents $a_{ij}x_j$, i.e. the validity of the type (p) or $(p + 1)$ equations and inequations.

where x_j is the percentage of raw material included in the formulation. The term nutrient is used here for true nutrients, macronutrients, energy, ash, fibre, etc.

A last calculation relates to cost, which must be minimised and confers a fundamental role on this parameter. The content of a given nutrient can effectively be zero in a given ingredient and can also be unknown, in which case it is possible to assume it is zero (if this is likely) without altering the entire operation. In contrast, a single unknown cost blocks the calculating process. In fact, if the inequations $(1)\ldots(m)$ are replaced by equations, a system of $(m+1)$ equations with n unknowns would have to be resolved. This system (except if $m = n$) would either be impossible to resolve or have an infinite number of solutions. If the equations are replaced by inequations, the number of solutions is infinite, but the optimal solution is chosen, while minimising the function:

$$C = c_1x_1 + c_2x_2 + \cdots + c_mx_m$$

where C is the cost, i.e. retaining the least costly solution.

The inequations are not all related to minima corresponding to requirements; in certain cases it is necessary to impose maxima, such as when the content of a nutrient needs to be limited because excesses are harmful, for example essential amino acids (EAA) or certain minerals. For certain nutrients, the formula is therefore as follows:

$$a_{i1}x_1 + a_{i2} + \cdots + a_{ij}x_j + \cdots + a_{in}x_n \geq b_i$$

$$a_{i1}x_1 + a_{i2} + \cdots + a_{ij}x_j + \cdots + a_{in}x_n \leq B_i$$

where B is the maximum allowed.

In practice, this type of inequation is little used for nutrients, since excesses are mostly expensive and consequently excluded from least cost formulae. However, a minimum level b_i and a maximum level B_i can be imposed in order to maintain the content within a given range. The use of an upper limit only is common for compounds such as fibres or total minerals. In certain cases, equations and not inequations are used: the energy level can be fixed, for example, since the recommended EAAs used are valid for a given energy level or even because the aforementioned level is that for which the energy is least costly. Data such as b_i or B_i constitute nutritional constraints.

Other specifications can be imposed on values of x_{ij} if the manufacturer wishes to limit or, conversely, boost the percentage of a given ingredient. This involves technical constraints. These set a minimum value for a given ingredient, even if it is not profitable for its nutrient supplies alone. In contrast, for technical reasons its use can be limited even where it appears to be economical. It can be seen how these constraints, whose use can only increase the cost of formulae established solely on the basis of nutritional requirements, allow a series of factors to be taken into consideration, which could not be introduced into the first set of inequations.

20.3 THEORETICAL LIMITS OF LINEAR FORMULATION

20.3.1 Linearity and additivity in nutrition

The principle of linear formulation by itself is based on the calculation of nutrient content using a simple mathematical formula which is not theoretically a subject for discussion, at least as far as crude nutrients are concerned. However, this is not always true for available nutrients, in the broad sense of the term. For example, as a first approximation it is assumed that the digestible energy of a diet is a linear function of its components. This hypothesis is accepted by scientists measuring digestible energy, but has rarely been perfectly verified: the energy value of an ingredient can decrease greatly when the level of incorporation (x) is high, as is observed in the saturation of digestion-absorption phenomena. In other words, a_{ij} is no longer a constant, but a function of x and the equations listed above should not be used. Among the classic examples of this type of non-linearity is the energy value of starch which, in fish, decreases at high levels of incorporation; this effect is even more pronounced when the starch is raw and the fish has low amylase activity (Chapter 8). Under these conditions, the principle of linear formulation is only valid when transposed into more complex models which allow the estimation of the dietary content of nutrients.

A more serious criticism should also be mentioned: the digestible energy of one ingredient can very easily depend on the presence of other ingredients, a phenomenon qualified as an interaction and which, in theory, invalidates the very principle of linear formulation. Interaction is frequent (Chapter 4). For example, the digestibility of saturated fatty acids is improved by the presence of unsaturated fatty acids added in another ingredient. With regard to requirements for essential amino acids, those for methionine depend on cystine input (Chapter 6). There are also many interactions between vitamins: choline – vitamin B_{12} – folic acid, and even choline – vitamin B_{12} – folic acid – methionine (Chapter 9).

20.3.2 Uncertainties and unknowns

The power of computing tools can often be used for linear programming under conditions where numerous parameters are missing. This is the case in the formulation of feeds destined for consumption by fish whose requirements are poorly understood, or in the use of primary constituents whose composition is not entirely known. It is also tempting to introduce many specifications even when they are not justified either scientifically or technically. It goes without saying that under such circumstances the power of the tool rapidly becomes an illusion. However, each gap does not have the same practical consequences: gaps concerning animal characteristics are even more numerous than the list of farmed species and tend to increase more rapidly than the acquisition of knowledge. Thus one is forced to borrow missing data on digestibility, requirements or even tolerance to certain antinutritional factors from related species. The risks taken in such a step are variable.

- For certain nutrients, such as vitamins, considerable margins of security are the rule; there is no reason, *a priori*, for not using the same supplements, in excessive amounts, for an unknown species as for a known species.
- Some requirements, such as those for essential amino acids expressed by 16 g N, appear to be very similar not only within a group of similar species but also for all fish. Therefore, only a minimal risk is taken in choosing the same essential amino acid profile for an unknown species as that determined for another species (Chapter 6).
- In many cases, it is tempting to replace an unknown value for requirements by a mean for requirements determined in related species, or even, if no closely related species has been studied, by the mean determined for fish in general. In this case the risk is always very difficult to evaluate. In the vast group of tilapias, for example, nutritional requirements appear to vary little between species. In the salmonids, however, where carnivorous diets predominate, optimal protein contents of foods and the sparing effect of proteins by lipids differ significantly between species.
- The digestibility of a nutrient appears to be relatively stable within a given type, for example in fish with carnivorous, omnivorous or herbivorous tendencies. A large error is thus not being made in estimating that the digestibility of a given ingredient is the same in sea bass as in trout. In contrast, a significant risk is being taken by considering that raw starch is equally digested by marine carnivorous fish as it is by carp or tilapia.

It is thus impossible to come to a clear conclusion about the repercussions caused by gaps in knowledge on the result of the linear programme. The unfortunately frequent lack of data necessitates interpolations or extrapolations which can greatly alter the validity of 'optimised' formulae which are apparently extremely precise.

20.3.3 Characteristics not easily quantified

The palatibility of an ingredient or a feed is a function of the concentration of certain substances that have defined gustatory characteristics, for example acidic, bitter or sweet flavours. However, although some of these substances can be measured, it is not possible to calculate the flavour of a mixture. In addition, very many other factors come into play, such as the hardness, solubility or viscosity of an ingredient or even those linked to the animal, such as age, physiological state and how accustomed it is to previous diets that might affect perception of taste. In practice, quantitative evaluation of organoleptic qualities is not possible by linear programming. Many other features such as the physical properties of the mixture, which are nevertheless very important in feeding technology, can also not be taken into account directly.

20.4 METHODS FOR COUNTERACTING DIFFICULTIES

20.4.1 Interactions – various solutions

In certain cases, interactions can be taken into account very simply using linear programming, for example with the interaction between methionine and cystine:

$$\text{supply of methionine} \geq b_m$$

$$\text{supply of methionine} + \text{cystine} \geq b_{SAA}$$

where b_m and b_{SAA} are the total requirements for methionine and sulphur amino acids (SAAs) respectively.

The first inequation shows that methionine requirements can only be met by this amino acid. The second indicates that when the requirement for methionine is met, one or other amino acid is effective in meeting the total SAA requirements.

In other cases, as with the interactions between vitamins or amino acids already described, the phenomena cannot be described using simple inequations, because of their complexity. Nevertheless, it is possible to have conditions where a nutrient is present in 'saturating' quantities (supply at its minimal quantity, equal to the standard, because of its cost) which allows simplification of the problem. Some interactions, such as those existing between phosphorus and calcium, or lysine and arginine, although they are often cited, exist in terrestrial animals but not in fish. Linear programming can then be used without any hesitation.

While interactions between nutrients alter the estimation of $\sum a_{ij}$, the principles of linear formulation itself are not respected. It must therefore be ensured that the biases are not excessive; this is outwith the field of this chapter.

20.4.2 Requirements, recommendations and safety margins

The simplest method of compensating uncertainty is to use a safety margin which corresponds to the difference between the recommendations (standards) and requirements (Chapter 2). The margin can be justified for very many reasons linked to the imprecision of determination itself, to the heterogeneity of the animals or even to uncertainties about the composition of the raw ingredients. It is often the nutritionists who choose the margin in relation to statistical calculations, or other considerations relative to their experiments. However, the manufacturer frequently modifies the recommendation, adding a new margin, taking into account the variability which may exist within the primary constituents being used. In certain cases (at least in terrestrial farming), the chosen margins are inversely proportional to the standard deviation of the content of a given ingredient. However, more often, a much more simple step is taken, each firm having its own buying and analysis policy and choosing its own margin.

When it is necessary to formulate a feed for a species whose requirements are poorly or little understood, most manufacturers tend to resolve the problem by borrowing the recommendations for another known species and increasing them. This new precaution is just as difficult to justify as the multiplication of margins, since it leads to a formulation which drifts further and further away from the 'least cost' concept.

20.4.3 Formulation using non-constant nutritional characteristics: utilisation of imaginary raw materials

The phenomenon of variation of digestibility of an ingredient with its level of incorporation has been described above. When this occurs, the manufacturers use a trick which, although not perfect, has the advantage of being simple: for example, instead of considering that they have a starch whose energy value decreases from 14 to 10.8 $MJ.kg^{-1}$ when its incorporation level increases from 10–30%, they consider that they have a series of four, whose use they set at between 0 and 10%, 10 and 20%, 20 and 30%, and 30 and 40%, and to which they attribute energetic values of 14, 12.4, 10.8 and 9.2 $MJ.kg^{-1}$ respectively; all these starches cost the same amount. This method can be used for values which increase or decrease in relation to their incorporation level; it does not matter which way as long as it is known. This involves the use of technical specifications, the use of which also involves many other considerations (see Box 20.2).

Box 20.2 Examples of use of technical specifications

1. $x_j = p$	Requirements met (for example for vitamins or trace elements) by the input of a premixture included as an ingredient.
2. $x_j \leq q$	Limited use of an ingredient – with certain toxicity; – decreasing appetibility; – with physical properties which are unfavourable for manufacturing processes; – too little known by the manufacturer; – too indigestible and thus too pollutant; – etc.
3. $x_j \geq r$	'forced' use of an ingredient – which improves appetite; – which contains 'unknown factors'; – containing nutrients, the requirements for which are poorly understood; – etc.
4. $r \leq x_j \leq s$	Imposition of a margin – ingredient has both type 2 and 3 properties; – linked specification (limited temporal changes to formulae). In this case the double inequation is applied to the main ingredients; – translation of a real primary ingredient whose nutritional characteristics vary with level of incorporation, into a series of imaginary primary ingredients in which this characteristic varies discretely (discontinuously).

20.4.4 Taking account of various 'nutritional' factors using technical constraints

There are many nutrients for which not only requirements are still imprecise, but analyses of raw materials are largely missing; this is the case for vitamins and mineral oligo-elements. In this case, where analyses are long and costly, the quasi-universal solution consists of incorporating a premix containing vitamins and possibly minerals in sufficient quantities to meet requirements, with large safety margins, without taking into account the supply of other ingredients. A technical specification imposing, for example, 1% of a premix, considered as a raw material, would thus replace more than 10 inequations which could only be used ill-advisedly.

Technical constraints also allow account to be taken of compounds that are difficult to measure or for which there is no standard. They can also be used to regroup, empirically, several advantages or disadvantages. In the chapter on primary constituents, for example, we have described several advantages of fish meals. Manufacturers often refuse (for foods destined for carnivorous species at least) to decrease the percentage of fish meal below a certain threshold in order to avoid the risk of unpalatibility, or deficiency in n-3 series essential fatty acids or EAAs, etc. Conversely, antinutritional factors, fibre-rich repellant compounds or simply factors which are repulsive to fish tend to limit the use of raw materials such as brewer's yeast, oilcakes or other plant products. It would certainly be possible to use analogous inequations to those of nutritional specifications in order to limit the antinutritional factors themselves, but this is not yet possible due to the lack of analysis and knowledge about the reactions of animals to these compounds. The practical solution thus consists of imposing a minimum on fish meal and a maximum on plant products which are sources of antinutritional factors and fibre.

20.4.5 Limited temporal changes

Over time, variations in costs of raw materials can decrease the profitability of certain ingredients and, conversely, render other products more competitive. The main aim of least cost linear formulation is to allow manufacturers to make immediate use of these changes in terms of formula. However, if these changes are too extreme, they can for example replace a formula rich in carbohydrates with one rich in lipids or substitute large amounts of plant protein for animal protein. Various adverse effects can result: a fall in the animals' appetite, temporary decrease in digestibility, or even loss of confidence by the fish farmer who might have the impression that they have bought two different feeds under the same name. The normal solution consists of making these changes more progressive using another category of technical specifications: limited temporal changes..

Technical specifications can therefore provide solutions to a wide range of problems and, despite their often empirical nature, constitute an important part of the know-how or 'secrets' of manufacturers.

20.5 FACTORS LINKED TO TECHNOLOGY OR THE ENVIRONMENT

20.5.1 Formulation and the manufacturing process

In order to be utilisable, a fish feed must have a certain stability and even a certain shape, both characteristics which are dependent on the physico-chemical properties of the raw materials. In other words, ingredients cannot be included in the formulae which render the final product too powdery, too doughy or too hard, too buoyant or conversely, too dense in relation to water. It is possible to modify the texture of the final agglomerate by choosing an adapted technological process, or by adding a binding agent – which may or may not have nutritional properties – which can be incorporated at a given rate (Chapter 21). In practice, however, the choice of ingredients, addition of binding agents and the choice of manufacturing process are directly linked.

Firstly, one can choose cooking-extrusion rather than pelleting because it is the only method of obtaining floating feed or even of incorporating the percentages of lipids required in salmonid culture which allows conservation of the environment (25–32% total lipids). However, even with this process it is necessary to ensure that the pellet have a solid texture (or 'matrix') and the choice of products such as pregelatinised starch or native proteins is essential. In this situation, no physical or physico-chemical feature of additives can be taken into account in the same way as a protein, mineral or fibre content. The solution consists of limiting the ingredients that make the pellet fragile and imposing a minimum level on those that provide texture, binding agents being one of these. The manufacturer thus uses technical specifications that are only valid for a process or for certain combinations of processes, species and main raw materials.

In the case of foods for shrimps, where water stability is important, manufacturers generally have a wider range of processes to choose from. They can opt either for cooking-extrusion or pressing. They thus use very distinct specifications for each process, the use of non-nutritive binding agents only being systematic for pelleting (Chapter 21).

20.5.2 Formulation, manufacture and distribution

Feed distribution methods and fish behaviour determine the physical characteristics of the food. It is clear that feed projected by a pressurised distributor or those delicately distributed by hand do not require the same texture which could, in theory at least, lead to different techniques or formulae.

20.5.3 Formulation and protection of the environment

The necessity for protecting the environment and, more particularly, limiting water pollution now constitutes a major constraint, although awareness of it is relatively recent. Pollution from fish farms depends on many factors which can be divided into those related to rationing and feed distribution and those related to digestibility and nutritional equilibrium (Chapters 4 to 7).

- The physical properties of the feed acquire increasing importance. A low-polluting diet must firstly break down as little as possible in water and be ingested as rapidly as possible without the risk of being rejected (food taken in and then spat out).
- The same feeds must also, as indicated above, result in minimal faecal emissions and excretion.

The concept of low-polluting diets therefore has several other consequences for formulation. In the traditional formulation method, account is taken only of the utilisable or digestible nutrients. Assume that two (theoretical) available ingredients have raw energy contents of 20 and 30 MJ.kg^{-1} and ADC of 90% and 60% respectively. The digestible energy of these two primary constituents will be the same (18MJ). However, the energy lost in the form of faeces is 2 and 12 MJ using ingredients 1 and 2 respectively. In order to limit pollution it is therefore clear that the first ingredient should be preferred. Several methods can be used to achieve this goal. It is possible to introduce non-digestible energy into the calculation and to minimise it (complex mathematical solution); it is also possible to fix a limit on non-digestible energy, i.e. to choose a minimum ADC for energy, organic matter, nitrogen or other nutrients. An even simpler method is to eliminate ingredients with low digestibility and replace them with more digestible ones (for example classic fish meals replaced with LT meals) or even impose a limit on the incorporation of raw materials. This is another use for technical specifications.

To summarise, manufacture of low-pollution foods requires both technological innovations and simultaneous changes in formulation methods.

20.5.4 Formulation and quality of the final product

The quality of the final product, i.e. the fish flesh, has for a long time only received minimal attention in comparison to growth rate, FCR or cost of production per kg of fish flesh. It is difficult to ignore this aspect when production reaches a ceiling, competition intensifies, the consumers' demands increase and prices fall. Whether the product is destined for the marketplace in its fresh form or after processing, quality must be taken into account.

Quality, with its technological, organoleptic and nutritional components, has no simple definition. Far from being an objective characteristic, it is perceived in very different ways depending on country, habits and food preferences of the consumers. There are some attributes or characteristics whose determinism is essentially genetic (conformation, coloration by pigments). Others are mainly linked to rearing practices and state of health (absence of malformations, food safety). Yet others are nutritional in origin (lipid, carotenoid or fatty acid content). In most cases, the determining factors are complex and nutrition, in a broad sense, exerts a variable and as yet poorly understood influence on quality. Two cases in which the role of nutrition is well demonstrated are those of carotenoids (Chapter 11) and lipids

(Chapter 7). In the latter, the influence of nutrition is exerted not only via formulation (fatty acid balance, lipid content, energy level of diet) but also via ration, energy intake determining the fat content.

Limiting or, conversely, imposing a minimum threshold on certain ingredients for quality reasons thus appears to be difficult, even for raw materials such as soya oilcake which is supposed to create abnormal flavours in certain cases. The main specifications which the manufacturer must impose in order to preserve quality involve carotenoids and fatty acids. Nevertheless, in the future, these specifications may multiply and become increasingly important.

20.6 OTHER ASPECTS AND AREAS OF APPLICATION OF THE LINEAR FORMULATION

The various software programmes for linear formulation currently available (for example MIXIT and PORFAL) do not have exactly the same advantages. Nevertheless, they allow similar secondary or supplementary calculations to be carried out that are very significant for manufacturers.

20.6.1 Calculation of threshold price

When an ingredient does not figure in a formula because it is too expensive, it is possible to determine at what threshold it could be retained. This threshold, or threshold price, is the maximal price which an ingredient can have in order to figure in the formula, the costs of other ingredients being fixed. These data can help the manufacturer considerably in conducting his buying policy.

The problem can also be approached in a more complex manner by taking into account the price variation of two raw materials at a time. This approach corresponds more to reality than to a simple mathematical approach, since the price of a significant ingredient (for example fish meal) usually depends on those of competing products (for example soya oilcake). This results, therefore, in two-dimensional representations that allow the visualisation of the incorporation of the ingredients studied into the formulae. A practical restriction to the above concerns raw materials which are rare on the market: if the threshold price of these ingredients attracts buyers, their true price is at risk – following the law of supply and demand – of increasing so rapidly that the manufacturer is practically obliged to include it immediately.

20.6.2 Range of tolerance

When the price of an ingredient decreases, its level of incorporation in least cost formulae increases, but this increase occurs discontinuously in stages, which results in margins of tolerance. In other words, a decrease in market price can lead to an immediate increase in the theoretical level of incorporation if the price is at the lower limit of the margin, while it has no effect if it is at the upper limit of the margin. This is a valid consideration for a given economic and methodological context.

20.6.3 Cost of saturated specifications

In practice, certain constraints – either nutritional or technical – have no effect on the price of the formula, always being satisfied after other specifications or the low cost of some nutrients (for example the requirement for phosphorus in formulae rich in fish meals). In contrast, others are costly (this is usually true for protein or limiting EAAs). For the most expensive nutrients, the amounts included in the formula never exceed the specifications i.e. the standard levels; these specifications are said to be saturated. Now the specifications, even nutritional ones, rest on bases that are inequally solid and tangible. Thus a protein standard includes a safety margin which results from a series of several precautions. If, in a given context, this specification is not only always saturated, but generates notably high costs, the manufacturer will naturally be led to re-examine this margin and question it. Examination of the cost of specifications can orient applied research towards sectors in which the greatest economies appear to be possible.

20.6.4 Multiformulation

We shall mention only briefly those processes of calculation which allow not only the optimisation of formulae one by one, but a global approach to the formulation of several very different foods using a limited group of primary constituents. It could be said that multiformulation allows various primary constituents to be aimed towards the species or physiological stages which can make best use of them. It is important for the manufacturers who are, in terms of practical considerations, limited in their possibilities for storage of raw materials and who know the relative tonnages of the feeds that they produce. While fish feeds only constitute a marginal part of the manufacturing process, their composition will hardly be modified by multiformulation.

20.7 CONCLUSION

The problems which are specific to the formulation of feeds destined for aquaculture are sometimes related to the nutritional peculiarities of fish and crustaceans, but more often arise from gaps in knowledge, which render specifications imprecise, as well as the requirement to take into account certain aspects which are not directly related to nutrition in the true sense. Within these aspects, the protection of the environment is of particular importance. It is probable that, in the future, the quality of the final products will create comparable problems which the manufacturers of today are still relatively powerless to answer.

20.8 BIBLIOGRAPHY

Garcia Gallego, M. (1987) Formulacion de dietas experimentales y piensos comerciales. In: Espinosa de los Monteros, J., Labarta, U. (eds) *Alimentacion en Acuicultura.* Industrias Grafica Espana, Madrid.

Houser, H. and Akiyama, D. M. (1997) Feed formulation principles. In: D'Abramo, L. R., Conklin, D. E. and Akiyama, D. M. (eds) *Crustacean Nutrition: Advances in World Aquaculture, Volume 6.* World Aquaculture Society, Baton Rouge, LA, 493–519.

Lapierre, D. (1979) *La Formulation des Aliments des Animaux. Cycles Approfondi d'Alimentation Animale.* INA-PG, Paris.

Larbier, M. and Leclerq, B. (1992) Modélisation des besoins et formulation des aliments. In: *Nutrition et Alimentation des Volailles.* INRA, Paris.

21

Feed manufacture

The manufacture of a compound feed consists of a series of operations, the end goal being to combine several raw materials in previously-determined proportions for a precise nutritional objective. This association is achieved by mixing components in their solid form (animal meals, oilcakes, cereal products, minerals and vitamins) or liquid form (fish oils, lecithins and certain vitamins and binding agents). Grinding of the largest solid components reduces the heterogeneity of the product and increases digestive utilisation to a certain extent.

The feed is formed into the required shape, and is then simpler to transport and handle, easier for the animals to take and allows pollution of ponds to be limited. Texturisation is the key operation that allows the feed to be adapted to the feeding behaviour of the animal.

Feeds destined for fish or crustaceans include raw materials which are by-products of other industries (oil, starch and cornflour works) or specially developed products (fishmeals, oils). All these raw materials have, to varying degrees, been subjected to technological treatments before being combined in a compound feed. Only the operations involved in the structuring of the feed are described here.

The diversity of forms (crumbs and different sizes of granules) and properties (physical resistance to handling and to leaching in water, susceptibility to rehydration, sinking and floating) required for feeds for aquatic animals impose significant adaptations to existing manufacturing processes. Most of the time, manufacturers prefer to develop specialised production lines which include adaptations of traditional processes (grinding, dosing, mixing) and specific processes (pressing, extrusion, drying, coating and crumbling). Examples are shown in Figures 21.1 and 21.2.

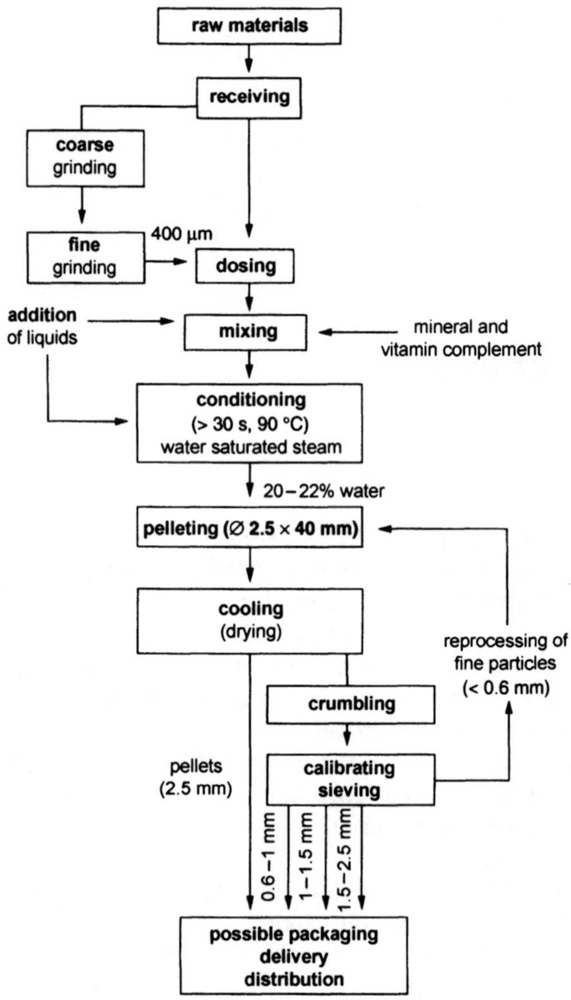

Figure 21.1. Example of manufacturing process for agglomerated water-stable feeds (for shrimps).

21.1 COMMON PROCESSES IN ANIMAL FEED MANUFACTURE

21.1.1 Grinding

Grinding consists of reducing a raw material to finer particles, resulting in a more homogeneous and stable mixture and more regular shaping of the feed. Particle size analysis by analytical sieving allows the degree of fineness of the meal to be determined. Because the temperature increases owing to the energy released by processing, some physico-chemical characteristics can also be modified.

Their versatility and robustness mean that hammer mills are the most widely-used industrial machinery. The product to be ground is tipped into a chamber where

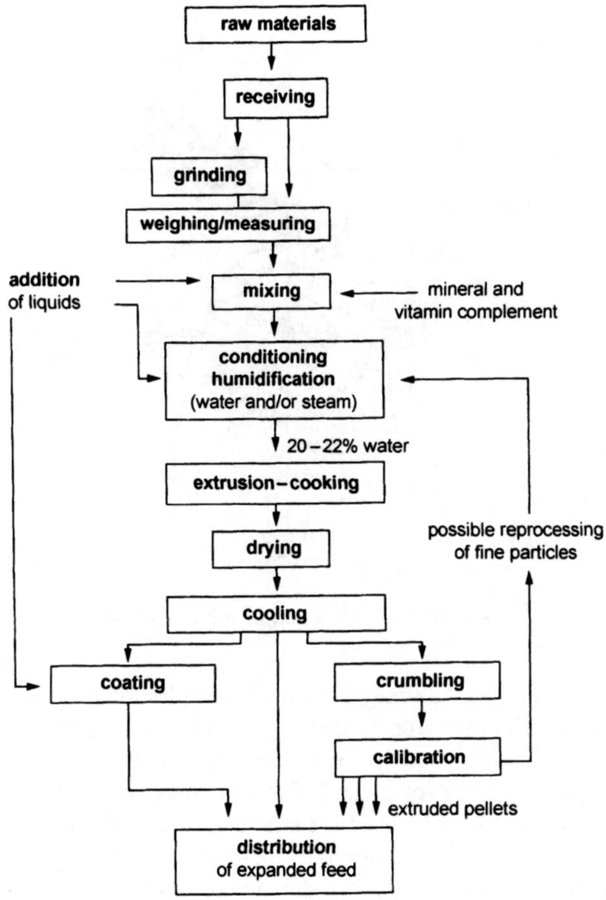

Figure 21.2. Example of manufacturing process for floating extruded feeds (for salmon, catfish).

it is smashed by hammers fixed to a rotor. The particles produced go round again until they are small enough to pass through a screen or sieve made of thick, perforated sheet metal, situated in the bottom part of the apparatus (Figure 21.3). Negative air pressure helps the particles pass through the holes in the screen.

The speed of impact is between 70 and 105 ms^{-1}, depending on the machine. It is usually constant for a given mill. The diameter of the holes ranges from 0.5 mm (feeds for shrimps and young fry) to 3.5 mm (feed for large fish). Particle size is small when the holes in the screen are small and when the peripheral speed of the hammers is high. The moisture content and texture of the raw material, linked in part to its composition (oil and fibre content), may also be involved. The diameter of the holes in the screen constitutes an upper limit on the desired granulometry, but does not describe the fineness of the meal, which must be measured separately.

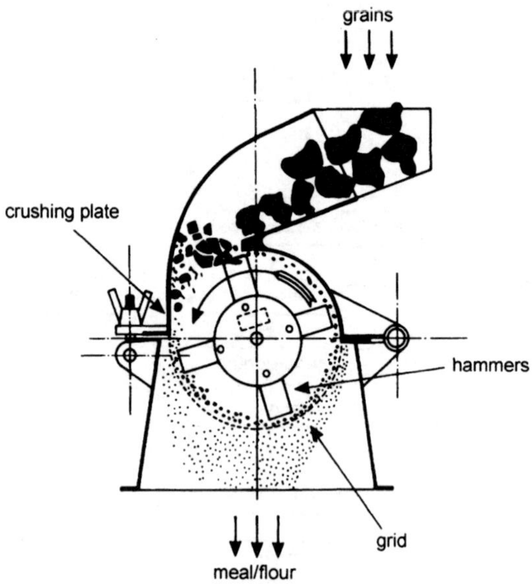

Figure 21.3. Diagram of a hammer mill.

It is assumed that the maximal size of the particles must not exceed one-third of the diameter of the dies used during their agglomeration. A mean particle size of less than or equal to 500 μm is usually recommended for fish feeds. It must be smaller (250 μm to 180 μm) for feeds for juvenile fish and crustaceans; it must be even smaller (<50 μm) when the feed is to be used for larvae and then milling is carried out in two stages. Supplementary sieving may be required to remove flexible fragments of material such as fish bones, which are difficult to reduce to powder and may block the dies or break the pelleted feed. A sieve mesh of 800 μm is sufficient to obtain a mean particle size of about 400–500 μm.

21.1.2 Dosing or proportioning or weighing

Measuring ensures the input of the different ingredients of the formula in well-defined proportions. It can be measured as mass (weight), which is most usual, or as volume. Precise measurement is vital when expensive active substances are incorporated in very small amounts into the formula: any excess or deficiency of an element constitutes an extra expenditure for the manufacturer or a risk for the fish farmer.

In practice, the manufacturer uses scales or a weighing machine, the range of which is adapted to the quantities being weighed. It is necessary to use two sets of scales, one for the raw materials used in large quantities, the other with a smaller range (1/5 to 1/10 of the former) for minor components. The scales are either mechanical or electronic with a strain gauge. The latter type now predominates in most factories where it is linked to a central computer which also manages the stocks of raw materials.

21.1.3 Homogenisation (mixing)

Homogenisation is an essential process in the manufacture of compound feeds and consists of bringing together the pre-crushed and measured raw materials while spreading them uniformly through the mixture.

A mixture is characterised by its homogeneity, but also by the maintenance of this homogeneity in time and space. The word 'homogeneous' is not a physically exact term for a mixture of meals, but indicates that each nutrient of the mixture is present in a given mass of product at a percentage which corresponds approximately to its theoretical value. It therefore involves the size of the sample: the animal must have all the nutrients it requires for its maintenance and growth, however small its daily intake of feed may be.

Homogeneity can be evaluated using tracers, various anions or cations (Cl^-, Fe^{++}) which are already present in the feed, colourings (methylene blue, fluorescin), or microscopic coloured crystals specially introduced for this purpose.

The effectiveness of the process depends greatly on the granulometry of the ingredients and the mixer used.

- In a homogeneous mixture, the large particles (and the lightest) are stuck together by the smaller ones, but if their proportion is too high, the fine fractions (often the heaviest) tend to separate out each time the produce is moved (mixing, transfers); this phenomenon is called separation. The test of pouring-out volumes is simple, allowing the risk of separation to be evaluated. Avoiding the risk of separation involves the grinding of the largest particles (larger than 400 μm) or adding liquids (oil) which slow the percolation of the finest particles.
- Horizontal mixers with double belts are the most usual machines used for powdered primary constituents: they can also be used for the incorporation of fatty materials in liquid form. The machine must be full to function and the mixing time is adjusted in relation to the formula: it is usually about 4–5 minutes. After mixing, total and fast emptying is recommended. Mixers with paddles or blades use more electrical energy, but are better adapted to the incorporation of liquids.

The mineral and vitamin complement or 'premix' is incorporated at this stage. It contains all the micro-components which are present at levels of 0.5–0.005% of the feed. The constituents must be homogenised separately using successive dilutions in a finely-ground base (often soybean meal). Their chemical nature must be taken into account when hygroscopic products (bicalcium phosphate, iron or copper sulphate) or electrostatic properties are involved. In the latter, it is vital that the mixer is earthed. The recommended final level of incorporation of pre-mix is between 1% and 4%. The lower this level, the greater the risk of heterogeneity.

21.2 STRUCTURING OF THE FEED

The feed can be given to fish as a paste after the addition of water (eels), but it is usually structured using two main methods: pelleting and extrusion-cooking. A third method, 'moist' pellet formation, used until the 1970s mainly in the United States, is now only used in the laboratory. Its principle lies between the two listed above.

The mixture, in the form of meal, is forced by rollers across perforations in a flat or annular die (Figure 21.4). It comes out of the press in the form of small cylinders (pellets) of certain length and diameter. The diameter varies from 2.5 to 6 mm depending on the size and type of the animal. The length is adjusted using knives situated at the edge of the die and is usually 1.5–2 times the diameter of the pellet. Water or steam are added beforehand using a conditioner.

The agglomeration process is delicate. The result depends on the composition of the formula, the efficiency of certain physical factors, humidity and heat, added

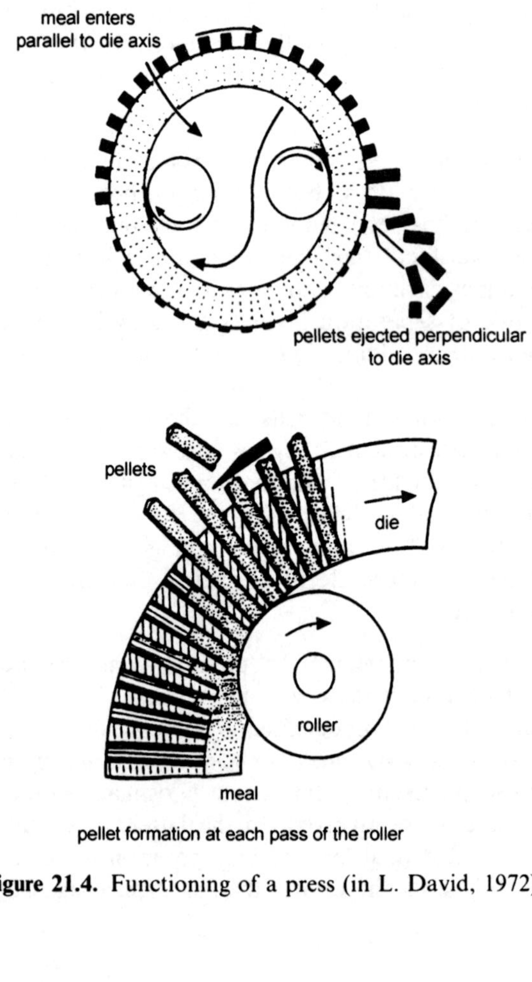

Figure 21.4. Functioning of a press (in L. David, 1972).

binding agents and the characteristics of the die being used. The pellet must stick together sufficiently in air and have a certain resistance to leaching in water. These properties can be obtained by:

- reducing the level of fatty materials in the feed (3–4% maximum) and/or incorporating wheat gluten or pre-gelatinised starches which help to form a continuous matrix binding the particles to each other;
- using more 'compressing' dies, in which the length/diameter ratio is >12. The higher this ratio, the harder the pellets;
- adding relatively large amounts of water or steam (2–6%) followed by a 'long conditioning' process which consists of maintaining the product at a relatively high temperature (60–80°C) for 2–5 minutes (instead of the 15–20 seconds in classic conditioning). If this temperature reaches or exceeds 100°C, the term 'thermo-conditioning' is used. The aim is to increase the homogeneity of water distribution and the plasticity of the meal particles;
- pre-densification or pre-compaction of the mixture which, instead of acting on the durability factor as above, uses the mechanical pressure to favour the penetration of liquids (fats) into the micropores in the meal.

Factory pelleting equipment uses a significant amount of energy (16–48 $kWh.t^{-1}$ of heat energy and 19–37 $kWh.t^{-1}$ of electrical energy). Energy expenditure is even greater when the dies are thicker and additional processes such as pre-compaction (+25–40% of electricity consumption) are added, which in practical terms involves two successive compressions. Agglomeration uses about two-thirds of the total energy required for manufacture. It leads to warming of the feed, the temperature of which is 70 to 85°C when it comes out of the presses. The pellets must therefore be cooled and dried to avoid any mould development, especially in conditions of high temperature and humidity.

21.2.1 Extrusion-cooking

This process consists of subjecting a mixture to the combined effects of pressure (30 to 120 bars) and temperature (90 to 180°C) over a short period of time (less than 30s) and shaping it by forcing through either one or several dies (Figure 21.5). Because of the low thermal diffusion of the ingredients and their high 'viscosity', it is necessary to work with a thin layer using one or two screws (single or twin screw extruder) which are encased in a barrel. Friction is obtained between screw and barrel casing in the first case, and between one screw and the other in the second. Depending on the machine, various forms of restriction to flow are situated along the channel in order to increase shearings. Water contained in the product is in a liquid state at high pressure and temperature. A large amount vaporises when it passes out of the dies and to the atmospheric pressure. This vaporisation creates a characteristic alveolar structure (expansion), at least when the composition of the feed allows. At the exit, the extruded product is cut by a knife, cooled and dried to about 10–12% humidity. In practice, extruded and expanded feeds are sometimes distinguished; the latter are only one type of the former.

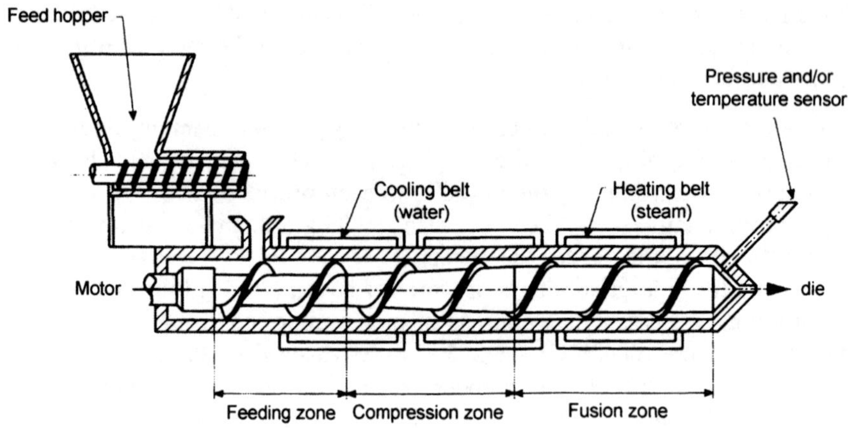

Figure 21.5. Diagram of a single screw extruder (drawn by Alain SIRE, INRA, Nantes).

When the feed is pre-conditioned with water and steam simultaneously before extrusion (single screw extruder), the extrusion is termed 'moist'. The humidity of the mixture is brought from 7–8% to 18–19% during a residence time in the conditioner of about a minute. This time needs to be longer when fat content is higher; for example about 3 min for 7% oil. In contrast, some machines function without the addition of water. This is called 'dry' extrusion and is usually more drastic. Intermediate solutions consist of injecting water or steam directly into the barrel.

Extrusion-cooking is a complex process, in which there is simultaneously intense mixing, cooking and texturising. These three functions together form a very versatile method although the mechanisms remain poorly understood because of the number of parameters involved: nature and structure of the mixtures, geometric characteristics and extruder controls.

Single or twin screw extrusion-cooking is applicable both to feeds with intermediate moisture content (38–40% moisture) which may need to be dried afterwards, and to dry feeds. The product completely loses its particulate structure and acquires new properties. Extruding in the presence of wheat gluten at high humidity and low temperature (<80–90°C) avoids expansion and allows feeds which sink and remain stable at the bottom of the water for several hours to be obtained (food for penaeid shrimps). A variation consists of sudden cooling the last part of the extrusion process in order to suppress expansion. In contrast, lower levels of hydration, higher temperatures (140–160°C) and higher shearing rates allow expanded, floating (or slow-sinking) feeds to be obtained, which rehydrate at varying rates (feeds for catfish, salmon or trout reared in the sea). The type of starch incorporated at 8–12% and its possible pre-treatment allows the degree of expansion and homogeneity of rehydration to be adjusted. Due to their functioning principle, twin screw machines can also be used for the incorporation of high proportions of oil (22% in practice), before extrusion.

Extruders constitute a very heterogeneous family of machines, the products of which are sometimes difficult to compare. In contrast to agglomeration, they involve a drastic process during which certain components can be favourably modified (starch) or profoundly altered (proteins, vitamins, in particular vitamin C and unsaturated fatty acids). The process is expensive, in terms of investment, man-hours and running costs. For example, the power consumption can be 2.5 times that for agglomeration, with the equivalent hourly output. This must be optimised for a given food. Nevertheless, its versatility makes it a method which is increasingly used for feeds destined for marine animals.

21.2.2 Complementary processes

21.2.2.1 Drying – cooling

Cooling is carried out by a flow of air through a layer of immobile pellets (expended or not) (vertical cooler) or laid out on a conveyor belt (horizontal and 'carousel' cooler). At the exit of the cooler, the feed is usually sieved to separate the non-pelleted fine particles, which are then recycled through the press, if agglomeration is involved.

The moisture content of the product is brought from 15–16% to about 9–10% in 9–14 minutes if the air is at ambient temperature. Water loss is greater if the air is hot (40–60°C). By forming a sort of 'shell' around the pellets, drying contributes significantly to the development of their mechanical characteristics. Intense mixing of the product during drying is to be avoided in order not to break up the pellets so that they remain very stable in water.

21.2.2.2 Crumbling

The extruded or pelleted product, usually small in diameter, is crushed on grooved cylinders (3 mm apart) after cooling and drying, then sieved in order to adapt particle size to that of the animal's mouth. For example, crumb size is 0.6–2.4 mm for shrimps and 1.5–3 mm for juvenile fish. It is approximately proportional to the size of the fish, for example 0.025 times the body length in trout. Sieving leaves behind a very fine residue (up to 25–35%), in a state that cannot be used as is and must therefore be recycled.

21.2.2.3 Coating

Pelleted or extruded products can be coated using fats in order to incorporate essential fatty acids, liposoluble vitamins and even hydrosoluble vitamins in emulsion (vitamin C) into the diet. In addition, coating forms an external hydrophobic layer that slows down the dissolving and leaching processes into the water. Fats, which are usually liquid at normal temperature, are sprayed onto products which have been cooled and especially dried and mixed in screw or drum coating machines. The fats thus penetrate into the pores left free by the departure of water. Some feeds for shrimps are subjected to two successive coatings: one with hydrosoluble additives (vitamins and attractants) in very small amounts, and the other with oil and liposoluble additives. The absorption capacity of the regular (not expended) pellets for oil is only 3–4%, while that of expanded products is much higher (>10%). It can be increased slightly by incorporation in stages. It is also possible to adsorb oil onto precipitated hydrophobic silicas.

21.2.3 Use of binding agents

Because of their high animal protein content and the almost complete absence of native plant raw material, feeds for carnivorous species contain few natural binding agents. It is thus inevitable that 'external' binding agents are incorporated as additives. Some, such as wheat gluten or pregelatinised starches, can have a nutritive value while others are inert raw materials used as gelling agents or thickeners in the feed industry (carboxy-methyl-cellulose, alginates, carragheens, agar and various gums). Synthetic binding agents are sold by the industry, but some (formaldehyde, for example) are not to be recommended because, although they are efficient, they could be harmful to target species and consumers.

The efficiency of binding agents reputed to be inert is often doubted when the feed is manufactured using the industrial methods described above. Thermo-mechanical treatment tends to destroy the structure of the gels formed if the moisture content of the product is insufficient. At higher moisture levels (45–50%), the feed would be distributed in a state in which the excess water would have to be eliminated by drying which is always expensive. In contrast, gelling-binding agents are used if the formulation contains high proportions of fresh meat by-products. Sodium alginate (1.5%) with 15% water in the presence of calcium (in the form of plaster) acting as an adhesive is used successfully in shrimp feeds. Processing is carried out at the rearing site, on an artisanal scale using simplified machines (press and meat mincer).

21.3 METHODS OF CHARACTERISING TEXTURED FEEDS

The quality of a feed is characterised in different ways depending on its destination and utilisation; certain criteria are associated with the manufacturing process while others are linked more to the feeding behaviour of the animal that will consume the product.

21.3.1 Geometrical and physical features

21.3.1.1 Particle size

The size of crumbs destined for juveniles is evaluated by analytical sieving. The sample (about 100 g) is tipped into the top of a set of sieves, the mesh dimensions of which are spread out according to a geometrical progression (AFNOR NF X 11-501 standard). After 'shaking' in controlled conditions (15 min), the amount present on each sieve is weighed and the results expressed as percentages. The resultant histograms are of little direct use, but it is possible to calculate a mean particle size and a standard deviation which is representative of the heterogeneity of the granulometric distribution, after adjustment to statistical laws (AFNOR NF X 11-635 and 636 standards).

21.3.1.2 Mass/volume or 'densities'

The unit density is the density of a pelleted or extruded feed particle taken individually. It influences the feed intake of the animal. It is measured by the volume of a granular solid (sand or calibrated glass beads) or a liquid (mercury or water) displaced by a known mass of product. This characteristic is usually expressed in $g.cm^{-3}$.

When non-porous products are involved, it is possible to use mercury or water after coating the sample in a hydrophobic substance which is solid at room temperature (tallow or paraffin wax). When the product is extruded and alveolar, the liquid can be replaced by homogeneous-sized sand or glass beads (1–2 mm). This is then packed down under controlled conditions before and after introduction of the product sample (AFNOR NF A 95-112 standard).

The apparent density (or bulk) is the density of a group of individual particles. This is essentially a practical criterion, which becomes significant during packaging, transport and distribution operations, since it is related to the degree of filling of sacks, hoppers and measuring systems. The apparent density is evaluated by weighing a known volume of a feed (2 litres), poured via a funnel, into a container. It is often expressed in $kg.m^{-3}$, sometimes in $g.l^{-1}$.

21.3.2 Mechanical characteristics in air

21.3.2.1 Resistance to abrasion (durability)

Friability is evaluated from the amount of fine particles broken off the food when it is subjected to a series of shocks and friction in a revolving box or in a machine with pneumatic circulation (Holmen method). By convention, fine particles are defined as all feed particles which pass through a mesh equal to 0.8 times the nominal diameter of the texturised product. There is a direct relationship between the results provided by these two techniques; the second is quicker and slightly more discriminatory in the case of compact, less friable products. Crushing in a small grinder in the laboratory over a short time period (10 or 20s) has also been proposed recently (*Quick test*). Durability is the reciprocal of the friability (in per cent).

21.3.2.2 Resistance to crushing (hardness)

Hardness is defined by the maximal resistance to radial compression of a group of pelleted or extruded products of the same diameter, if they are round in shape, and of known length. It is expressed in MPa (megapascals).

Crushing force is measured using a traction-compression machine. Recording the curve in relation to displacement allows the crush energy to be calculated; this is useful when the products are not regular in shape. Simplified machines allow a hardness index to be obtained by the application of a point force.

When feeds are very rich in oils or semi-moist or hydrated, no hardness threshold can be described. Measures of durability and especially hardness lose their significance. By analogy with certain practical tests on meat or fruit, the product can be subjected to a double compression at fixed speed and deformation. The two peaks obtained can be used to define a hardness (height of the first peak), elastic recovery and cohesion (ratio of the areas of the two peaks) often called 'plasticity'.

21.3.3 Behavioural characteristics in the aquatic environment

21.3.3.1 Aptitude for rehydration

A simple test consists of plunging the available sample in wire mesh baskets into an excess of water at 20°C and, after blotting on filter paper, measuring either the quantity of water absorbed or the rate of rehydration (ratio: mass of rehydrated product/initial mass of dry product) in relation to the immersion period. A simpler method is to express the quantity of water absorbed over a fixed period of time, conventionally between 8 and 15 minutes. This time can be reduced by placing the sample in a partial vacuum (3 min). The method thus has a fault which must be corrected in relation to possible leaching.

Oil absorption can be evaluated according to the same principle. The soaking period is shorter (1 min at 40°C).

21.3.3.2 Water stability

Resistance to dissolution or stability in water is very important for feeds destined for hunting predators or browsers or detritivores such as sole, penaeid shrimps and, to a lesser extent, carp.

21.3.3.3 Static methods

Neither feed nor liquid is moved: the swelling and leaching of the feed submerged in the water are estimated visually in relation to time. A complementary method consists of tipping the feed onto a sieve (2.4 mm). The residue is then retained and dried. The results are expressed as a percentage of dry matter.

21.3.3.4 Dynamic methods

Displacement of liquid in relation to solid The sample is placed on a sieve and subjected to an ascending water current that gathers the particles and dissolved substances. The opacity of the liquid, which is related to the proportion of material trapped, is measured by nephelometry. The method has the advantage of not requiring desiccation in order to measure the amount of dry matter; nevertheless, the significance of the opacity of the liquid remains to be determined.

Table 21.1. Main features of feeds for aquatic animals (examples).

Characteristics	'floating' feed	slow sinking feed	sinking feed which is water stable
diametric expansion	1.2–1.6		0.98–1.02
unit density (g.cm^{-3})			0.78–0.95
apparent density (kg.m^{-3})	350–400	390–450	520–820
hardness (MPa)			0.4–0.9
durability (%)	96–97		98–99
rehydration after 8 min (%)	140–220		60–70
stability in water (%)			82–85

Displacement of solid in relation to liquid The feed is placed in wire mesh baskets with a mesh of 2 mm and alternately plunged into water and pulled out into the air at a rate of 10 movements per minute. When the product passes through the liquid's surface, it is mixed: the soluble substances and particles less than 2 mm in size are trapped. After 40 or 60 minutes, the residue is collected and dried in an oven until its mass is constant. Stability in water is expressed in relation to the dry matter of the control feed under the same conditions. A stability of 80–84% is considered satisfactory for feeds destined for penaeid shrimps: this corresponds to an apparent maintenance of the integrity of the product *in situ* of about three hours.

Examples of characteristics are given in Table 21.1. Relationships between criteria are often complex.

Aptitude for rehydration, which renders the product more easily consumable, is inversely proportional to stability in water. In the case of non-expanded feeds for penaeid shrimps, there is no relationship between durability in air and stability in water, just as there is none for pelleted or extruded feeds.

21.4 CONCLUSION

Obtaining a feed for marine animals is similar to a classic feed but with more requirements: it demands the perfect management of grinding, structuring methods (pelleting and cooking-extrusion) and drying-cooling, operations during which care must be taken to avoid altering the biological value of the proteins in the feed.

The choice of method must be related to the type of animal, but is also a function of the cost of the operation and its sometimes risky nature. Thus, cooking-extrusion, often favoured because of its versatility and its ease of use, is not the answer to every problem. It is a costly operation, currently not properly managed, which can yield very variable results. In certain cases it can be replaced by pelleting, with possible adaptations such as heat treatment or pre-compaction. It is the only method where floating feed is required.

Lastly, there are several methods for the physical characterisation of feeds which, as a first step, allow a method to be optimised for a fixed objective.

21.5 BIBLIOGRAPHY

Bouroche, A. and Le Bars, M. (1996) *La Cuisson-Extrusion*. Vocabulaire français-anglais-allemand. INRA, Paris.

Colonna, P. and Della Valle, G. (1994) *Cuisson-Extrusion*. Tec & Doc Lavoisier, Paris.

Csavas, I. (1991) Physical and nutritional properties of aquafeeds. In: *Feed Production Tomorrow*. VICTAM, Asia, 2–19.

David, L. (1972) *Conduite et Entretien des Presses*. La Jarrie.

David, L. and Lefumeux, J. (1972) *Pratique de la Compression et Techniques Nouvelles*. La Jarrie.

Kearns, J. P. (1988) Preparation of fish and shrimp feeds by extrusion. In: Applewhite, T. H. (ed.) *Proceedings of World Congress on Vegetable Protein Utilization in Human Food and Animal Feedstuffs*. AOCS, Champaign, Ill, 152—161.

Kearns, J. P., Huber, G. H., Dominy, W. G. and Freeman, D. W. (1988) Properties of extrusion cooked shrimp feeds containing various commercial binders. Presented at World Aquaculture Society 19th Annual Conference, 4–8 January 1988, Waikiki, Hawaii.

Melcion, J-P., Guillaume, J., Méhu, J., Métailler, R. Cuzon, G. (1983) Préparation d'aliments pour animaux marins par cuisson-extrusion. *Rev. Alim. Anim.*, **371**, 26–31.

Tan, R. K. H. (1988) Pelleting for aquaculture feeds. In: Applewhite, T. H. (ed.) *Proceedings of World Congress on Vegetable Protein Utilization in Human Food and Animal Feedstuffs*. AOCS, Champaign, Il, 162–166.

22

Feeding practices

In practice, feeding fish presents particular difficulties for three main reasons: firstly, whatever the fish's behaviour, the farmer comes up against the problem of ensuring that all the feed distributed in the water is effectively eaten. Secondly, feed not consumed can lead to deterioration in the quality of the rearing environment or even pollution of the neighbouring aquatic environment, as well as economic loss. Lastly, the feeding requirements of fish vary in relation to numerous parameters, notably environmental temperature, which can require the farmer to make a permanent change to the amount of food distributed. In practice, the farmer must adjust the amount of feed allowed for the supposed requirements of the fish at a given time as well as possible (this is called the feed ration), and also choose appropriate food distribution of 'feeding' methods.

22.1 BASIS FOR RATIONS

21.1.1 Influence of rationing on growth and feed conversion

For a given food, species and conditions, specific growth rate (SGR) and feed conversion ratio (FCR) depend on the feeding rate (FR) of the fish. Maximal growth and feeding efficiency are not obtained for the same level of feeding (Chapter 2). The gap between the feeding levels corresponding to these two significant values delimits the margins for manoeuvre available for the farmer. In the range of feeding, it is possible either to maximise growth or to minimise feeding costs in relation to economic or other constraints. Obtaining maximal growth implies overspending on feeding. For example, in a trial where brown trout were reared in the sea, this additional cost amounted to 12% for feeding to satiation in contrast to a ration of 85% of this; the difference in the gain in mass obtained with the two feeding levels was 12.3%, representing 10 days of growth, i.e. 8% of the duration of the experiment. However, this change in weight gain[1] and utilisation of feed in relation to FR is only an outline applicable to feeding conditions and is rather controversial. The minimum feeding cost and maximum growth rate are obtained using the same feeding level in trout if they are fed by controlled 'self-feeding', a method of feeding that takes into account the feeding rhythms and demands of the fish (Chapter 3).

It is also important to consider if the quantity of food taken into account in the calculation of FCR may considerably exceed the quantity which is actually consumed by the fish. The risks of waste increase with ration size, which occurs generally when maximal growth is sought.

22.1.2 Feeding to satiation

Satiation is much more difficult to assess in fish than in terrestrial animals since it is not easy to have either direct or indirect visual evaluation of the quantity ingested in the aquatic environment. Two different approaches can be used.

- The first consists of giving the fish all the feed they are capable of eating while monitoring their behaviour. The end of the activity linked to food intake indicates a momentary halt in the meal. A brief period of satiation can be obtained by distributing a certain amount of food by hand until it is refused. However, true satiation is only obtained with a sufficient number of meals (two or more per day) according to size, species and environmental temperature. These factors, as well as the nature of the diet, determine the rate of intestinal transit.
- The second method consists of leaving an excess of food available to the fish (using a 'self-feeding' system), the animals taking what they need as they please. Management of this method is simpler than that of the previous methods.

As a general rule, management of distribution to satiation is complex and frequently leads to losses of food, reduced feed efficiency and degradation of the rearing environment, which can be sufficient to alter biological performance. The problems encountered can also be aggravated by the energy density. It is sometimes possible to limit the losses by waste control. Remaining feed can be collected using screens placed at the pond outflows, when there is a sufficient current flowing across them, or using 'plates' placed below the feeding stations (a method frequently used in shrimp rearing). However, these corrective measures are rarely satisfactory.

22.1.3 Energy density of feeds and regulation of feed intake by fish

Regulation of intake occurs, in fish as in terrestrial animals, via the energy content of the food. In the medium term, the animal adjusts its consumption in such a way that

[1] In this chapter which, more than the preceding ones, relates to the practicalities of farming rather than laboratory rearing, the term mass (body mass) has been replaced by W, for weight, a term which is incorrect but widely used in practice.

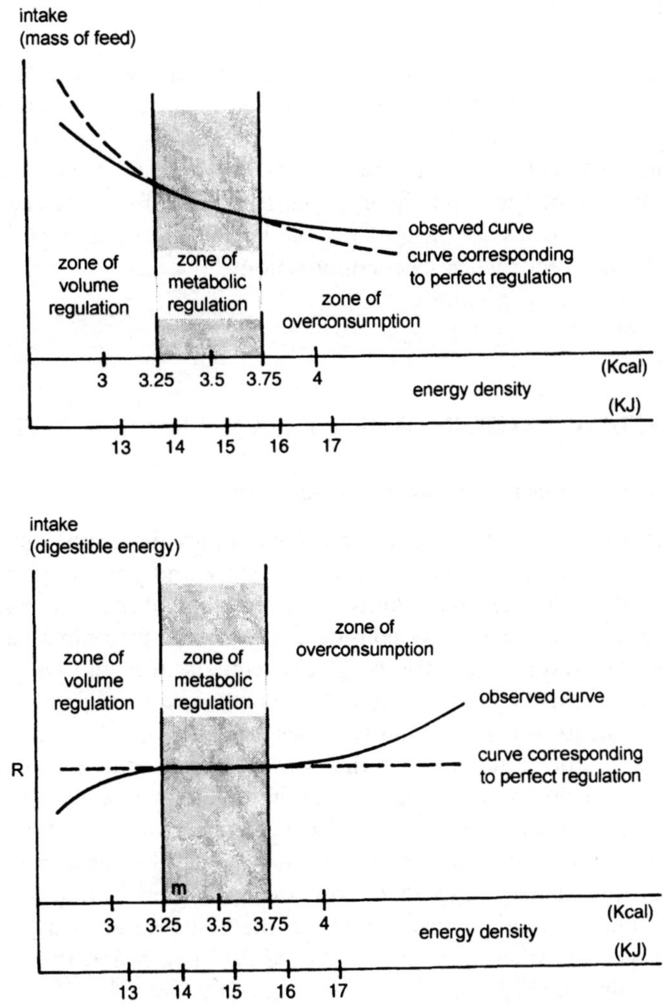

Figure 22.1. Regulation of intake in relation to energetic density of the feed. R: energy requirements; m: minimal energetic level.

its energy requirements are met. With low-energy diets, the amount of food intake may be limited by stomach volume and, in certain cases, this may be insufficient for good performance (Figure 22.1). In contrast, with the high-energy foods now available, the amount of energy taken in is not limited by the intake capacity of the animal and can exceed requirements; it is as if the short duration of the meal does not allow the physiological mechanisms of satiation to come into play sufficiently quickly. In this case, the excess fraction is stored in the form of fats in parts of the body which are of variable significance depending on the species: muscle, peri-visceral adipose tissue, liver and subcutaneous tissue, etc. (Chapter 6). The ration must therefore be limited in order to minimise losses when large fish destined for processing are gutted. Ration levels must therefore take into account the characteristics of the food.

In practice, the fish farmer only needs to be concerned with the energy density of the feed, since the feed manufacturer is supposed to ensure a balance between the energy level and other nutrients (proteins or amino acids, vitamins, minerals, etc.). Similarly, the manufacturer is supposed to take into account important factors such as the digestibility of the food. Energy density is usually expressed in $kJ.g^{-1}$ or $MJ.kg^{-1}$ of digestible energy (in practice, the kcal is still used, see Chapter 4). It is of particular importance when calculating rations.

22.2 PRINCIPAL FACTORS WHEN ESTABLISHING A FEEDING RATE

22.2.1 Fish size, temperature and predicted growth

Fish's requirements for nutrients, and thus for feed, are determined by many factors linked either to the animal (species, stock, size, stage and physiological status) or to the environment (temperature, salinity, water quality, dissolved oxygen content). Two of these factors are usually considered as dominant: animal size and water temperature. However, even for a given temperature, size is insufficient to determine requirements. In addition to maintenance requirements, there are growth requirements which are, among other things, a function of the growth rate and hence of genetic potential. In other words, rather than the body mass, it is growth kinetics or potential that determine the requirements.

The general principle of the effect of water temperature is well known; requirements increase with temperature up to a certain threshold, considered as the thermal optimum for the species, and then decreases (Chapter 2 and Table 5.5). This optimum also corresponds approximately to the best feed efficiency since the relationship between the fraction of the food used for growth and that used for maintenance is maximised (Figure 22.2 and Table 5.5). However, the effect of temperature

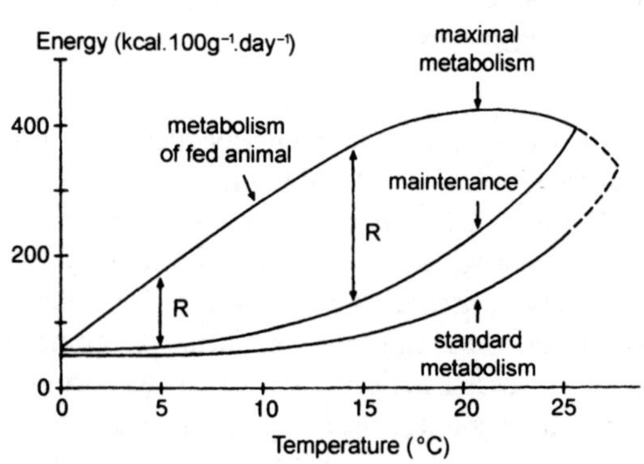

Figure 22.2. Effect of temperature on intake ($kcal.100\,g^{-1}.day^{-1}$) and energy retained (R).

has still not been quantified for all farmed species and the large number of species involved renders this type of modelling difficult.

In addition, many other factors influence requirements: state of health, physiological status, rearing environment, season (for a given temperature), etc. Exact evaluation of the energy requirements of fish is thus particularly complex. In the absence of a scientific solution, it is only by carrying out successive observations and re-evaluations that the farmer can arrive at an empirical solution. The efficiency of the feeding level in terms of economic performance of the fish farm will depend greatly on following the methods set in place by the fish farmer.

22.2.2 Limitations owing to rearing environment

Many factors linked to the rearing environment can affect the feeding level. The most important of these is the concentration of dissolved oxygen. The fish's requirement for oxygen is directly linked to the amount of food distributed; trout, for example, require about 250–300 g of oxygen to metabolise 1 kg of food. If it is difficult to find the oxygen they require in the environment, they tend to utilise their food less efficiently because of the energetic cost of oxygen uptake (muscular expenditure linked to respiration). Below a certain threshold, (4–5 ppm for trout), food intake decreases and growth slows down. Above this threshold, even if the quantity of oxygen available is sufficient, the fish reaches its potential and transforms its food increasingly well as the dissolved oxygen concentration is increased, since the energetic cost of obtaining oxygen decreases. The fish farmer must therefore reduce the ration if the oxygen content is not optimal for the species. This is less important for semi-extensive farms and for species which are capable of surviving on lower oxygen levels (for example carp). However, methods of oxygenation are increasingly used to allow the farmer to rear livestock in an environment which is not limited by oxygen and to maintain a biomass per unit flow at a much higher level than under natural conditions.

Other environmental factors must also be taken into account, in particular those that are likely to limit oxygen availability for the fish or affect its utilisation: agents affecting gill integrity, either by clogging (suspended matter, colloidal iron), or by necrosis or hyperplasia (ammonia) or by hypersecretion of mucus (acidic pH, irritant substances). Similarly, the level of nitrite must be monitored, as this ion limits the use of oxygen in the organism by the formation of methaemaglobin.

22.2.3 Regulation of waste discharge

Most intensive aquaculture units are or will be subject to waste standards, expressed either as concentration for fish farms with flowing water, or as flux (quantity discharged per unit of time) for marine farms. The pollutants covered by these regulations are, at the moment, residues from digestion (suspended matter) and metabolic excretions. Biological oxygen demand (BOD) which results from the metabolism of bacteria transforming waste, is directly linked to the discharge of suspended matter. It is usually measured over 5 h (BOD5). Other criteria, such as the discharge of medical or therapeutic products, may be taken into account in the future. The total quantity of feed distributed per unit of time (24 h for example) can be limited by these standards. This is sometimes the case in the summer for salmon farms.

Under these conditions, the fish farmer must adjust the total quantity of food Q, which he can distribute per 24 h, in relation to river flow and the waste standards applied, using a formula of the type:

$$Q_{kg/j} = D^3_{m/j} \times N_{mg/L}/E_{g/kg}$$

where D is the flow in the river through the fish farm
N is the standard applied to the system[2]
E is the quantity of product produced by digestion or excretion, discharged per kg of food.

Note that the initial oxygen level in the water is not taken into account, since it can be modified by the use of appropriate equipment.

22.3 CHOOSING A FEEDING LEVEL

The feeding rate is chosen in relation to a growth objective that can be a 'standard' growth or a performance level dictated by a particular imperative. It can be taken from a feeding table or calculated from previous results.

22.3.1 Feeding tables

Feeding tables for commercial feeds usually only take into account the species, size or mass of the fish and water temperature. They apply only to the feed concerned and to a fish farm where 'best performances' are sought (usually imprecisely based on the compromise between optimisation of growth and FCR). These tables must therefore be considered as guides and be adapted in relation to the unique characteristics both of the fish and the rearing environment in the fish farm, and even the season.

For an average mass and a given temperature, a table of energy requirements provides the quantity of energy required by a fish. The use of tables sometimes requires interpolations for body mass, temperature or both factors simultaneously (Box 22.1).

Energy requirements of fish are calculated from maintenance requirements (fasted animals), growth rate of the animals, temperature and losses during food utilisation by the fish.

Cho (1992) proposed a simplified method which is summarised in Box 22.2. An example of the application of this method is given in Table 22.1. From the calculations of Cho (1992), it is also possible to determine the theoretical FCR that can be

[2] This standard is set by legislation. In France it varies between regions, for example from 0.1 to 1 ppm for ammonia nitrogen ($N.NH_4$).

Box 22.1 Calculation of ration size by double interpolation of body mass and temperature

Although variation in ration size in relation to body mass and temperature is not linear, a linear interpolation is sufficient for an approximation. It is a rapid process.

Interpolation of masses
assuming t_1 and t_2 are the temperatures and P_1 and P_2 the masses found in the tables of real temperatures and masses, respectively, and assuming at temperature t_1, TP_1t_1 is the ration size for mass P_1 and TP_2t_1 for the mass P_2, the ration size TPt_1 for mass P will be:

$$TPt_1 = TP_1t_1 + (TP_2t_1 - TP_1t_1) \times \frac{P - P_1}{P_2 - P_1}$$

TPt_2 can be calculated in the same way, for temperature t_2.

$$TPt_2 = TP_1t_2 + (TP_2t_2 - TP_1t_2) \times \frac{P - P_1}{P_2 - P_1}$$

Interpolation of temperatures
assuming TPt is the ration size found for mass P and temperature t.

$$TPt = TPt_1 + (TPt_2 - TPt_1) \times \frac{t - t_1}{t_2 - t_1}$$

obtained with the recommended FR, which can constitute a reference point for the farmer.

The data required to establish these tables are only available for a limited number of species (certain salmonids). Energy requirements of fish can be affected by many factors and so the sufficiency of the energy content of the diet distributed for the fish's requirements must be verified by regular monitoring of the livestock's performance. The CI must be monitored regularly. In order to calculate it, weights must be carefully measured, possibly adding the total mass of dead fish and taking account of the numbers of individuals. This index must be within the limits usually found for the species and size and also the energy density of the diet. A correction factor (multiplying factor) must be applied to the ration if FCR is beyond these limits. In general, this coefficient must be less than 1 (FCR above the upper value of the margin). In addition, FCR can reflect different characteristics of the diet and not only the amount distributed.

22.3.2 Feeding levels from previous data

Feeding levels can be established so that a batch of fish reaches an average body mass at the end of a certain period of time. This method requires *a priori* knowledge

Box 22.2 Calculation of energy requirements of fish (Cho, 1995)

Determination of growth coefficient of a species (or stock)

$$GC = \frac{M_f^{1/3} - M_i^{1/3}}{t \times \Sigma\theta}$$

This very empirical coefficient is known for several salmonid species (trout, brown trout, Atlantic and chinook salmon) in which it is relatively constant in value. It is necessary to determine and to verify its constancy for other species.

Prediction of body mass at time *t* from mass at time 0

$$M_t = M_0^{1/3} + (GC \times t \times \Sigma\theta)]^3$$

Calculation of gain in energy GE from gain in biomass

$$GE = (M_t - M_0) \times dm \times ev$$

where dm = dry matter content
ev = energy content

Estimation of energy requirements for maintenance (EM)

$$EM = (-1.04 + 3.26\theta + 0.05\theta^2) \times M^{0.824} \text{ in kJ.day}^{-1}$$

Estimation of heat increment of feeding (HiF)

$$HiF = 0.6\,EM$$

Calculation of energy losses in urine and through the gills

$$ELUG = 0.06\,(GE + EM + HiF)$$

Calculation of total energy requirement (ER)

$$ER = GE + EM + HiF + ELUG$$

of the growth potential of the animals as well as the change in FCR in relation to size. It also assumes that the fish will be reared under optimal conditions or, failing that, that the fixed objective will be inferior or equal to the potential of the stock. In the case where a given mass has already been attained with a given diet, two parameters are required, growth rate and feed : gain ratio during the preceding period.

22.3.2.1 Utilisation of the specific growth rate from previous results

If P_2 is the final mass required, then:

$$P_2 = P_1(1 + SGRp)^D$$

Table 22.1. Predictions, at three temperatures, of body mass (g), energy requirements ($kJDE.100g^{-1}.day^{-1}$) and feeding rate (%) in trout, using the method of Cho (1995).

	5°C			10°C			15°C		
Week	g	kJ	%	g	kJ	%	g	kJ	%
0	1			1			1		
1	1.2	0.33	1.93	1.4	0.68	3.98	1.7	1.04	6.14
2	1.4	0.37	1.82	1.9	0.85	3.56	2.5	1.45	5.17
3	1.7	0.42	1.73	2.5	1.05	3.22	3.7	1.94	4.47
4	1.9	0.46	1.65	3.3	1.28	2.95	5.2	2.50	3.96
5	2.2	0.52	1.58	4.2	1.52	2.72	7.0	3.14	3.55
6	2.5	0.57	1.51	5.2	1.79	2.53	9.2	3.85	3.23
7	2.9	0.63	1.45	6.4	2.09	2.36	11.8	4.65	2.97
8	3.3	0.69	1.39	7.7	2.40	2.22	14.9	5.53	2.75
9	3.7	0.75	1.34	9.2	2.75	2.10	18.5	6.50	2.56
10	4.2	0.82	1.29	10.9	3.12	1.99	22.6	7.55	2.40
11	4.7	0.89	1.25	12.8	3.51	1.89	27.3	8.69	2.26
12	5.2	0.96	1.21	14.9	3.93	1.80	32.6	9.92	2.14
13	5.8	1.03	1.17	17.3	4.37	1.72	38.5	11.23	2.03
14	6.4	1.11	1.14	19.8	4.85	1.65	45.1	12.64	1.93
15	7.0	1.19	1.10	22.6	5.35	1.59	52.4	14.14	1.84
16	7.7	1.28	1.07	25.7	5.87	1.53	60.5	15.73	1.77
17	8.4	1.37	1.04	29.0	6.42	1.47	69.3	17.42	1.69
18	9.2	1.46	1.02	32.6	7.00	1.42	79.0	19.20	1.63
19	10.0	1.55	0.99	36.4	7.61	1.37	89.5	21.08	1.57
20	10.9	1.65	0.97	40.6	8.25	1.33	100.9	23.05	1.51
21	11.8	1.75	0.94	45.1	8.91	1.29	113.3	25.13	1.46
22	12.8	1.86	0.92	49.9	9.61	1.25	126.6	27.30	1.42
23	13.9	1.96	0.90	55.0	10.33	1.22	141.0	29.57	1.37
24	14.9	2.08	0.88	60.5	11.08	1.18	156.4	31.95	1.33
25	16.1	2.19	0.86	66.3	11.86	1.15	172.8	34.42	1.30
26	17.3	2.31	0.85	72.4	12.67	1.12	190.4	37.00	1.26
27	18.5	2.43	0.83	79.0	13.51	1.10	209.1	39.69	1.23
28	19.8	2.55	0.81	85.9	14.38	1.07	229.1	42.48	1.19
29	21.2	2.68	0.80	93.2	15.28	1.05	250.2	45.37	1.17
30	22.6	2.81	0.78	100.9	16.21	1.02	272.6	48.37	1.14
31	24.1	2.95	0.77	109.1	17.17	1.00	296.3	51.48	1.11
32	25.7	3.09	0.75	117.6	18.16	0.98	321.4	54.69	1.09
33	27.3	3.23	0.74	126.6	19.18	0.96	347.8	58.02	1.06
34	29.0	3.38	0.73	136.1	20.24	0.94	375.6	61.45	1.04
35	30.7	3.52	0.72	146.0	21.32	0.92	404.9	65.00	1.02
36	32.6	3.68	0.70	156.4	22.44	0.90	435.7	68.66	1.00
37	34.5	3.83	0.69	167.2	23.59	0.89	468.0	72.43	0.98
38	36.4	3.99	0.68	178.6	24.77	0.87	501.8	76.31	0.96
39	38.5	4.16	0.67	190.4	25.98	0.86	537.2	80.30	0.94
40	40.6	4.32	0.66	202.8	27.22	0.84	574.3	84.41	0.92

where P_1 = mass at last weighing
SGRp = specific growth rate during previous period
D = time in days to change from P_1 to P_2.

N.B. depending on case, the unknown in the equation is P_2 or D.

The disadvantage of this method is that it consists of extrapolating growth observed in the previous period. If this has been poor, or poorly evaluated, growth for the predicted period risks being underestimated. The method can no longer be used if the difference in temperature between the two periods is large. Note also that SGR does not constitute the best parameter for assessment of growth and that it can give rise to significant errors in estimation for high values of D. However, the same principle can be applied to any other growth parameter.

22.3.2.2 Calculation or estimation of the feed consumption ratio

CI is an important parameter in the calculation of feeding rate (FCR). It can be set by the farmer, either in relation to results usually observed, or as a fixed objective to be attained, or calculated in relation to mass using a mathematical model. The linear model below gives a good evaluation of this coefficient, for which experimental data can be used for 'validation':

$$\text{FCR} = a + b \times P_{\text{kg}}$$

where a is the FCR observed at the start of rearing
b is the increase in the coefficient per kg of mean mass
P is the mean mass of fish.

22.3.2.3 Calculation of feeding rate

Knowing the mass P_2, the objective of the rearing system, and thus the SGR allowing this to be attained from P_1 as well as the probable FCR, FR can be simply calculated from:

$$\text{FR} = \text{CI} \times \text{SGR}$$

22.4 FEED DISTRIBUTION METHODS

22.4.1 Manual distribution

Manual distribution is still widely used. It has the advantage of involving surveillance of the fishes' behaviour, especially if they are not free range and can be easily observed. However, the cost of this method is often high in relation to the margins cleared by fish farming.

22.4.2 Automatic distribution

Depending on the apparatus chosen, automatic distribution allows division of distribution or continuous distribution over the entire feeding period. Different systems can be utilised, from conveyor belt distributors that discharge food continuously throughout the day to automatic machines that calculate, weigh out and distribute the rations. The determining factor between these different systems is often cost.

In certain cases, automatic distribution allows the FCR to be improved, especially for species which are still barely domesticated, whose feeding behaviour is easily disturbed and so are difficult to observe (brown trout or Atlantic salmon, for example). The resulting fractionation of the ration is also important in limiting the risks inherent to distribution of too large amounts of food at inappropriate times. For example, excessive distribution can occur in the sea during low tides, periods during which water turnover in the rearing cages is insufficient to maintain a sufficient oxygen level.

It is possible to combine the different methods of feeding in order to profit from their advantages or minimise their disadvantages, respectively. Thus, automatic distribution of 80% of a calculated ration can be followed at the end of the day by feeding manually to satiation, while carefully observing behaviour. In most cases, however, quantitative evaluation of the fishes' requirements remains necessary to ensure rational feeding management.

22.4.3 Demand feeding

In this system, the fish can obtain the quantity of feed it desires using an appropriate device (tactile rod with mechanical or electronic control). The quantity of food available in the device can be determined by the fish farmer (calculated daily ration) but, as a general rule, it is much higher. An electronic device can control both the delay in response to demands by the fish on the tactile rod and the amount of food distributed at each activation, to limit the losses of feed linked to 'play' or external events (water movement, wind).

This feeding system is frequently blamed for exacerbating competitive phenomena and thus increasing growth disparities. This disadvantage can be reduced by increasing the number of distribution points or improving their distribution. A complementary method used in raceway-type ponds is to position feed distributors in the last third of the ponds in order to induce fed fish to move upstream into more oxygenated water and allow other fish access to the feeding points. Demand feeding is a relatively recent method under practical rearing conditions. It is subject to some uncertainties concerning the quality of the products obtained, notably their lipid content. In addition, the investment cost for the equipment, from which one hopes to optimise distribution (electronic control), can make farmers hesitate. Nevertheless, the significance of this method is that it allows the factors of voluntary intake variation in fish to be taken into account; thus it may well develop in the future.

22.5 CONCLUSION

Feeding methods are approached in very different ways by fish farmers. From manual methods using empirical feeding rates, automatic methods have developed where the use of precise programming is increasingly used. However, few of the necessary models required for automation are as yet available. Researchers are more interested in the determination of requirements than in the way in which the fish farmer can satisfy them via feed distribution. The use of the microcomputer is opening many avenues to the fish farmer, as a powerful tool for integrating the numerous parameters concerning the animals' requirements and consequently the ration which they should receive.

22.6 BIBLIOGRAPHY

Brett, J. R. (1976) *Feeding metabolic rates of young sockeye salmon,* Oncorhynchus nerka, *in relation to ration level and temperature.* Research Techniques 675, Ministry of Environment, Fisheries and Science, Ottawa, Canada.

Cho, C. Y. (1992) Feeding systems for rainbow trout and other salmonids with reference to current estimate of energy and protein requirement. *Aquaculture*, **100**, 107–123.

Cho, C. Y. and Bureau, D. P. (1995) Determination of the energy requirements of fish with particular reference to salmonids. *J. Appl. Ichtyol.*, **11**, 141–163.

Muller-Fuega, A. (1990) *Modélisation de la croissance des poissons en elevage.* Rapport Technique Ifremer No. 3.

Appendix A

Growth curves and length – body mass relationships for some fish species

The theoretical growth curves shown in Figures A.1 to A.8 were drawn using the model of Mueller-Feuga (Chapter 2, Boxes 2.1, 2.2). The values of the parameters of the somatic and thermal functions are indicated in the figure legends. The growth curves in Figures A.9 and A.10 were drawn using a much simpler model, while the curves in Figure A.11 correspond to real data with no modelling.

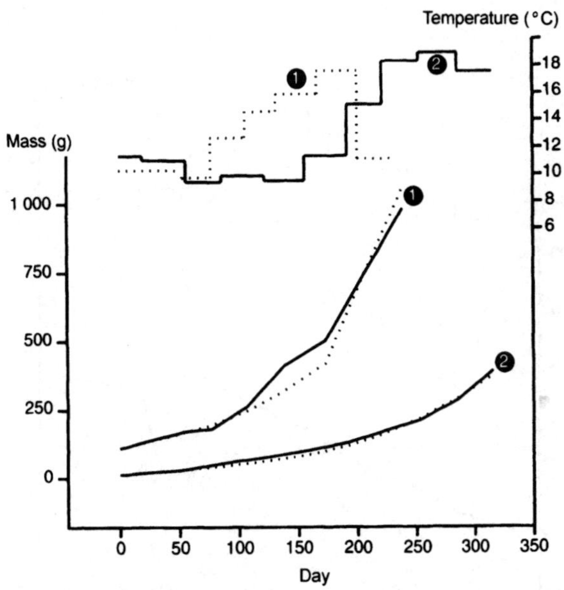

Figure A.1. Growth curve showing increase in mass of brown trout in freshwater. Effect of temperature (2 thermal cycles: 1 and 2); observed values (——); modelled curves (- - - -); $\alpha = 0.119, \beta = 0.140, \theta_M = 24$ and $\Gamma = 1$ (data from SEMii[1])

[1] Salmoniculture Experimentale INRA-IFREMER, 29450 Sizun.

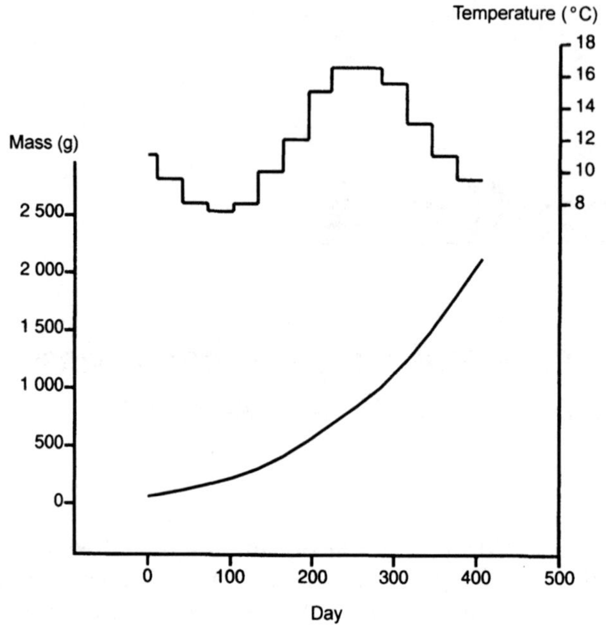

Figure A.2. Growth in mass of brown trout (marine phase) for the thermal regime described; $\alpha = 0.128, \beta = 0.156, \theta_M = 20$ and $\Gamma = 1$ (data from SEMII).

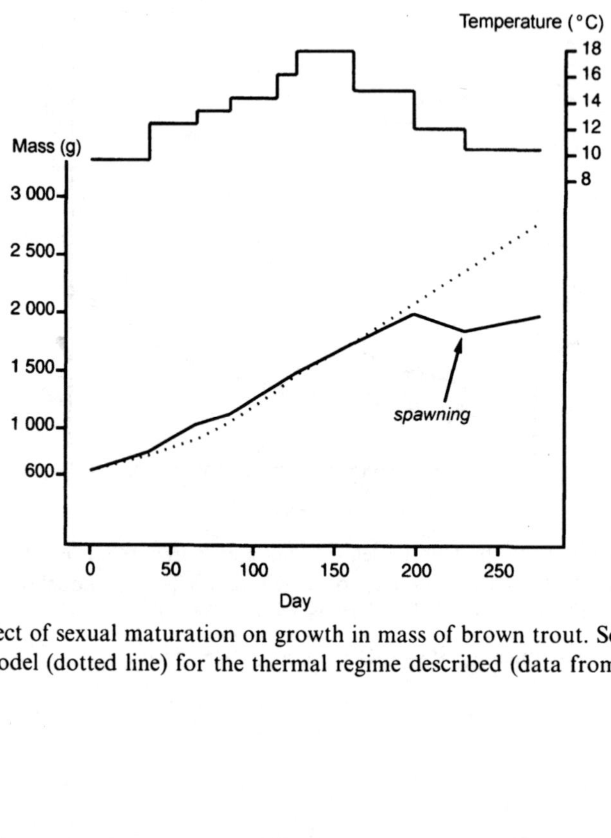

Figure A.3. Effect of sexual maturation on growth in mass of brown trout. Scaled in relation to calculated model (dotted line) for the thermal regime described (data from SEMII).

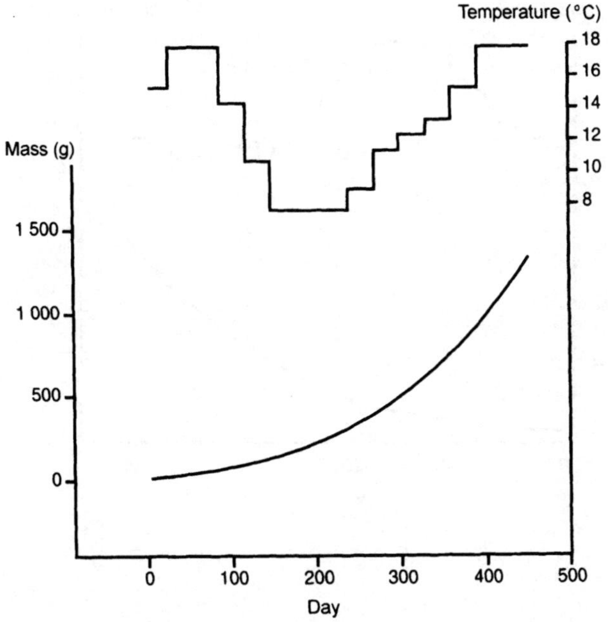

Figure A.4. Growth in mass of trout for the thermal range described; $\alpha = 0.115$, $\beta = 0.137, \theta_M = 22$ and $\Gamma = 1$ (data from SEMII).

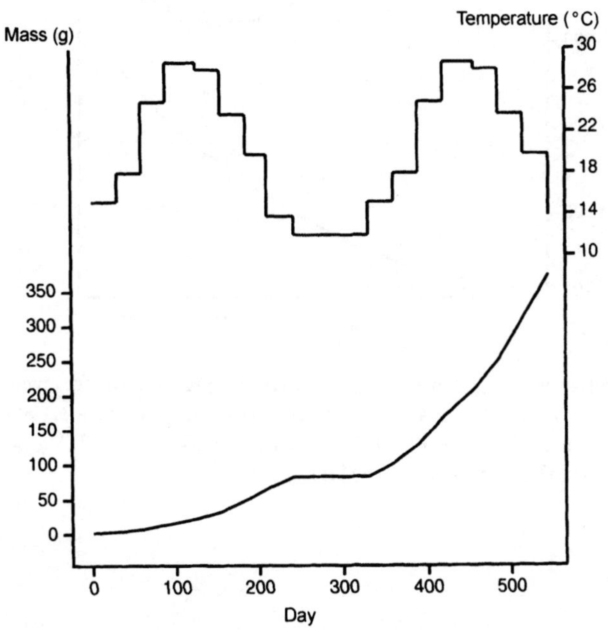

Figure A.5. Growth in mass of sea bass in the ongrowing phase and for the thermal regimes described; $\alpha = 0.123, \beta = 0.140, \theta_M = 31.5$ and $\Gamma = 1$ (data from IFREMER).

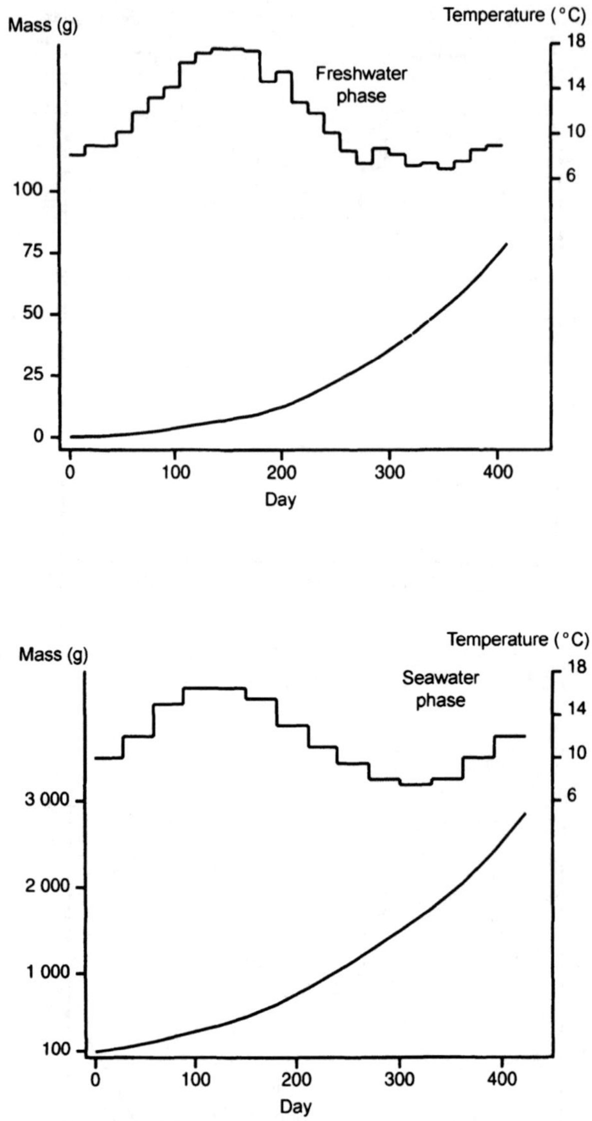

Figure A.6. Growth in mass for Norwegian stock Atlantic salmon for freshwater and seawater phases and for the thermal regimes described. Top: the freshwater phase: $\alpha = 0.129$, $\beta = 0.140, \theta_M = 20$ and $\Gamma = 1$; Bottom: the marine phase: $\alpha = 0.128, \beta = 0.157, \theta_M = 20$ and $\Gamma = 1$ (data from IFREMER and SEMII).

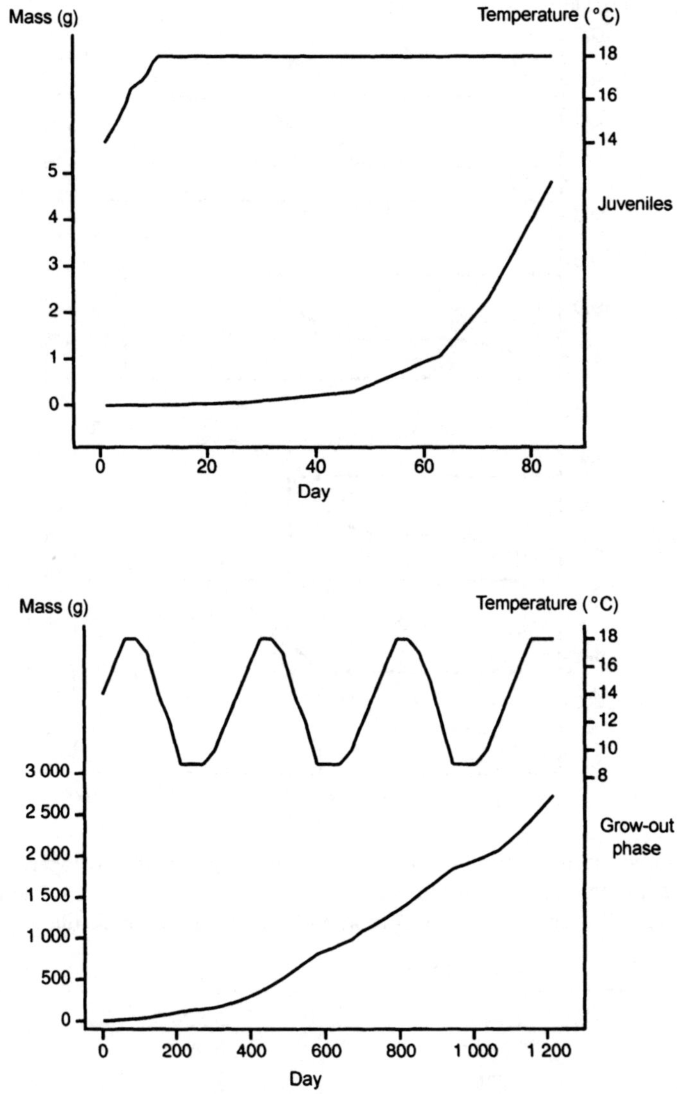

Figure A.7. Growth in mass for turbot for the production phase in juveniles, as well as the growing-out phase, for the thermal regimes described. The values of parameters are given in the table (data from IFREMER):

Mass	α	β	θ_M
< 1 kg	0.1235	0.140	24.8
> 1 kg	0.1305	0.140	24.8

The best adjustment to the observed values requires a modification of the model parameters above an individual mass of about 1kg, which corresponds to the start of sexual maturation in the males.

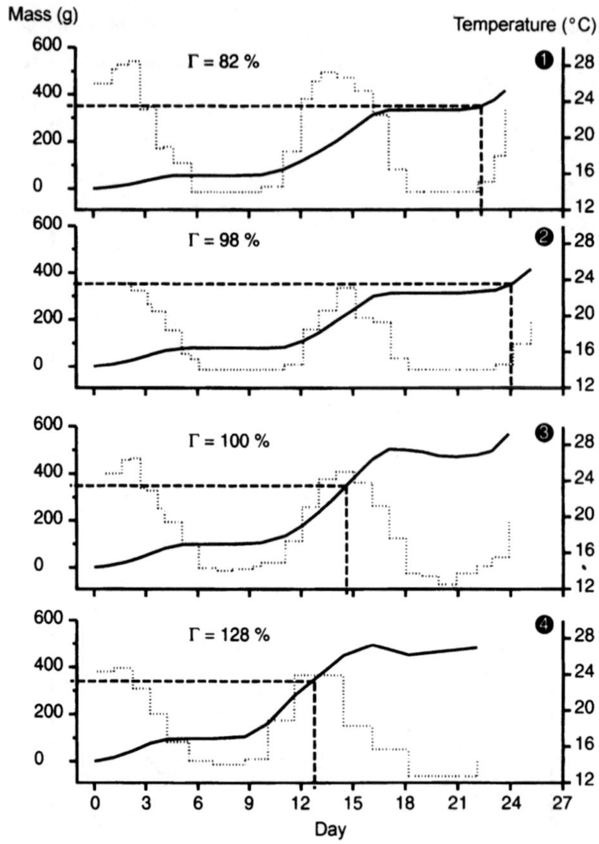

Figure A.8. Growth in mass of gilthead sea bream showing the effect of operation type and thermal regime. In each case, $m = 0.5, \alpha = 0.136, \beta = 0.1965$ and $\theta_M = 31.5$. The values of the performance index Γ found for each operation and type of site were as follows:

Operation	Site type	Γ
1	closed	0.82
2	closed	0.98
3	open	1
4	open	1.28

(Date from IFREMER).

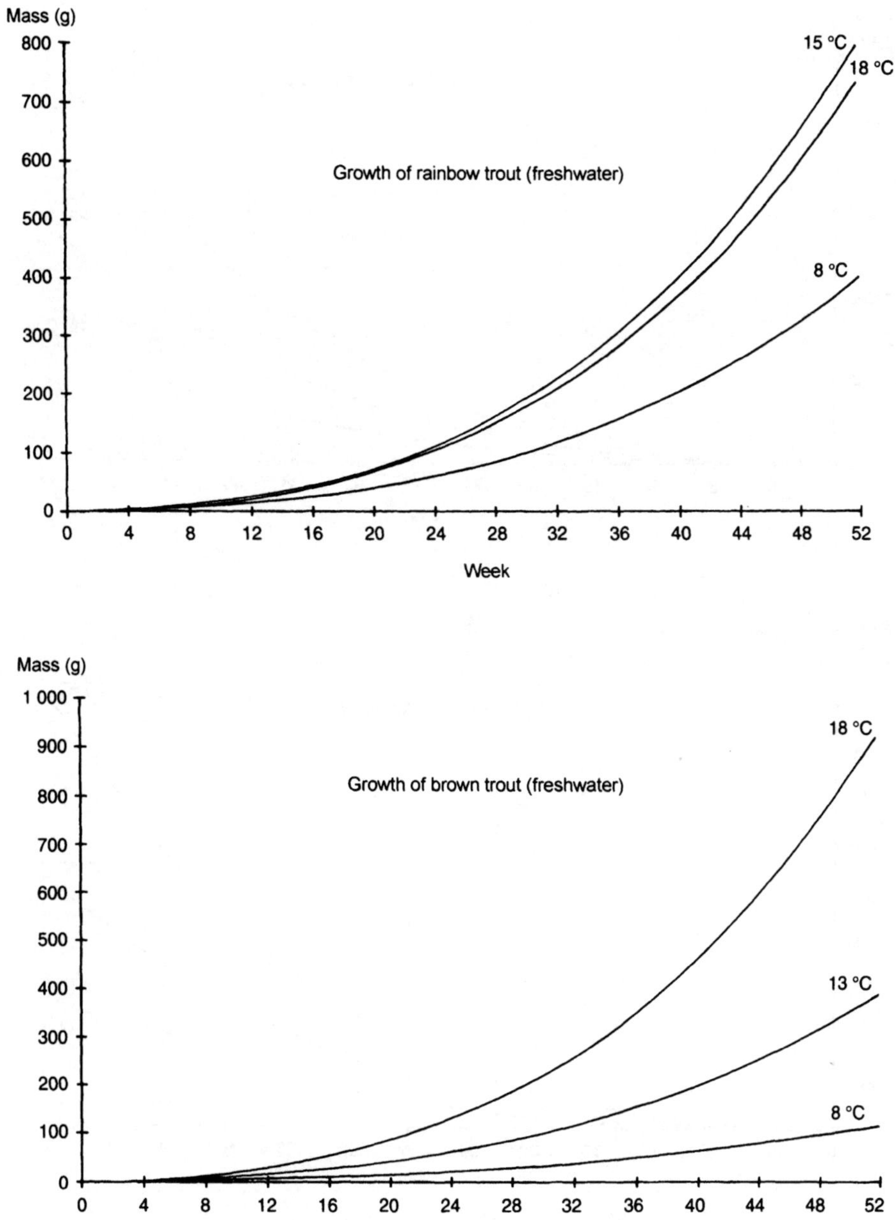

Figure A.9. Theoretical growth curves (mass) under ideal conditions for several temperatures, assumed to be constant throughout rearing (model from S. Kaushik).

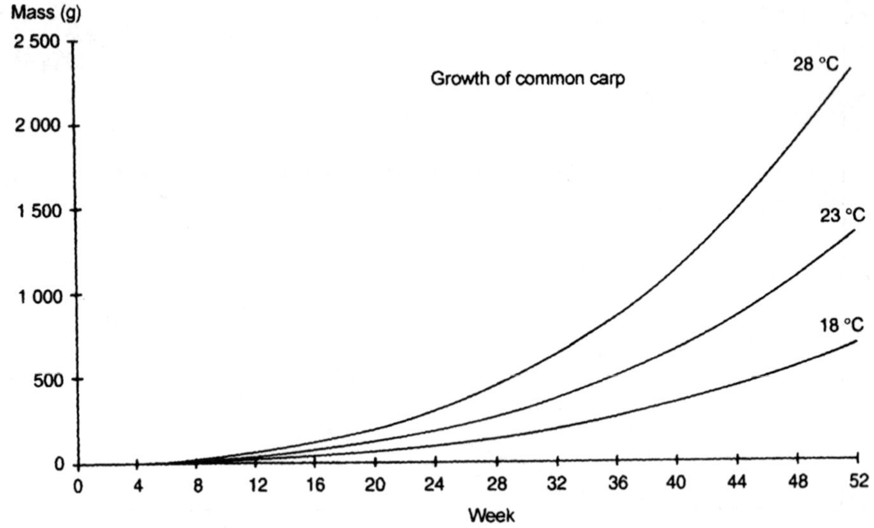

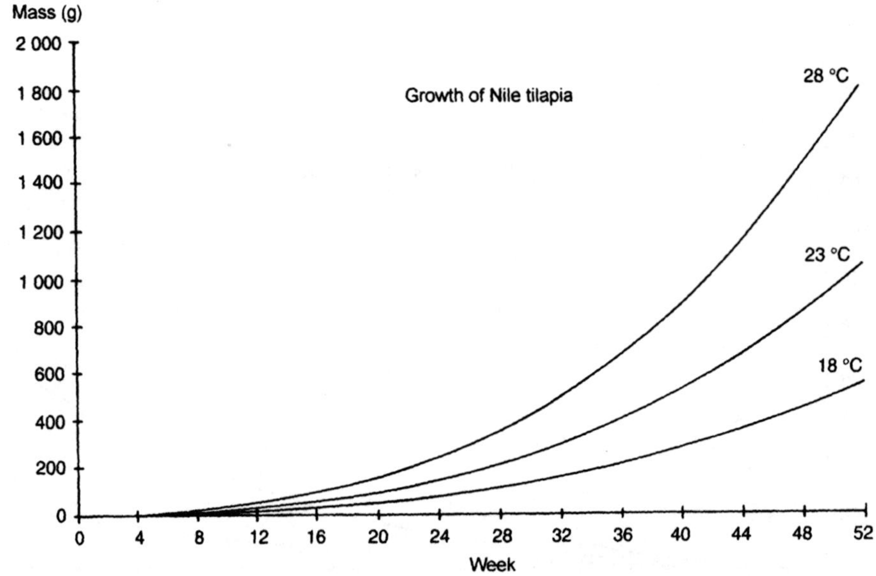

Figure A.9. (continued). Theoretical growth curves (mass) under ideal conditions for several temperatures, assumed to be constant throughout rearing (model from S. Kaushik).

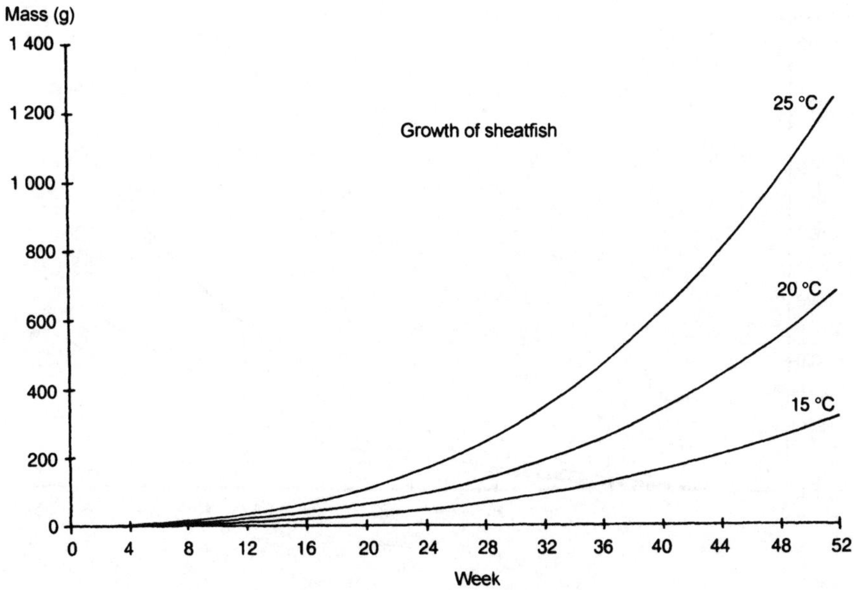

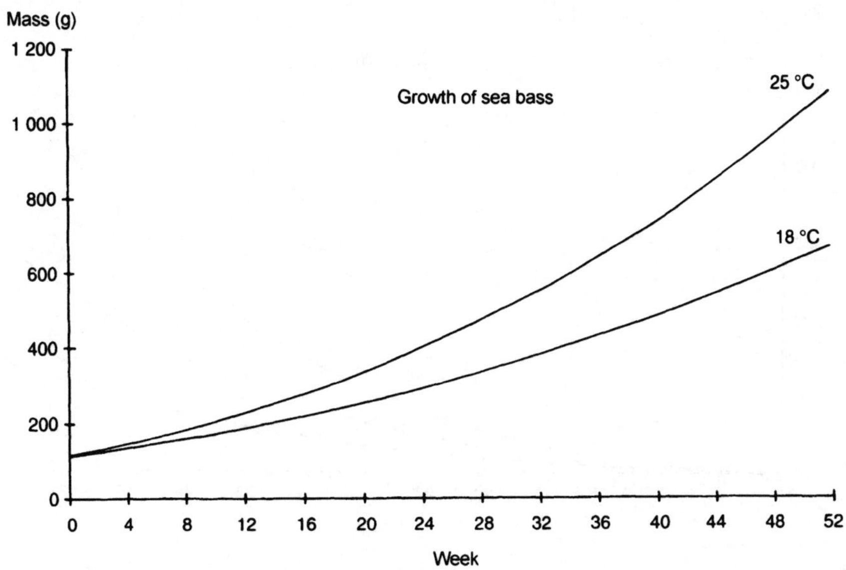

Figure A.9. (continued). Theoretical growth curves (mass) under ideal conditions for several temperatures, assumed to be constant throughout rearing (model from S. Kaushik).

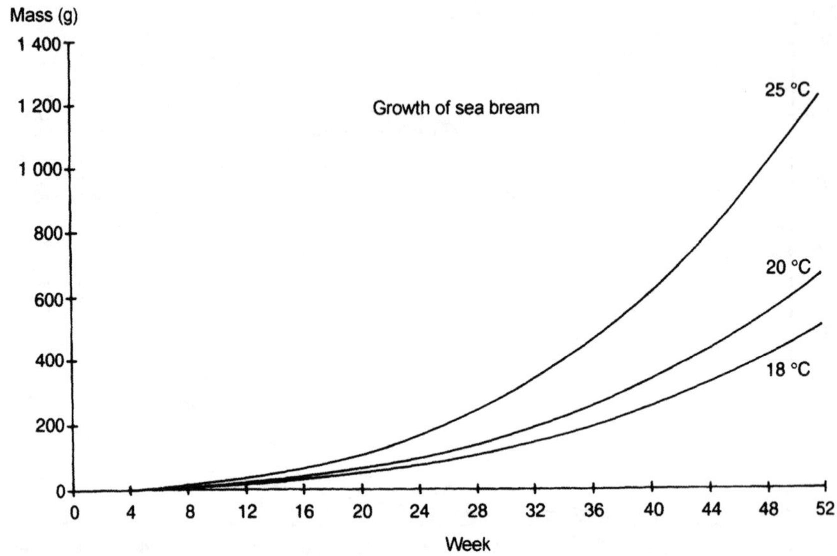

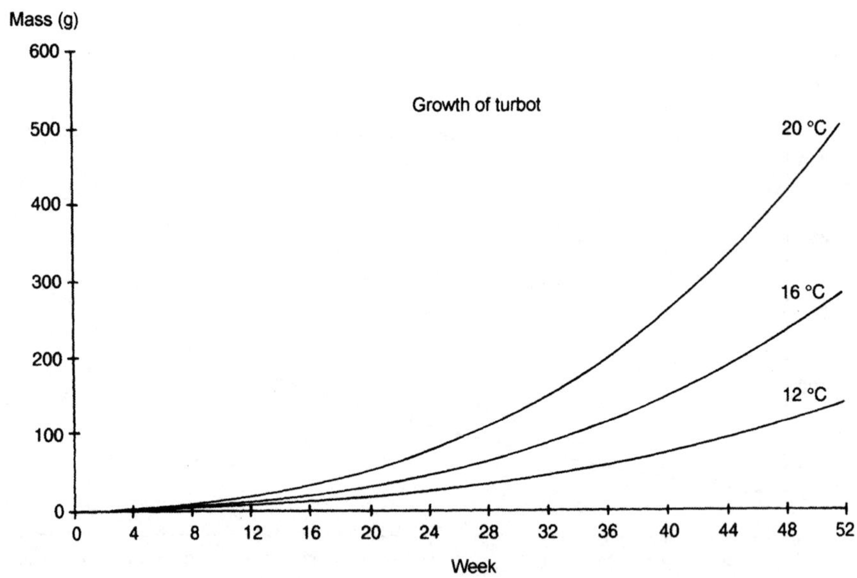

Figure A.9. (continued). Theoretical growth curves (mass) under ideal conditions for several temperatures, assumed to be constant throughout rearing (model from S. Kaushik).

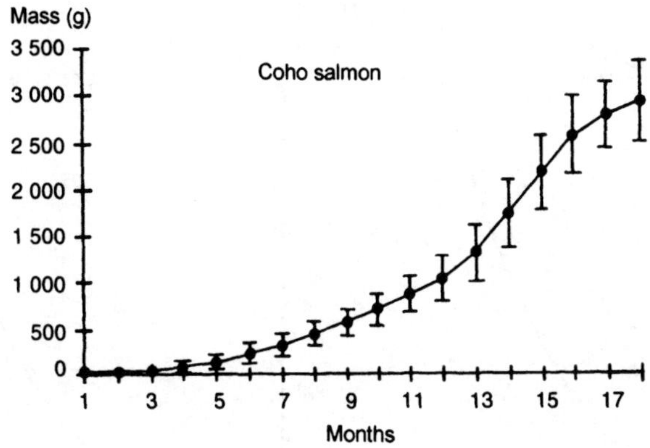

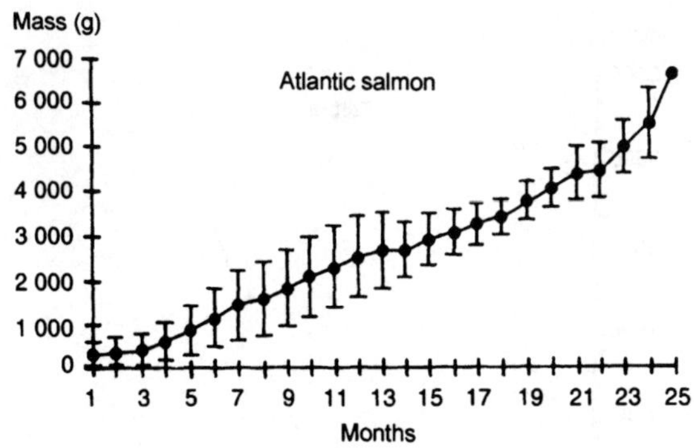

Figure A.10. Actual growth curves (mass) observed in Atlantic and coho salmon under good conditions.

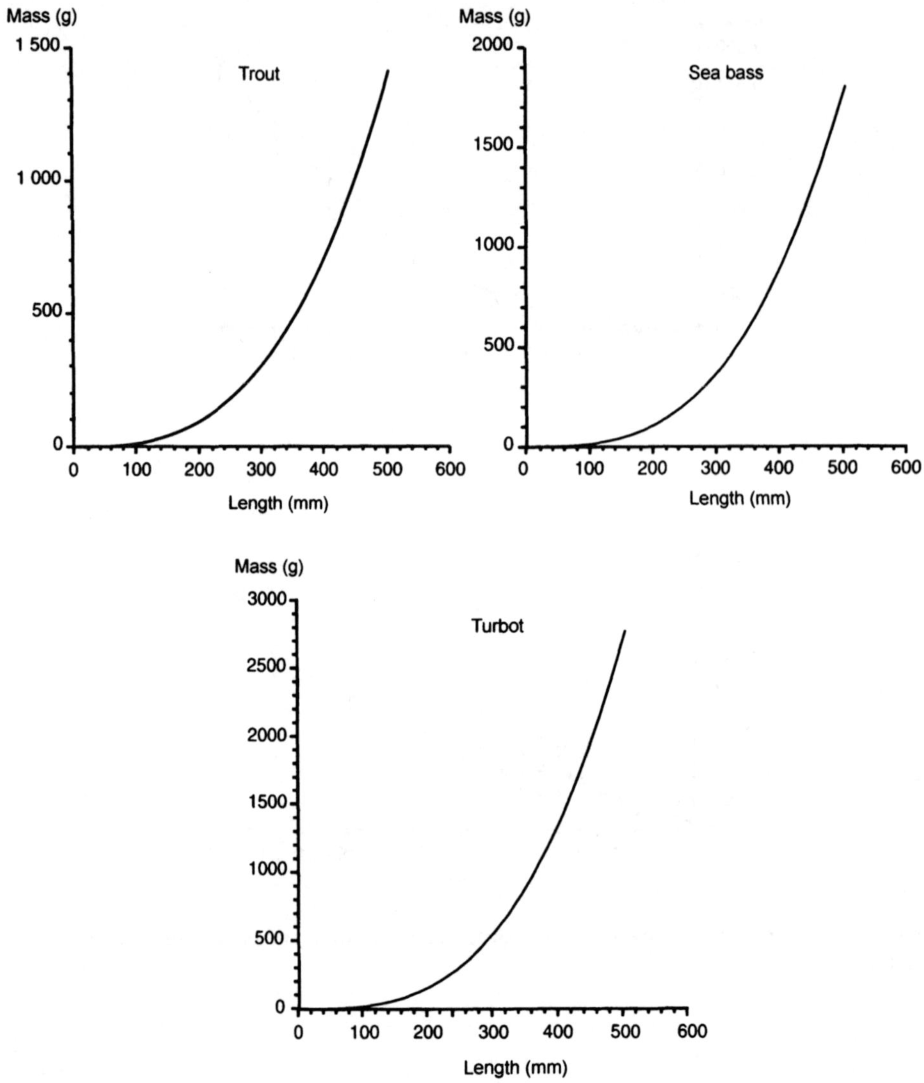

Figure A.11. Graphs showing the relationship between length and body mass in three species of fish.

Appendix B

Composition and nutritive value of raw materials

The tables give the chemical composition as well as other nutritional characteristics (digestible energy and proteins, available phosphorus) of the main ingredients used for feeding fish and crustaceans.

They are mainly extracted from the collective works published by INRA *The Feeding of Monogastric Animals, Pigs, Rabbits and Poultry* (INRA, 1984). Some primary ingredients that do not figure in the cited work have been added; these include products which are more specifically used in the feeding of aquatic animals (squid and shrimp meals, for example).

Another change concerns the n-3 and n-6 essential fatty acids which are particularly important in aquaculture nutrition, as well as cholesterol which is an essential nutrient for crustaceans. The following abbreviations have been adopted:

- Long chain polyunsaturated fatty acids (HUFA): n-3 or n-6 fatty acids with more than 18 carbon atoms.
- Polyunsaturated fatty acids (PUFA): n-3 or n-6 fatty acids, including those with 18 carbon atoms.

Note that, in general, PUFAs with 18 carbon atoms, C18:3, n-3 and C18:2, n-6, can only meet the requirements for essential fatty acids of freshwater fish (Chapter 7).

Table B.1. Tables of composition.

Characteristics % product (as fed)	Fish meal			
	65 high fat	72 high fat	65 lean	72 lean
Dry matter	**92**	**92**	**92**	**92**
Nitrogen-free extract	–	–	–	–
Starch (acid hydrolysis)	–	–	–	–
Starch (enzyme hydrolysis)	–	–	–	–
Sugars	–	–	–	–
Cellulose	–	–	–	–
ADF	–	–	–	–
NDF	–	–	–	–
Crude proteins	**66.2**	**71.6**	**64.6**	**71.3**
Lysine	5.03	5.48	5.04	5.42
Methionine	1.92	2.08	1.81	2.07
Methionine + cystine	2.52	2.80	2.39	2.71
Tryptophan	0.70	0.79	0.65	0.75
Threonine	2.88	3.06	2.73	3.10
Leucine	5.10	5.20	4.81	5.20
Isoleucine	3.11	3.42	3.04	3.35
Valine	3.67	3.88	3.55	3.80
Histidine	1.52	1.61	1.58	1.63
Arginine	3.77	3.92	3.71	4.06
Phenylalanine + tyrosine	4.83	4.85	4.81	5.00
Fats (ether extract)	**9.6**	**9.5**	**5.5**	**1.80**
saturated fatty acids	1.86	2.05	1.19	0.39
monosaturated fatty acids	4.82	5.30	3.07	1.00
n-6 HUFA	0.05	0.06	0.03	0.01
n-6 PUFA	0.17	0.19	0.11	0.04
n-3 HUFA	1.27	1.39	0.80	0.26
n-3 PUFA	1.37	1.50	0.87	0.28
Cholesterol	–	–	–	–
Minerals (ash)	**15.6**	**11.0**	**21.4**	**16.8**
Calcium	3.90	2.70	6.30	4.20
Total phosphorus	2.55	1.80	3.50	2.75
Sodium	0.90	0.77	1.00	0.95
Potassium	0.75	0.90	0.70	0.74
Chlorine	1.15	1.00	1.30	1.20
Magnesium	0.20	0.18	0.23	0.21
Gross energy				
$Mcal.kg^{-1}$	4.78	4.94	4.60	4.30
$MJ.kg^{-1}$	18.7	20.7	19.3	18.0
Digestible energy				
$Mcal.kg^{-1}$	4.14	4.41	3.73	3.77
$MJ.kg^{-1}$	17.3	18.5	15.6	15.8
Digestible proteins	59.6	64.4	58.1	64.2
Available phosphorus	2.20	1.53	3.00	2.35

Characteristics % product (as fed)	Low temperature fishmeal (LT)	Soluble protein composition of the fish	
		CPSP G	CPSP 90
Dry matter	**92**	**95**	**95**
Nitrogen-free extract	**–**	**–**	**–**
Starch (acid hydrolysis)	–	–	–
Starch (enzyme hydrolysis)	–	–	–
Sugars	–	–	–
Cellulose	–	–	–
ADF	–	–	–
NDF	–	–	–
Crude proteins	**74.2**	**70.0**	**83.3**
Lysine	6.94	4.67	5.67
Methionine	2.43	2.08	2.48
Methionine + cystine	2.93	2.23	3.25
Tryptophan	0.75	0.52	0.62
Threonine	2.65	2.66	3.16
Leucine	6.68	4.18	4.98
Isoleucine	4.04	2.52	3.00
Valine	4.38	3.00	3.56
Histidine	1.74	1.41	1.68
Arginine	5.55	4.52	5.38
Phenylalanine + tyrosine	5.94	4.40	5.24
Fats (ether extract)	**11.0**	**18–20**	**7.5**
saturated fatty acids	2.16	–	–
monosaturated fatty acids	5.94	–	–
n-6 HUFA	0.06	–	–
n-6 PUFA	0.20	–	–
n-3 HUFA	1.47	–	–
n-3 PUFA	1.58	–	–
Cholesterol	–	–	–
Minerals (ash)	11.7	5.3	7.1
Calcium	3.27	0.60	0.60
Total phosphorus	0.70	0.60	0.80
Sodium	–	–	–
Potassium	–	–	–
Chlorine	–	–	–
Magnesium	–	–	–
Gross energy	–	–	–
$Mcal.kg^{-1}$	5.23	5.90	5.38
$MJ.kg^{-1}$	21.9	24.6	22.5
Digestible energy			
$Mcal.kg^{-1}$	4.76	5.23	5.16
$MJ.kg^{-1}$	19.9	24.0	21.6
Digestible proteins	67.9	65.6	79.1
Available phosphorus	0.60	0.51	0.60

Characteristics % product (as fed)	Squid meal	Shrimp meal
Dry matter	**92.0**	**88.0**
Nitrogen-free extract	–	–
Starch (acid hydrolysis)	–	–
Starch (enzyme hydrolysis)	–	–
Sugars	–	–
Cellulose	–	–
ADF	–	–
NDF	–	–
Crude proteins	**78.0**	**39.5**
Lysine	4.85	2.17
Methionine	2.16	0.82
Methionine + cystine	3.17	1.41
Tryptophan	–	0.42
Threonine	3.17	1.42
Leucine	5.10	2.60
Isoleucine	3.00	1.46
Valine	2.82	1.83
Histidine	2.12	0.90
Arginine	5.20	2.35
Phenylalanine + tyrosine	–	1.59
Fats (ether extract)	**6.0**	**3.2**
saturated fatty acids	1.78	–
monosaturated fatty acids	0.56	–
n-6 HUFA	0.03	–
n-6 PUFA	0.04	–
n-3 HUFA	2.72	–
n-3 PUFA	2.72	–
Cholesterol	–	–
Minerals (ash)	**11.0**	**27.2**
Calcium	0.10	–
Total phosphorus	0.60	–
Sodium	1.10	–
Potassium	–	–
Chlorine	–	–
Magnesium	–	–
Gross energy		
$Mcal.kg^{-1}$	4.70	3.60
$MJ.kg^{-1}$	19.8	15.0
Digestible energy		
$Mcal.kg^{-1}$	4.0	3.30
$MJ.kg^{-1}$	16.8	13.8
Digestible proteins	70.0	36.3
Available phosphorus	–	–

Characteristics % product (as fed)	Meat meal 50 thin	 60 thin
Dry matter	**93.0**	**93.0**
Nitrogen-free extract	–	–
Starch (acid hydrolysis)	–	–
Starch (enzyme hydrolysis)	–	–
Sugars	–	–
Cellulose	–	–
ADF	–	–
NDF	–	–
Crude proteins	**50.0**	**60.0**
Lysine	2.80	3.46
Methionine	0.70	0.85
Methionine + cystine	1.20	1.39
Tryptophan	0.28	0.36
Threonine	1.79	2.15
Leucine	3.10	3.82
Isoleucine	1.40	1.94
Valine	2.30	2.80
Histidine	0.95	1.13
Arginine	3.48	3.89
Phenylalanine + tyrosine	2.75	3.36
Fats (ether extract)	**5.0**	**4.0**
saturated fatty acids	2.38	1.90
monosaturated fatty acids	2.01	1.61
n-6 HUFA	traces	traces
n-6 PUFA	0.16	0.12
n-3 HUFA	–	–
n-3 PUFA	0.03	0.024
Cholesterol	0.035	0.028
Minerals (ash)	**32.6**	**22.5**
Calcium	10.20	7.05
Total phosphorus	4.85	3.35
Sodium	0.76	0.52
Potassium	0.54	0.37
Chlorine	0.65	0.45
Magnesium	0.76	0.52
Gross energy		
$Mcal.kg^{-1}$	3.28	3.75
$MJ.kg^{-1}$	13.70	15.69
Digestible energy		
$Mcal.kg^{-1}$	2.81	3.21
$MJ.kg^{-1}$	11.7	13.4
Digestible proteins	42.5	51.0
Available phosphorus	3.90	2.70

Characteristic % product (as fed)	Connective tissue meal	Hydrolysed feather meal	Blood meal
Dry matter	**93.0**	**93.0**	**93.0**
Nitrogen-free extract	–	–	–
Starch (acid hydrolysis)	–	–	–
Starch (enzyme hydrolysis)	–	–	–
Sugars	–	–	–
Cellulose	–	–	–
ADF	–	–	–
NDF	–	–	–
Crude proteins	**67.2**	**85.8**	**84.0**
Lysine	3.72	1.84	7.62
Methionine	0.95	0.53	0.93
Methionine + cystine	1.57	4.08	1.68
Tryptophan	0.38	0.43	1.06
Threonine	2.27	3.91	4.00
Leucine	4.00	7.14	11.58
Isoleucine	1.96	3.95	0.79
Valine	3.10	6.88	7.02
Histidine	1.12	0.58	5.18
Arginine	4.42	5.66	3.63
Phenylalanine + tyrosine	3.74	7.01	6.97
Fats (ether extract)	**10.1**	**3.5**	**1.1**
saturated fatty acids	4.80	0.93	–
monosaturated fatty acids	4.06	1.51	–
n-6 HUFA	traces	traces	–
n-6 PUFA	0.31	0.68	–
n-3 HUFA	–	–	–
n-3 PUFA	0.06	0.04	–
Cholesterol	0.07	0.01	–
Minerals (ash)	**15.0**	**3.2**	**4.5**
Calcium	4.10	0.20	0.30
Total phosphorus	2.10	0.70	0.25
Sodium	1.00	–	0.32
Potassium	0.40	0.24	0.10
Chlorine	–	–	0.40
Magnesium	0.12	0.18	0.22
Gross energy			
Mcal.kg^{-1}	4.71	5.16	4.85
MJ.kg^{-1}	19.69	21.60	20.27
Digestible energy			
Mcal.kg^{-1}	4.08	4.50	4.31
MJ.kg^{-1}	17.1	18.8	18.1
Digestible proteins	57.8	74.6	74.8
Available phosphorus	1.80	0.60	0.22

Characteristics % product (as fed)	Lacteroserum	Lactic casein
Dry matter	**95.0**	**91.0**
Nitrogen-free extract	**71.6**	**4.8**
Starch (acid hydrolysis)	–	–
Starch (enzyme hydrolysis)	–	–
Sugars	66.5	–
Cellulose	–	–
ADF	–	–
NDF	–	–
Crude proteins	**13.3**	**82.0**
Lysine	1.12	6.72
Methionine	0.22	2.29
Methionine + cystine	0.50	2.57
Tryptophan	0.15	1.39
Threonine	0.72	4.02
Leucine	1.14	7.54
Isoleucine	0.72	5.00
Valine	0.67	5.91
Histidine	0.24	2.54
Arginine	0.33	3.36
Phenylalanine + tyrosine	0.77	9.26
Fats (ether extract)	**1.3**	**1.5**
saturated fatty acids	–	0.81
monosaturated fatty acids	–	0.37
n-6 HUFA	–	traces
n-6 PUFA	–	0.03
n-3 HUFA	–	0.00
n-3 PUFA	–	0.00
Cholesterol	–	0.01
Minerals (ash)	**8.8**	**2.7**
Calcium	0.94	0.12
Total phosphorus	0.70	0.41
Sodium	0.80	0.004
Potassium	2.23	0.006
Chlorine	1.20	0.04
Magnesium	0.12	0.002
Gross energy		
$Mcal.kg^{-1}$	3.72	5.01
$MJ.kg^{-1}$	15.58	20.95
Digestible energy		
$Mcal.kg^{-1}$	3.20	4.78
$MJ.kg^{-1}$	13.40	20.0
Digestible proteins	11.5	78.7
Available phosphorus	–	–

Characteristics % product (as fed)	Soya oilcake 44	 50	Double zero rape oilcake
Dry matter	**88.0**	**88.0**	**89.0**
Nitrogen-free extract	**30.3**	**28.5**	**33.3**
Starch (acid hydrolysis)	3.3	3.1	5.2
Starch (enzyme hydrolysis)	–	–	–
Sugars	10.0	8.7	8.3
Cellulose	7.4	3.4	11.7
ADF	9.6	5.3	18.5
NDF	13.5	7.5	25.5
Crude proteins	**42.5**	**48.0**	**38.7**
Lysine	2.70	3.05	1.97
Methionine	0.59	0.66	0.76
Methionine + cystine	1.27	1.43	1.73
Tryptophan	0.57	0.65	0.43
Threonine	1.67	1.88	1.57
Leucine	3.26	3.68	2.43
Isoleucine	2.14	2.42	1.44
Valine	2.18	2.46	1.84
Histidine	1.05	1.19	0.92
Arginine	3.18	3.59	2.19
Phenylalanine + tyrosine	3.65	4.12	2.45
Fats (ether extract)	**1.8**	**1.9**	**1.8**
saturated fatty acids	0.23	0.24	0.11
monosaturated fatty acids	0.31	0.35	0.94
n-6 HUFA	0.00	0.00	0.00
n-6 PUFA	0.85	0.90	0.34
n-3 HUFA	0.00	0.00	0.00
n-3 PUFA	0.12	0.12	0.14
Cholesterol	0.00	0.00	0.00
Minerals (ash)	**6.0**	**6.2**	**7.0**
Calcium	0.30	0.27	0.75
Total phosphorus	0.62	0.69	1.10
Sodium	0.01	0.01	0.07
Potassium	1.70	2.20	1.25
Chlorine	traces	traces	–
Magnesium	0.25	0.28	0.45
Gross energy			
Mcal.kg^{-1}	4.15	4.20	4.11
MJ.kg^{-1}	17.36	17.57	17.18
Digestible energy			
Mcal.kg^{-1}	3.11	3.23	3.40
MJ.kg^{-1}	13.0	13.5	14.2
Digestible proteins	34.1	41.1	36.0
Available phosphorus	0.10	0.10	0.22

Characteristics % product (as fed)	Copra cake	Cotton cake
Dry matter	**90.0**	**91.0**
Nitrogen-free extract	**43.9**	**29.1**
Starch (acid hydrolysis)	–	–
Starch (enzyme hydrolysis)	–	–
Sugars	9.0	–
Cellulose	16.0	13.0
ADF	30.0	21.1
NDF	54.0	29.8
Crude proteins	**21.5**	**41.0**
Lysine	0.66	1.72
Methionine	0.32	0.59
Methionine + cystine	0.63	1.24
Tryptophan	0.15	0.49
Threonine	0.67	1.39
Leucine	1.33	2.44
Isoleucine	0.76	1.39
Valine	1.18	1.97
Histidine	0.47	1.11
Arginine	2.62	4.33
Phenylalanine + tyrosine	1.50	3.30
Fats (ether extract)	**1.6**	**1.4**
saturated fatty acids	0.99	0.36
monosaturated fatty acids	0.99	0.25
n-6 HUFA	0.00	0.00
n-6 PUFA	0.03	0.72
n-3 HUFA	0.00	0.00
n-3 PUFA	–	–
Cholesterol	0.00	0.00
Minerals (ash)	**7.0**	**6.5**
Calcium	0.18	0.20
Total phosphorus	0.60	1.00
Sodium	0.04	0.05
Potassium	1.20	1.25
Chlorine	0.10	0.04
Magnesium	0.30	0.50
Gross energy		
$Mcal.kg^{-1}$	3.85	4.28
$MJ.kg^{-1}$	16.11	17.91
Digestible energy		
$Mcal.kg^{-1}$	–	2.16*
$MJ.kg^{-1}$	–	9.1*
Digestible proteins	18.5	32.8
Available phosphorus	0.09	0.10

* Calculations according to National Research Council.

Characteristics % product (as fed)	Wheat gluten	Maize gluten
Dry matter	**94.0**	**90.0**
Nitrogen-free extract	**–**	**37.7**
Starch (acid hydrolysis)	10	–
Starch (enzyme hydrolysis)	–	–
Sugars	–	–
Cellulose	–	4.0
ADF	–	5.0
NDF	–	14.4
Crude proteins	**79.0**	**60.0**
Lysine	1.44	0.77
Methionine	1.21	1.02
Methionine + cystine	2.65	1.65
Tryptophan	0.76	0.21
Threonine	2.05	1.48
Leucine	5.39	7.20
Isoleucine	2.81	2.12
Valine	3.19	2.22
Histidine	1.59	1.01
Arginine	2.81	1.41
Phenylalanine + tyrosine	6.22	4.69
Fats (ether extract)	**5.0**	**2.8**
saturated fatty acids	0.95	0.31
monosaturated fatty acids	0.48	0.62
n-6 HUFA	0.00	0.00
n-6 PUFA	2.74	1.41
n-3 HUFA	0.00	0.00
n-3 PUFA	0.16	–
Cholesterol	0.00	0.00
Minerals (ash)	**1.0**	**2.8**
Calcium	–	0.15
Total phosphorus	–	0.43
Sodium	–	0.008
Potassium	–	0.03
Chlorine	–	0.10
Magnesium	–	0.08
Gross energy		
$Mcal.kg^{-1}$	5.33	4.54
$MJ.kg^{-1}$	22.32	19.00
Digestible energy		
$Mcal.kg^{-1}$	4.73	4.21
$MJ.kg^{-1}$	19.8	17.6
Digestible proteins	75.1	57.6
Available phosphorus	–	0.13

Characteristics % product (as fed)	Young wheat	Maize
Dry matter	**86.0**	**86.0**
Nitrogen-free extract	**68.8**	**69.0**
Starch (acid hydrolysis)	56.0	60.5
Starch (enzyme hydrolysis)	–	–
Sugars	–	2.1
Cellulose	2.3	2.2
ADF	3.3	3.0
NDF	10.5	9.0
Crude proteins	**11.3**	**9.0**
Lysine	0.32	0.25
Methionine	0.19	0.19
Methionine + cystine	0.37	0.39
Tryptophan	0.2	0.06
Threonine	0.5	0.32
Leucine	0.8	1.13
Isoleucine	0.4	0.35
Valine	0.5	0.46
Histidine	0.24	0.26
Arginine	0.49	0.43
Phenylalanine + tyrosine	0.82	0.85
Fats (ether extract)	**1.9**	**4.20**
saturated fatty acids	0.25	0.46
monosaturated fatty acids	0.26	0.98
n-6 HUFA	0.00	0.00
n-6 PUFA	0.77	2.11
n-3 HUFA	0.00	0.00
n-3 PUFA	0.12	–
Cholesterol	0.00	0.00
Minerals (ash)	**1.7**	**1.6**
Calcium	0.06	0.01
Total phosphorus	0.33	0.27
Sodium	0.05	0.01
Potassium	0.40	0.33
Chlorine	0.06	0.05
Magnesium	0.12	0.11
Gross energy		
$Mcal.kg^{-1}$	**3.78**	**3.86**
$MJ.kg^{-1}$	15.83	16.13
Digestible energy		
$Mcal.kg^{-1}$	3.03	3.59
$MJ.kg^{-1}$	12.7	15.0
Digestible proteins	10.3	7.8
Available phosphorus	0.18	0.05

Characteristics % product (as fed)	White lupin, shelled and extruded	Pea, shelled and extruded
Dry matter	**87.0**	**86.0**
Nitrogen-free extract	**27.6**	**52.7**
Starch (acid hydrolysis)	–	42.4
Starch (enzyme hydrolysis)	0.3	–
Sugars	–	2.1
Cellulose	–	6.3
ADF	–	8.7
NDF	–	12.0
Crude proteins	**47.1**	**22.0**
Lysine	1.68	1.60
Methionine	0.28	0.25
Methionine + cystine	0.80	0.59
Tryptophan	0.28	0.20
Threonine	1.29	0.87
Leucine	2.50	1.53
Isoleucine	1.57	0.97
Valine	1.50	1.01
Histidine	0.82	0.52
Arginine	3.75	2.12
Phenylalanine + tyrosine	3.00	1.73
Fats (ether extract)	**9.6**	**1.6**
saturated fatty acids	1.19	–
monosaturated fatty acids	4.96	–
n-6 HUFA	0.00	0.00
n-6 PUFA	1.12	–
n-3 HUFA	0.00	0.00
n-3 PUFA	0.44	–
Cholesterol	0.00	0.00
Minerals (ash)	**3.4**	**3.4**
Calcium	0.18	0.08
Total phosphorus	0.40	0.40
Sodium	0.01	0.01
Potassium	0.90	1.10
Chlorine	0.02	0.03
Magnesium	0.15	0.12
Gross energy		
$Mcal.kg^{-1}$	4.35	3.57
$MJ.kg^{-1}$	18.2	14.9
Digestible energy		
$Mcal.kg^{-1}$	3.14	3.18
$MJ.kg^{-1}$	13.14	13.3
Digestible proteins	54.9	21.2
Available phosphorus	0.08	0.15

Characteristics % product (as fed)	Wheat bran	Rice bran	Wheat remillings
Dry matter	**88.0**	**90.0**	**88.0**
Nitrogen-free extract	**53.0**	**41.0**	**54.6**
Starch (acid hydrolysis)	53.0	41.0	54.6
Starch (enzyme hydrolysis)	19.0	–	–
Sugars	–	–	–
Cellulose	10.0	11.6	8.2
ADF	12.1	–	9.9
NDF	36.9	–	32.8
Crude proteins	**15.6**	**12.8**	**16.4**
Lysine	0.65	0.56	0.74
Methionine	0.25	0.22	0.24
Methionine + cystine	0.62	0.42	0.56
Tryptophan	0.19	0.13	0.20
Threonine	0.54	0.44	0.52
Leucine	0.99	0.90	1.06
Isoleucine	0.53	0.51	0.56
Valine	0.73	0.77	0.77
Histidine	0.38	0.34	0.39
Arginine	1.04	0.97	1.03
Phenylalanine + tyrosine	1.10	1.19	1.08
Fats (ether extract)	**6.5**	**13.8**	**4.8**
saturated fatty acids	–	0.81	–
monosaturated fatty acids	–	6.00	–
n-6 HUFA	0.00	0.00	0.00
n-6 PUFA	–	5.50	–
n-3 HUFA	0.00	0.00	0.00
n-3 PUFA	–	0.11	–
Cholesterol	0.00	0.00	0.00
Minerals (ash)	**4.4**	**10.7**	**4.0**
Calcium	0.15	0.07	0.13
Total phosphorus	0.93	1.40	0.90
Sodium	–	0.05	–
Potassium	1.00	1.50	–
Chlorine	0.06	0.06	–
Magnesium	0.35	0.85	–
Gross energy			
Mcal.kg^{-1}	4.00	1.97	4.2
MJ.kg^{-1}	16.7	8.2	17.6
Digestible energy			
Mcal.kg^{-1}	3.7	–	3.8
MJ.kg^{-1}	15.5	5.7	15.9
Digestible proteins	13.9	11.9	13.9
Available phosphorus	–	0.14	0.1

Characteristics % product (as fed)	Brewer's yeast	Lactic yeast
Dry matter	**93.0**	**95.0**
Nitrogen-free extract	**32.7**	**30.3**
Starch (acid hydrolysis)	4.9	0.5
Starch (enzyme hydrolysis)	–	–
Sugars	6.9	–
Cellulose	2.8	–
ADF	–	–
NDF	–	–
Crude proteins	**48.4**	**50.7**
Lysine	3.38	3.70
Methionine	0.67	0.66
Methionine + cystine	1.19	1.16
Tryptophan	0.55	0.55
Threonine	2.21	2.99
Leucine	3.11	3.80
Isoleucine	2.15	2.28
Valine	2.48	2.68
Histidine	1.12	0.96
Arginine	2.27	2.43
Phenylalanine + tyrosine	3.36	3.74
Fats (ether extract)	**1.8**	**6.2**
saturated fatty acids	–	2.13
monosaturated fatty acids	–	4.40
n-6 HUFA	–	0.00
n-6 PUFA	–	1.40
n-3 HUFA	–	0.00
n-3 PUFA	–	–
Cholesterol	0.00	0.00
Minerals (ash)	**7.3**	**7.8**
Calcium	0.14	0.23
Total phosphorus	1.40	1.80
Sodium	0.07	0.12
Potassium	1.70	2.20
Chlorine	0.15	0.22
Magnesium	0.20	0.14
Gross energy		
$Mcal.kg^{-1}$	4.30	–
$MJ.kg^{-1}$	18.00	–
Digestible energy		
$Mcal.kg^{-1}$	3.32	3.39
$MJ.kg^{-1}$	13.9	14.2
Digestible proteins	40.4	44.8
Available phosphorus	0.91	1.20

Table B.2. Variability in the chemical composition of fishmeals. Data in relation to dry matter (after Anderson *et al.*, 1993).[1]

	Range
Crude protein (%)	77.8–83.7
Crude fats (%)	9.4–15.0
Minerals (%)	11.0–14.6
Gross energy ($kJ.g^{-1}$)	21.4–22.9
Ca (%)	2.16–3.94
P (%)	1.69–2.44
Mg (%)	0.20–0.30
K (%)	0.43–1.05
Cu (ppm)	3.20–6.50
Fe (ppm)	76–220
Mn (ppm)	7–16
Zn (ppm)	78–120
Available lysine ($g.16\,gN^{-1}$)	7.13–9.00
Volatile base nitrogen (mg%)	46–156
Sulphydryl groups (SH) ($mM.16\,gN^{-1}$)	0.48–4.19
Bisulphates ($mM.16\,gN^{-1}$)	0.65–3.34
Sulphydayl ($mM.16\,gN^{-1}$)	1.86–5.62
Cysteine ($mM.16\,gN^{-1}$)	1.98–5.98

Table B.3. Availability (%) of phosphorus from various sources in carp, trout and catfish (after NRC, 1983, McVey, 1991).[2]

Source	Common carp	Catfish	Trout
Inorganic salts			
Monosodium phosphate	94	90	98
Monopotassium phosphate	94	–	98
Calcium phosphates			
Monocalcium	94	94	94
Bicalcium	46	65	71
Tricalcium	13	–	64
Raw materials			
Fishmeals	18–24	39–40	66-74
Casein	97	90	90
Brewers' yeast	93	–	91
Wheat germ	57	–	58
Maize	–	25	–
Soybean meal	–	29–54	–
Phytates	8–38	–	19

[1] Anderson, J. S., Lall, S. P., Anderson, D. M. and Mivray, M. A. (1999) Evaluation of protein quality in fish meals by chemical and biological assays. *Aquaculture*, **115**, 305–325.

[2] NRC (National Research Council) (1983) Nutrient requirements of warm water fishes and shellfishes. In: *Nutrient Requirements of Domestic Animals VIII*. National Academy Press, Washington D.C.

Appendix C

Nutritional requirements, types of formulae, feeding tables and various data

Table C.1. Optimal protein level for some fish.

Species	Optimal content (% DM)
Atlantic salmon	45
Chinook salmon	40
Coho salmon	40
Sockeye salmon	45
Trout (rainbow)	40
Channel catfish	32–36
Common carp	31–38
Grass carp	41–43
Japanese eel	44
Sturgeons	40
Nile tilapia	30
Mozambique tilapia	40
Zilli tilapia	35
Blue tilapia	34
Milk fish	40
Plaice	50
Japanese puffer	50
Sea bass	45–50
Turbot	55
Grouper	40–50
Gilthead sea bream	40
Japanese sea bream	45–55
Smallmouth bass	45
Largemouth bass	40
Striped bass	47
Yellowtail	55

Table C.2. Quantitative requirements for essential amino acids of some teleost fish (g/16g/N).

Amino acid	Salmonids	Catfish	Carps	Tilapias	Eels	Milk fish	Sea bream	Sea bass	Tropical sea bass
Arginine	4.4	4.3	4.4	4.1	4.5	5.6	<6.0	4.1	3.6
Histidine	1.6	1.5	2.4	1.7	2.1	2.0	–	–	–
Isoleucine	2.0	2.6	3.0	3.1	4.0	4.0	–	–	–
Leucine	3.6	3.5	4.7	3.4	5.3	5.1	–	–	–
Lysine	4.8	5.0	6.0	4.6	5.3	4.0	5.0	4.8	4.5
Methionine + cystine	3.3	2.3	3.5	3.2	3.2	4.8	4.0	4.4	3.4
Phenylalanine + tyrosine	5.3	4.8	8.2	5.6	5.8	5.2	–	–	–
Threonine	2.0	2.1	4.2	3.8	4.0	4.9	–	–	–
Tryptophan	0.6	0.5	0.8	1.0	1.1	0.6	0.6	–	0.5
Valine	5.3	4.8	8.2	5.6	5.8	5.2	–	–	–

N.B. These values, expressed as % crude protein, are valid for the protein levels used by the scientists who determined them. While they apply to the levels appearing in Table C.1. certain under- or over-estimations may occur (Chapter 6).

Table C.3. Essential fatty acid requirements of some teleost fish.

Species	Essential fatty acid	Requirement (% of ration)
Trout	18:3n-3	0.8–1.7
Chum salmon	{ 18:2 n-6 or 20:4 n-6	1
(freshwater or seawater)	{ 18:3 n-3	1
Coho salmon	18:3 n-3	1.0–2.5
Common carp	{ 18:2 n-6	1
	{ 18:3 n-3	1
Eel	{ 18:2 n-6	0.5
	{ 18:3 n-3	0.5
Zilli tilapia	18:2 n-6 or 20:4 n-6	1
Nile tilapia	18:2 n-6	0.5
Catfish	{ 18:3 n-3 or	1.0–2.0
	{ AGPI n-3	0.5–0.75
Ayu	18:3 n-3 or 20:5 n-3	1
Yellowtail	20:5 n-3 or 22:6 n-3	0.5
Turbot	AGPI n-3	0.8
Red sea bream	AGLPI n-3	0.4
Striped bass	AGLPI n-3	1.7

Table C.4. Vitamin requirements of trout and common carp (IU or mg.kg^{-1} food).

Vitamin	Trout	Common carp
Vitamin A (IU)	2000	1000–2000
Vitamin D (IU)	2400	–
Vitamin E	50	80–300
Vitamin K	10	–
Thiamine (B_1)	1.5	2–3
Riboflavin (B_2)	4	4–5
Pyridoxine (B_6)	3	4–10
Pantothenic acid	12	25–40
Niacin (PP)	10	30–50
Folic acid	5	35
Biotin	0.15	1–1.5
Vitamin B_{12}	0.015	–
Vitamin C	10–50	100–150
Choline	3000	500–3000
Inositol	–	200–440

Table C.5. Mineral and trace element requirements (% DM) of some teleost fish.

	P	Ca	Mg	K	Zn	Mn
Atlantic salmon*	0.6	–	–	–	–	20
Pacific salmon	0.6	–	–	0.8	R	R
Trout	0.5–0.8	0.03–0.24	0.05	–	15–2	12
Channel catfish	0.45	0.45**	0.04	0.26	2	2.4
Common carp	0.65	0.3	0.05	–	25	13
Tilapias	0.9	0.65**	0.06	–	1	12
Eels	0.3	–	0.14	–	–	–
Red drum	0.86	–	–	–	20	–
Red sea bream	–	0.34	–	–	–	–

* in freshwater and seawater
** in water containing no Ca
R: required (existing requirement)

Table C.6. Feeds for intensively farmed freshwater fish (g/kg).

Ingredients	Trout	Carp	Catfish	Tilapias	Eels
Fishmeal	300	270	80	150	650
Lactoserum	100	–	–	–	–
Soya cake	130	60	482	200	–
Maize gluten	170	50	–	–	–
Wheat germ	165	410	100	–	–
Maize	–	–	312	–	–
Wheat	–	–	–	200	–
Sorghum	–	–	–	450	–
Rice bran	–	55	–	–	–
Extruded/gelatinised starch	–	–	–	–	210
Fish oil	115	–	15	–	50
Vegetable oil	–	50	–	–	20
Tallow	–	50	–	–	–
Vitamin mixture	10	5	0.5	–	20
Mineral mixture	10	5	0.5	20	–
Bicalcium phosphate	–	25	10	–	–
Binding agents	–	–	–	–	30
Protein ($g.kg^{-1}$ DM)	380	310	320	290	360–400
Lipid ($g.kg^{-1}$ DM)	150	125	40	80	120–150
Digestible energy ($MJ.kg^{-1}$ DM)	17.2	16.4	13.0	14.5	18.0

Table C.7. Feeds for intensively farmed marine fish ($g.kg^{-1}$).

Ingredients	Atlantic salmon	Brown trout	Sea-bass	Sea-bream	Turbot
Fishmeal	550	340	360	430	300
CPSP	–	150	100	100	200
Soya cake	–	–	50	100	100
Yeast	–	50	–	–	–
Maize gluten	–	–	90	100	–
Extruded wheat	250	90	–	–	145
Wheat remillings	–	–	–	–	150
Gelatinised/extruded starch	–	120	230	170	–
Fish oil	170	120	120	50	50
Vegetable oil	–	50	–	–	20
Soya lecithin	10	10	–	–	–
Vitamin mixture	10	20	20	20	20
Mineral mixture	10	10	10	10	10
Binding agents	–	20	20	20	25
Protein ($g.kg^{-1}$ DM)	420	500	400	520	480–520
Lipid ($g.kg^{-1}$ DM)	210	150	180	120	120
Digestible energy ($MJ.kg^{-1}$ DM)	19	19	21	19	18

Table C.8. Semi-purified feeds for freshwater fish (g/kg).

Ingredients	Salmonids	Catfish	Sturgeon	Tilapias, carps
Casein	400	320	310	330
Gelatine	40	80	–	–
Wheat gluten	–	–	150	–
Egg white	–	–	40	–
Starch	115	–	–	200
Dextrin	90	330	280	270
Glucose	50	–	–	–
Fish oil	150	30	40	–
Vegetable oil	–	30	80	80
DL-methionine	5	–	–	–
L-arginine	10	–	–	–
Vitamin mixture	30	10	40	20
Mineral mixture	80	40	30	50
Carboxymethyl cellulose*	–	20	–	–
Cellulose	30	140	30	50

* binding agent

Table C.9. Semi-purified feeds for marine fish ($g.kg^{-1}$).

Ingredients	Sea bream	Turbot	Tropical sea bass	Yellowtail
Casein	500	250	500	640
CPSP	–	313	–	–
Gelatine	–	–	100	–
Starch	–	–	50	–
Dextrin	100	50	–	30
Glucose	50	50	–	–
Fish oil	120	80	60	140
Vegetable oil	–	–	30	–
DL-methionine	–	10	–	5
L-arginine	–	7	–	20
Mixture of amino acids	–	–	70	27.17
Vitamin mixture	10	28	20	30
Mineral mixture	40	5	40	80
Attractants	20	50	–	1.53
Carboxymethyl cellulose*	50	–	50	26.3
Cellulose	110	157	80	–

* binding agent

Table C.10. Very simplified feeds for extensively farmed fish in tropical ponds ($g.kg^{-1}$).

Ingredients	Tilapias	Carps	Shrimps
Fishmeal	150	–	400
Shrimp waste	–	–	100
Soybean meal	200	–	–
Oil-producing plant meal	–	200–500	400
Whole wheat	200	–	–
Sorghum	450	–	–
Rice bran	–	800–500	–
Cassava	–	–	90
Vitamin and mineral mixtures	–	–	10

Table C.11. Feeds for intensively and extensively farmed penaeid shrimps (g.kg^{-1}) (*Penaeus stylirostris, P. monodon, P. chinensis*).

	Type of farming	
Ingredients	intensive	extensive
Fishmeal	400	100
Shrimp meal	60	60
Soybean meal	100	190
Whole wheat	250	–
Wheat middlings	150	600
Fish oil	20	20
Soya lecithin	5	–
Vitamin mixture	10	5
Mineral mixture	5	1
Binding agents	–	24
Protein (g.kg^{-1} DM)	400	150
Lipid (g.kg^{-1} DM)	80	50

N.B. For optimal protein values, refer to values indicated in Table 17.1.

Table C.12. Feeds for shrimps of the genus *Penaeus* (g.kg^{-1}).

Ingredients	*P. monodon*	*P. stylirostris*	*P. vannamei*	*P. chinesis*
Fishmeal	360	200	100	200
Shrimp meal	70	100	100	100
Squid waste	30	–	–	50
CPSP G	10	–	20	–
Yeast	50	50	–	50
Soybean meal	100	150	250	150
Wheat gluten	50	50	–	–
Wheat flour	260	370	450	390
Fish oil	20	30	25	20
Soya lecithin	10	10	10	5
Cholesterol	–	–	5	5
Vitamin mixture	10	10	10	10
Mineral mixture	20	20	20	10
Binding agents	10	10	10	10

Table C.13. Inert foods used for weaning fish (g.kg^{-1}).

Diets	I	II	III	IV	V
Fishmeal	386	–	445	–	–
Fish proteolysate	180	40	–	–	–
Mussel meal	–	40	–	–	–
Scallop meal	–	–	83	–	–
Squid meal	–	140	53	–	–
Norway lobster meal	–	140	–	–	–
Fish gonad	–	190	–	–	–
Connective tissue meal	180	–	–	–	–
Beef liver	–	–	–	350	–
Hen's egg yolk	–	140	–	–	–
Casein	–	40	–	–	–
Basic casein	–	–	–	–	940
Protein from pulses	–	–	94	–	–
Lactic yeast	–	40	–	–	–
Brewers' yeast	60	–	–	–	–
Petroleum grown yeast	–	–	–	500	–
Fish oil	38	50	60	–	–
Peanut oil	–	–	–	–	40
Soya lecithin	10	30	30	–	–
Vitamin mixture	10	25	60	100	100
Vitamin C	2.8	2.5	–	–	–
Vitamin E	0.35	–	–	–	–
Choline chloride	6	–	–	–	–
Egg phosphatidylcholine	–	–	–	–	20
Cholic acid	–	0.02	–	–	–
Mineral mixture	10	40	50	50	50
Attractants	–	–	36	–	–
BHT	0.1	–	–	–	–
Pregelatinised starch	117	–	–	–	–
Zein	–	60	80	–	–
Cellulose	–	–	3	–	–
Protein (g.kg^{-1} DM)	380	310	320	290	360–40
Lipid (g.kg^{-1} DM)	150	125	40	80	120–15
Digestible energy (g.kg^{-1} DM)	17.2	16.4	13.0	14.5	18.0

I. Person-Le Ruyet, J., Baudin-Laurencin, F., Devauchelle, N., Métailler, R., Nicholas, J. L., Robin, J. and Guillaume, J. (1991) Culture of turbot (*Scophthalmus maximus*). In: McVey, J. P. (ed). *CRC Handbook of Mariculture*. CRC Press, Boca Raton, FL. 21–41.

II. IFREMER, 1990. Internal report. Unpublished data.

III. Kanazawa, A., Koshio, S. and Teshima, S.-L. (1989) Growth and survival of larval red sea bream *Pagrus major* and Japanese flounder *Paralichthys olivaceus* fed microbound diets. *J. World Aquac. Soc.*, **20**, 31–37.

IV. Szlaminska, N., Escaffre, A. M., Alami-Durante, H., Charlon, N. and Bergot, P. (1990) Casein in the place of beef liver in artificial diets for common carp (*Cyprinus carpio* L.) larvae. *Aquat. Living Resour.*, **3**, 229–234.

V. Radünz-Neto, J., Corraze, G., Bergot, P. and Kaushik, S. J. (1996) Estimation of essential fatty acid requirements of common carp larvae using semi-purified artificial diets. *Arch. Anim. Nutr.* **49**, 41–48.

Table C.14. Alginated squid-based feeds for giant freshwater shrimp larvae: 'Acal' microparticles.

Ingredients	$g.kg^{-1}$
Ground squid	400
Ground shrimp	150
Mussels	100
Roe	50
Fresh hen egg	30
Vitamins and minerals mixture	120
Sodium alginate	150
Moisture content	10
Protein ($g.kg^{-1}$ DM)	40
Lipid ($g.kg^{-1}$ DM)	23

Table C.15. Example of a vitamin mixture for fish.

Vitamins	IU or $mg.kg^{-1}$ feed
Vitamin A acetate (IU)	2500
Vitamin D3 (IU)	2400
Vitamin E acetate	50
Vitamin K3	10
Thiamine	1
Riboflavin	4
Pyridoxine	3
Calcium pantothenate	20
Niacin	10
Biotin	0.15
Folic acid	1
Vitamin B12	0.01
Ascorbyl monophosphate	50
Choline	1000

Recommended level of incorporation: 1% in feeds based on common ingredients (fishmeal, oilcakes ...); 2–3% in semi-purified feeds.

Table C.16. Example of a mineral mixture for fish.

Ingredients	$g.kg^{-1}$ of premixture
Bicalcium phosphate ($CaHPO_4.2H_2O$)	500.00
Calcium carbonate or chloride ($CaCO_3/CaCl_2$)	215.00
Sea salt (NaCl)	40.00
Potassium chloride (KCl)	90.00
Magnesium hydroxide or magnesium carbonate* ($Mg(OH)_2$)	124.00
Iron citrate or sulphate ($FeSO_4.7H_2O$)	20.00
Zinc sulphate ($ZnSO_4.7H_2O$)	4.00
Copper sulphate ($CuSO_4.5H_2O$)	3.00
Potassium iodide (KI)	0.004
Cobalt sulphate ($CoSO_4$)	0.02
Manganese sulphate ($MnSO_4.H_2O$)	3.00
Sodium fluoride (NaF)	1.00

All ingredients must be finely ground (<300 μm) before mixing. Level of incorporation in feed: 1–2%
* Can be replaced with magnesium sulphate ($MgSO_4.7H_2O$), 62 g instead of 124 g.

Table C.17. Example of simplified mineral mixture for feeds containing natural ingredients (fishmeal, oilcakes etc.).

Ingredients	$g.kg^{-1}$ of premixture
Sea salt (NaCl, 39% Na)	300.00
Iron sulphate ($FeSO_4.7H_2O$, 21% Fe)	6.30
Zinc sulphate ($ZnSO_4.7H_2O$, 36% Fe)	14.40
Copper sulphate ($CuSO_4.5H_2O$, 25% Cu)	2.50
Potassium iodide (KI, 76% I)	0.80
Manganese sulphate ($MnSO_4.H_2O$, 33% Mn)	8.60
Cellulose or glucose	1000.00

Level of incorporation in feed: 1–2%.

Table C.18. Mean amino acid composition of different species of fish (g.16gN^{-1}).

	Muscle		Whole fish	
Amino acid	mean	standard deviation	mean	standard deviation
Alanine	5.74	0.14	5.60	0.96
Arginine	5.70	0.47	5.25	1.31
Aspartic acid	10.18	0.26	8.83	1.10
Cystine	0.94	0.37	0.93	0.32
Glutamic acid	13.87	0.29	13.03	1.26
Glycine	4.60	0.13	5.93	2.57
Histidine	3.15	0.36	2.36	0.82
Isoleucine	4.60	0.56	4.08	0.49
Leucine	8.00	0.58	7.06	0.37
Lysine	9.11	0.45	7.82	0.85
Methionine	2.28	0.81	2.90	0.31
Phenylalanine	4.33	0.41	4.14	0.71
Proline	3.37	0.37	4.13	1.16
Serine	3.92	0.18	3.97	0.75
Threonine	4.71	0.30	4.38	0.85
Tryptophan	1.18	0.15	0.93	0.05
Tyrosine	3.29	0.37	3.04	0.39
Valine	5.33	0.52	5.10	0.64
Methionine + cystine	3.21	1.07	3.46	0.72
Phenylalanine + tyrosine	7.61	0.74	7.18	1.06

Table C.19. Fatty acid composition of muscle in some species of fish (% of fatty acids).

Fatty acids	Trout	Common carp	Sea bass	Turbot
C14:0	6.40	2.3	3.72	4.30
C15:0	0.44	–	0.26	0.30
C16:0	23.87	19.6	18.33	13.50
C18:0	3.89	4.5	0.15	2.50
C20:0	0.11	–	6.65	–
C16:1	8.93	–	3.38	7.90
C17:1	0.49	–	6.05	–
C18:1	24.45	23.4	16.69	20.90
C20:1	3.52	0.8	3.82	7.50
C22:1	2.16	–	8.24	0.55
C24:1	–	–	2.04	0.45
18:2n-6	4.33	3.9	0.48	7.70
20:2n-6	0.26	–	0.10	0.20
20:3n-6	0.13	–	0.58	–
20:4n-6	0.56	3.5	0.95	0.70
18:3n-3	1.02	6.0	0.37	1.40
18:4n-3	1.16	0.2	4.01	2.23
20:3n-3	0.11	–	1.71	–
20:4n-3	0.46	–	0.56	0.60
20:5n-3	3.51	0.60	13.95	7.60
22:5n-3	0.91	1.2	1.07	2.25
22:6n-3	8.60	5.1	3.70	11.20
Saturated	35.83	36.3	29.22	20.60
Mono-unsaturated	39.76	35.6	43.28	36.80
n-6	5.55	9.4	2.12	8.60
n-3	15.77	18.5	25.38	25.40
PUFA	21.32	27.9	27.49	34.00
PUFA/saturated	0.59	0.78	0.94	1.65
n-3/n-6	2.84	2.0	11.99	2.95

N.B. It must be noted that this composition is greatly influenced by dietary lipids.

Table C.20. Ration size and meal frequency in tilapia reared in ponds (25–28°C) expressed in quantity of dry feed (g.100 g biomass^{-1}day^{-1}).

Live weight (g)	Ration (per 100 g)	Number of meals
< 1 g	30–10	> 8
1–5	10–6	6
5–20	6–4	4
20–100	4–3	3–4
> 100	3	3

Table C.21. Recommended ration size for shrimps of the genus *Penaeus* (*P. vannamei*, *P. stylirostris*, *P.monodon*) in relation to body mass: quantities of dry food (g.100 g biomass^{-1}day^{-1}).

Body mass (g)	Ration
1	11–16
2	8–11.7
3	4.7–8.6
4	4.2–7.2
5	3.9–6.2
6	3.6–6.0
7	3.2–5.0
8	3.0–4.4
9	2.9–4.0
10	2.8–3.9
11	2.6–3.6
12	2.5–3.4
13	2.3–3.2
14	2.2–3.0
15	2.1–2.9
16	2.0–2.8
17	2.0–2.7
18	1.9–2.7
19	1.8–2.6
20	1.8–2.6

Table C.22. Recommended ration size for trout in relation to fish size and water temperature : quantities of dry food (g.100 g biomass^{-1}day^{-1}).

	Temperature (°C)								
Length (cm)	4	6	8	10	12	14	16	18	20
2.5–3.5	2.6	3.0	3.5	4.1	4.8	5.6	6.5	4.6	2.8
3.5–5	2.4	2.8	3.3	3.9	4.6	5.4	6.3	4.4	2.6
5–5.7	2.1	2.5	2.9	3.4	4.0	4.6	5.3	3.7	2.3
7.5–10	1.8	2.1	2.4	2.8	3.2	3.7	4.3	3.0	2.0
10–12.5	1.5	1.7	1.9	2.2	2.5	2.9	3.5	2.4	1.6
12.5–15	1.3	1.5	1.7	2.0	2.3	2.6	3.0	2.2	1.4
15–17.5	1.1	1.3	1.5	1.7	2.0	2.3	2.6	1.9	1.2
17.5–20	0.9	1.1	1.3	1.5	1.7	2.0	2.3	1.6	1.0
22–22.5	0.8	1.0	1.2	1.4	1.6	1.8	2.1	1.5	0.9
22.5–25	0.7	0.9	1.1	1.3	1.5	1.7	2.0	1.4	0.8

Glossary of common and scientific names of species cited

Common name	Scientific name
South Pacific anchovy, South American anchovy	*Engraulis ringens*
European eel	*Anguilla anguilla*
Japanese eel	*Anguilla japonica*
Artemia, brine shrimp	*Artemia salina*
Atipa	*Hoplostermum littorale*
Ayu	*Plecoglossus altivelis*
Sea bass	*Dicentrarchus labrax*
Striped bass	*Morone* sp.
Tropical sea bass	*Lates calcarifer*
Brill	*Scophthalmus rhombus*
Smallmouth bass	*Micropterus dolomieui*
Largemouth bass	*Micropterus salmoides*
Pike	*Esox* sp.
Capelin	*Mallotus villosus*
Goldfish	*Carassius carassius*
Silver carp	*Hypophtalmichthys molitrix*
Common carp	*Cyprinus carpio*
Grass carp	*Ctenopharyngodon idella*
Indian major carp	*Catla catla*
Sculpin	*Cottus* sp.
Giant fresh water shrimp (prawn)	*Macrobrachium rosenbergii*
White fish	*Coregonus* sp.
Fiddler crab	*Uca pugilator*
Blue shrimp (prawn)	*Penaeus vannamei*
Common prawn	*Palaemon serratus*
Chinese shrimp (prawn)	*Penaeus chinensis (orientalis)*

Indian shrimp (prawn)	*Penaeus indicus*
Japanese shrimp (prawn)	*Penaeus japonicus*
Tiger shrimp (prawn)	*Penaeus monodon*
Red sea bream	*Chrysophrys major*
Gilthead sea bream	*Chrysophrys aurata*
European crayfish (crawfish)	*Astacus fluviatilis*
Haddock	*Gadus eglefinus*
Smelt	*Osmerus* sp.
Stickleback	*Gasterosteus aculeatus*
White sturgeon	*Acipenser transmontanus*
Siberian sturgeon	*Acipenser baeri*
Flounder	*Platichthys flexus*
Halibut	*Hypoglossus* sp.
(Japanese) puffer	*Fugu* sp.
Herring	*Clupea harengus*
Japanese flounder	*Paralichthys olivaceus*
American lobster	*Homarus americanus*
Krill	*Euphasia superba*
Sand eel	*Ammodytes tobianus*
Saithe, coalfish	*Pollachius virens*
Dab	*Limanda limanda*
Blue ling	*Mova dypterygia* or *M. byrkelange*
Mackerel	*Scomber* sp.
Medaka	*Oryzias latipes*
Menhaden	*Brevoortia* sp.
Hake	*Merluccius* sp.
Grouper	*Epinephelus* sp.
Cod	*Gadus morhua*
Mullet	*Mugil* sp.
Grayling	*Thymallus thymallus*
Red drum	*Scianops ocellatus*
Couch's sea bream	*Pagrus pagrus*
Perch	*Perca fluviatilis*
Plaice	*Pleuronectes platessa*
African catfish	*Clarias gariepinus*
Channel catfish	*Ictalurus punctatus*
Milk fish	*Chanos chanos*
Skate	*Raja* sp.
Yellow pickerel	*Stizostedion vitreum*
Pilchard, sardine	*Sardina pilchardus*
American, South African, Japan pilchard	*Sardinops* sp.
Pacific salmon	*Oncorhynchus* sp.
Sockeye salmon	*Oncorhynchus gorbuscha*
Coho salmon	*Oncorhynchus kisutch*

Atlantic salmon	*Salmo salar*
Chum salmon	*Oncorhynchus keta*
Chinook salmon	*Oncorhynchus tshawytscha*
Red fish	*Sebastes* sp.
Cuttlefish	*Sepia* sp.
Yellowtail	*Seriola quinqueradiata*
Sheatfish	*Silurus glanis*
Sprat	*Sprattus sprattus*
Dover sole	*Solea solea*
Pout	*Trisopterus luscus*
Zilli tilapia	*Tilapia zilli*
Blue tilapia	*Oreochromis aureus*
Mozambique tilapia	*Oreochromis mozambicus*
Nile tilapia	*Oreochromis niloticus*
Rainbow trout	*Oncorhynchus mykiss* (ex. *Salmo gairdneri*)
Brown trout	*Salmo trutta*
Turbot	*Scophthalmus maximus*

Glossary of technical terms and abbreviations

AA	amino acid
absolute requirements	requirements expressed in relation to the animal (for example in amount per day) (Chapter 2)
absorption	passage (of nutrients) across the intestinal wall, never ingestion or consumption (Chapter 4)
accretion	increase in mass of a component of an organism, for example proteins, during growth
active transport	one of the processes of nutrient absorption (Chapter 4)
ADF	acid detergent fibre (Chapter 8)
ADP	adenosine diphosphate
ammonotelic	pertaining to organisms that excrete ammonia as the principal nitrogenous waste product of protein metabolism
AMP	adenosine monophosphate
apical	in anatomy, close to the apex (head)
apparent digestibility coefficient (ADC)	one of the definitions of the coefficient of digestive utilisation (Chapter 4)
appetence	attraction of an animal towards a food
artemia	primitive crustacean (branchiopod) of the genus *Artemia* used as live prey in aquaculture (Chapter 13)
ATP	adenosine triphosphate
availability	similar to digestibility, but applied to a nutrient which is not digested (hydrolysed) (Chapter 2)
bioconversion	transformation of a molecule by an organism, often restricted, in nutrition, to the process of elongation-desaturation of PUFAs (Chapter 7)

biomass	any quantitative estimate of the total mass of organisms comprising all or part of a population or any other specified unit
body mass	term used in this book instead of the common expressions body weight or live weight
Brachionus	a genus of the Phylum Rotifera widely used as live prey in aquaculture (Chapter 13)
brush border	digestive tract cells on the surface of the lumen, with many microvilli (Chapter 4)
BV	biological value (of a protein): percentage of food nitrogen utilised for growth and body maintenance (Chapter 6)
coefficient of variation (CV)	in statistics, the relationship between the standard deviation and the mean as a percentage (Chapter 2)
compound feed	a feed composed of several ingredients
consumption	intake
copepod	Subclass of Crustacea, abundant in the zooplankton (Chapter 13)
crude fibre	fraction of fibre measured by Weende's method (Chapter 8)
crude protein	amount of protein estimated from nitrogen content ($N \times 6.25$)
daily growth rate	coefficient used to quantify a growth curve (Chapter 2)
DE	digestible energy (Chapter 5)
DGI	daily growth index (Chapter 2)
diet	food or feed (mainly used in experimentation)
digestibility	the relative rate and completeness at which a food is digested (Chapter 4)
digestible	that which can be digested
digestive gland	gland in many invertebrates, including crustaceans, performing various functions, in particular digestion and absorption (Chapter 15)
digestive utilisation	utilisation of food in the digestive tract (from ingestion to absorption) (Chapter 2)
distal	in anatomy, away from the anterior end, opposite of proximal
DP	digestible proteins
duodenum	anterior part of the small intestine
EAA	essential amino acids (Chapter 6)
ectothermy	(approximate synonym: poikilothermy) refers to animals that lack a mechanism for internal temperature

	regulation; internal temperature is close to that of the environment
elasmobranch	a subclass of the Chondrichthyes
endocytosis	penetration of molecules into a cell without previous hydrolysis (Chapter 4)
enterocyte	cell forming part of the intestinal wall (Chapter 4)
enthalpy	heat content (Chapter 5)
essential	substance (e.g. amino acid, fatty acid) (Chapter 2) which cannot be synthesised by an organism, but is required in the diet to avoid dietary deficiency (Chapter 2)
euryhaline	used of organisms which are capable of osmoregulating in a wide range of salinities
eurythermic	used of organisms which can adapt to a wide range of temperatures
excretion	the process of discharging waste matter arising from catabolism
FA	fatty acid (Chapter 7)
facilitated diffusion	one of the mechanisms involved in nutrient absorption (Chapter 4)
feed conversion efficiency	relationship between production food intake (usually gain in body mass) (Chapter 2) Often used to define the performance of formulated fish diets
feed conversion ratio (FCR)	feed ingested/production (usually gain in body mass) ratio
fibre	non-digestible carbohydrates (Chapter 8)
food chain	the sequence of organisms on successive trophic levels within a community, firm fixation by primary producers, passes to the herbivores and then the carnivores.
growing out	maintaining crops until they reach harvest size
hepatopancreas	the major digestive gland in crustaceans and molluscs
HUFA	highly unsaturated fatty acid (Chapter 7)
ileum	the mid-part of the small intestine
indirect calorimetry	indirect measure of the thermal energy released by an animal during respiratory exchanges (Chapter 4)
inert feed	all food other than live prey
ingredient	raw material or other compound included in the formulation of manufactured food
intake	consumption (not to be confused with absorption)

jejunum	mid-part of the small intestine, between ileum and duodenum
limiting factor	any environmental factor which exists at a suboptimal level and thereby prevents an organism from reaching its full biotic potential
macronutrient	molecule which, when digested, provides nutrients (Chapter 2)
maintenance	process ensuring maintenance of an organism without any accretion or loss
ME	metabolisable energy (Chapter 5)
metabolic utilisation	utilisation of food within the organism (retention, bioconversion, catabolism, etc.) (Chapter 2)
microencapsulation	technique to facilitate the rearing of fish or crustacean larvae on artificial diets; involves the production of a micro-diet of appropriate size consisting of dietary ingredients, such as yeast, encapsulated by a membrane or binder
National Research Council (NRC)	American organisation whose specialist commissions publish, among other things, standards for animal nutrition
nauplius	first larval stage of many crustaceans (Chapter 16)
NDF	neutral detergent fibre (Chapter 8)
NE	net energy (Chapter 5)
NEAA	non-essential amino acids (Chapter 6)
net protein value (NPV)	criterion describing total protein utilisation (Chapter 5), the proportion of protein (nitrogen) retained by an animal over the total protein (nitrogen) consumed
nitrogen-free extract	in the analysis using the Weende method, total amount of carbohydrates obtained from the difference between organic matter and the sum of proteins, lipids and crude fibres
nutrient	a substance that provides nourishment (Chapter 2)
nutritional requirements	level of a nutrient required for optimal performance of an animal (Chapter 2)
palatibility	characteristic of a food which leads to the appetence of the animal
passive diffusion	one of the mechanisms involved in nutrient absorption (Chapter 4)
peroxidation	process of oxidation particularly affecting certain fatty acids (Chapter 7)

pinocytosis	almost synonymous with endocytosis, but involves intake of liquid
PL	phospholipid
PL*n*	post-larvae, *n* days old (Chapter 16)
poikilothermy	almost synonymous with ectothermy
preferendum	range (of temperature, salinity, etc.) providing the best conditions for a species
pre-grow out	technical term devoted to the start of growing out prior to moving animals to true growing out units
production	the assimilation of energy into organic materials by an organism, population or community. Also used to describe products of value to man (flesh, eggs, etc.)
protein efficiency ratio (PER)	Body mass gain/protein ratio (Chapter 6)
proximal	in anatomy, close to the anterior extremity
PUFA	polyunsaturated fatty acid (Chapter 7)
purified diet	a feed made out of refined ingredients with specified analyses
Q_{10}	coefficient by which the rate of a physiological or biochemical process (for example, enzyme activity) is increased for a temperature increase of 10°C
ration	the amount of feed given
rationing	establishment of rations. Does not necessarily mean food is restricted (Chapter 22)
relative requirements	requirements expressed in relation to the food (for example as a percentage) (Chapter 2)
replication	in an experiment, the number of groups subjected to the same treatment
retention	difference between absorption and excretion in an animal with a positive budget
SDA	specific dynamic action: see heat increment (Chapter 5)
Selachii	one of the subclasses of the cartilagenous fish, consisting of sharks and rays
semi-essential	used to describe an important nutrient which can be synthesised by an organism, but only from an essential nutrient
specific growth rate (SGR)	coefficient used to describe a growth curve (Chapter 2)
stenohaline	organisms that can exist only within a narrow range of salinity
stenotherm	organisms that can exist only within a narrow range of temperature

suspended solid	term used to describe the presence of sediment in culture water
TAG	triacylglycerol (new name for triglyceride)
Teleostei	the largest group of bony fishes, belonging to the subclass Actinopterygii
urotelic	producing urea and/or uric acid as the major nitrogenous waste product of protein metabolism

Index

Lightning Source UK Ltd.
Milton Keynes UK
UKOW01n0140090813

215026UK00001B/22/A

9 781852 332419